② Collins International Primary Maths models of teaching and learning

Collins International Primary Maths is based on the constructivist model for teaching and learning developed by Piaget, Vygotsky, and later Bruner, and the importance of:

- starting from what learners know already and providing them with guidance that moves their thinking forwards
- students learning the fundamental principles of a subject, as well as the connections between ideas within the subject and across other subjects
- focusing on the process of learning, rather than the end product of it
- developing learners' intuitive thinking, by asking questions and providing opportunities for learners to ask questions

- learning through discovery and problem solving, which requires learners to make predictions, hypothesise, make generalisations, ask questions and discuss lines of enquiry
- using active methods that require rediscovering or reconstructing norms and truths
- using collaborative as well as individual activities, so that learners can learn from each other
- evaluating the level of each learner's development so that suitable tasks can be set.

At the core of the course are four of the major principles of Bruner's 'Theory of Instruction'* that characterise the organisation and content of Collins International Primary Maths.

Predisposition to learn

The concept of 'readiness for learning'.

A belief that any subject can be taught at any stage of development in a way that fits the learner's cognitive abilities.

Structure of knowledge

A body of knowledge can be structured so that it can be most readily grasped by the learner.

Effective sequencing

No one sequencing will fit every learner, but in general, curriculum content should be taught in increasing difficulty.

Modes of representation

Learning occurs through three modes of representation: *enactive* (action-based), *iconic* (image-based), and *symbolic* (language-based).

In particular, the teaching and learning opportunities in Collins International Primary Maths reflect Bruner's three modes of representation whereby learners develop an understanding of a concept through the three progressive steps (or representations) of concrete-pictorial-abstract, and that reinforcement of an idea or concept is achieved by going back and forth between these representations.

---

* Bruner, J. S. (1966) *Toward a Theory of Instruction*, Cambridge, MA: Belknap Press

Concrete Representation

The *enactive* stage

The learner is first introduced to a concept using physical objects. This 'hands on' approach is the foundation for conceptual understanding.

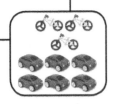

Pictorial Representation

The *iconic* stage

The learner has sufficiently understood the hands-on experiences and can now relate them to images, such as a picture, diagram or model of the concept.

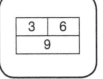

3	6
9	

Abstract Representation

The *symbolic* stage

The learner is now capable of using numbers, notation and mathematical symbols to represent the concept.

$3 + 6 = 9$

The model below illustrates the six proficiencies in mathematics that Collins International Primary Maths believe learners need to command in order to become mathematically literate and achieve mastery of the subject at each stage of learning.

All of the teaching and learning units throughout the Collins International Primary Maths course aim to achieve each of these six proficiencies, in order to help teachers to establish successful mathematics learning.

⅄ Knowledge and understanding of mathematical concepts, principles, fundamental operations and procedures.

⅄ Ability to carry out procedures (or mathematical operations) flexibly, accurately, efficiently and appropriately.

⅄ Capacity for making conjectures and generalisations, and explaining and justifying conclusions.

⅄ Skill in using and applying mathematics to understand, represent, solve and interpret and evaluate problems.

⅄ Facility to communicate clearly in the language of mathematics by understanding and using precise vocabulary and symbolic and diagrammatical representations.

⅄ Process of identifying relationships and making connections by linking thoughts and ideas together.

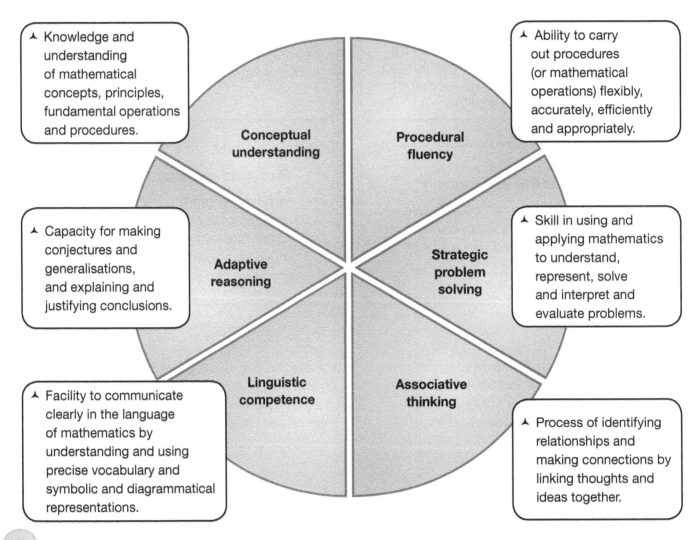

Collins
INTERNATIONAL
PRIMARY
MATHS

Teacher's Guide 1

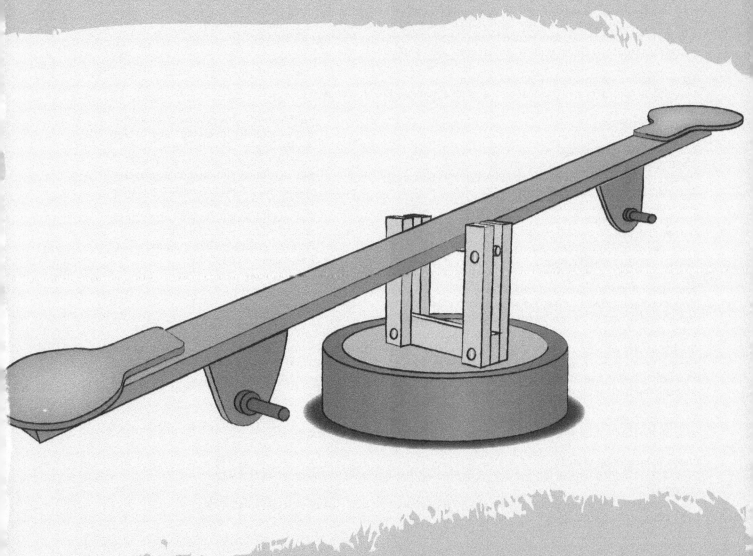

William Collins' dream of knowledge for all began with the publication of his first book in 1819. A self-educated mill worker, he not only enriched millions of lives, but also founded a flourishing publishing house. Today, staying true to this spirit, Collins books are packed with inspiration, innovation and practical expertise. They place you at the centre of a world of possibility and give you exactly what you need to explore it.

Collins. Freedom to teach.

Published by Collins
An imprint of HarperCollins*Publishers*
The News Building
1 London Bridge Street
London
SE1 9GF

HarperCollins*Publishers*
Macken House, 39/40 Mayor Street Upper
Dublin 1, D01 C9W8, Ireland

Browse the complete Collins catalogue at
www.collins.co.uk

© HarperCollins*Publishers* Limited 2021

10 9 8 7

ISBN 978-0-00-836951-4

British Library Cataloguing-in-Publication Data
A catalogue record for this publication is available from the British Library.

Author: Lisa Jarmin
Series editor: Peter Clarke
Publisher: Elaine Higgleton
Product developer: Holly Woolnough
Project manager: Mike Harman (Life Lines Editorial Services)
Development editor: Joan Miller
Copyeditor: Tanya Solomons
Proofreader: Catherine Dakin
Answer checker: Steven Matchett
Cover designer: Gordon MacGilp
Cover illustrator: Ann Paganuzzi
Typesetter: QBS Learning
Illustrators: Ann Paganuzzi and QBS Learning
Production controller: Lyndsey Rogers
Printed and bound in the UK by Ashford Colour Press Ltd

With thanks to the following teachers and schools for reviewing materials in development: Calcutta International School; Hawar International School; Melissa Brobst, International School of Budapest; Rafaella Alexandrou, Pascal Primary Lefkosia; Maria Biglikoudi, Georgia Keravnou, Sotiria Leonidou and Niki Tzorzis, Pascal Primary School Lemessos; Taman Rama Intercultural School, Bali.

The publishers gratefully acknowledge the permission granted to reproduce the copyright material in this book. Every effort has been made to trace copyright holders and to obtain their permission for the use of copyright material. The publishers will gladly receive any information enabling them to rectify any error or omission at the first opportunity.

Cambridge International copyright material in this publication is reproduced under licence and remains the intellectual property of Cambridge Assessment International Education.

Photo acknowledgements
Every effort has been made to trace copyright holders. Any omission will be rectified at the first opportunity.
p6t Tatiana Popova/Shutterstock; p6b Studio KIWI/Shutterstock; p280 ColinCramm/Shutterstock; p281 Wongstock/Shutterstock.

Contents

Introduction

Revise activities

Collins International Primary Maths Stage 1 units

Introduction

① Key features of Collins International Primary Maths

Collins International Primary Maths places the learner at the centre of the teaching and learning of mathematics. To this end, all of the components and features of the course are aimed at helping teachers to address the distinct learning needs of *all* learners.

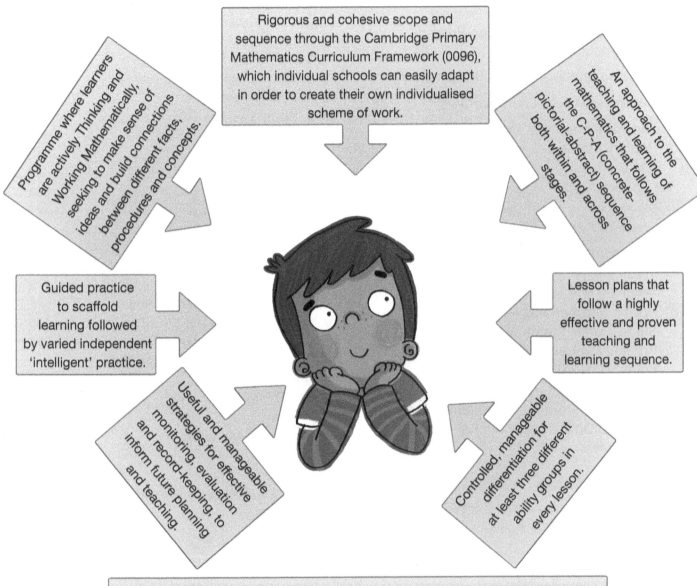

Programme where learners are actively Thinking and Working Mathematically, seeking to make sense of ideas and build connections between different facts, procedures and concepts.

Rigorous and cohesive scope and sequence through the Cambridge Primary Mathematics Curriculum Framework (0096), which individual schools can easily adapt in order to create their own individualised scheme of work.

An approach to the teaching and learning of mathematics that follows the C-P-A (concrete-pictorial-abstract) sequence both within and across stages.

Guided practice to scaffold learning followed by varied independent 'intelligent' practice.

Lesson plans that follow a highly effective and proven teaching and learning sequence.

Useful and manageable strategies for effective monitoring, evaluation and record-keeping, to inform future planning and teaching.

Controlled, manageable differentiation for at least three different ability groups in every lesson.

In addition, the course offers extensive teacher support through materials that:
- promote the most effective pedagogical methods in the teaching of mathematics
- are sufficiently detailed to aid confidence
- are rich enough to be varied and developed
- take into account issues of pace and classroom management
- give careful consideration to the key skill of appropriate and effective questioning
- provide a careful balance of teacher intervention and learner participation
- encourage communication of methods and foster mathematical rigour
- are aimed at raising levels of attainment for *every* learner.

③ How Collins International Primary Maths supports Cambridge Primary Mathematics

Cambridge Primary is typically for learners aged 5 to 11 years. It develops learner skills and understanding through the primary years in English, Mathematics and Science. It provides a flexible framework that can be used to tailor the curriculum to the needs of individual schools.

In Cambridge Primary Mathematics, learners:

- engage in creative mathematical thinking to generate elegant solutions
- improve numerical fluency and knowledge of key mathematical concepts to make sense of numbers, patterns, shapes, measurements and data
- develop a variety of mathematical skills, strategies and a way of thinking that will enable them to describe the world around them and play an active role in modern society
- communicate solutions and ideas logically in spoken and written language, using appropriate mathematical symbols, diagrams and representations
- understand that technology provides a powerful way of communicating mathematics, one that is particularly important in an increasingly technological and digital world.

Cambridge Primary Mathematics supports learners to become:

RESPONSIBLE	INNOVATIVE	ENGAGED
• Learners understand how principles of mathematics can be applied to real-life problems in a responsible way.	• Learners solve new and unfamiliar problems, using innovative mathematical thinking. They can select their own preferred mathematical strategies and can suggest alternative routes to develop efficient solutions.	• Learners are curious and engage intellectually to deepen their mathematical understanding. They are able to use mathematics to participate constructively in society and the economy by making informed mathematical choices.

CONFIDENT	REFLECTIVE
• Learners are confident and enthusiastic mathematical practitioners, able to use appropriate techniques without hesitation, uncertainty or fear. They are keen to ask mathematical questions in a structured, systematic, critical and analytical way. They are able to present their findings and defend their strategies and solutions as well as critique and improve the solutions of others.	• Learners reflect on the process of thinking and working mathematically as well as mastering mathematics concepts. They are keen to make conjectures by asking sophisticated questions and thus develop higher-order thinking skills.

Cambridge Primary is organised into six stages. Each stage reflects the teaching targets for a year group. Broadly speaking, Stage 1 covers the first year of primary teaching, when learners are approximately five years old. Stage 6 covers the final year of primary teaching, when learners are approximately 11 years old.

The Cambridge Primary Mathematics Curriculum Framework (0096) replaces the previous curriculum framework (0845), and is presented in three content areas (strands), with each strand divided into sub-strands. Thinking and Working Mathematically underpins all strands and sub-strands, while mental strategies are a key part of the Number content.

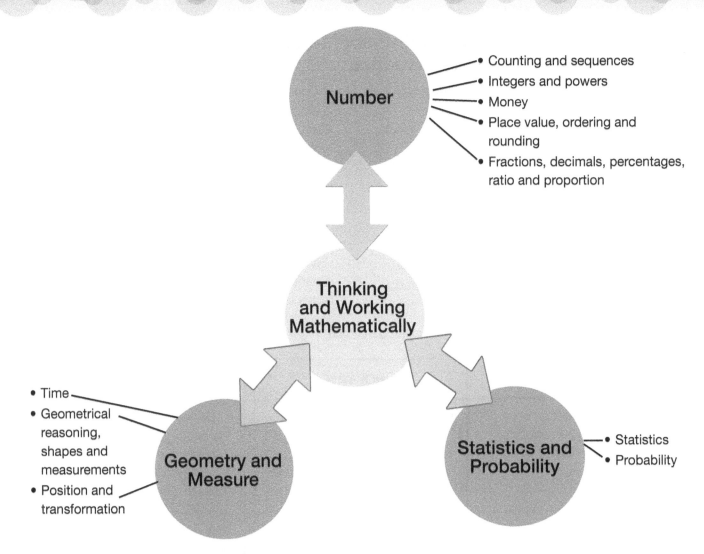

Number
- Counting and sequences
- Integers and powers
- Money
- Place value, ordering and rounding
- Fractions, decimals, percentages, ratio and proportion

Thinking and Working Mathematically

Geometry and Measure
- Time
- Geometrical reasoning, shapes and measurements
- Position and transformation

Statistics and Probability
- Statistics
- Probability

Planning, teaching and assessment

Effective planning, teaching and assessment are the three interconnected elements that contribute to promoting learning, raising learners' attainment and achieving end-of-stage expectations (mastery). All of the components of Collins International Primary Maths emphasise, and provide guidance on, the importance of this cyclical nature of teaching in order to ensure that learners reach the end-of-stage expectations of the Cambridge Primary Mathematics Curriculum Framework (0096). This teaching and learning cycle, and the important role that the teacher plays in this cycle, are at the heart of Collins International Primary Maths.

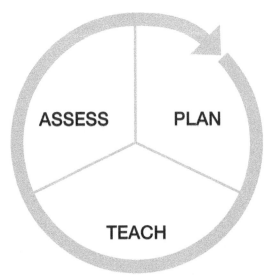

ASSESS | PLAN

TEACH

Collins International Primary Maths supports teachers in planning a successful mathematics programme for their unique teaching context and ensures:

- a clear understanding of learners' pre-requisite skills before they undertake particular tasks and learn new concepts
- considered progression from one lesson to another
- regular revisiting and extension of previous learning
- a judicious balance of objectives, and the time dedicated to each one
- the use of a consistent format and structure.

The elements of Collins International Primary Maths that form the basis for planning can be summarised as follows:

Long-term plans

The Cambridge Primary Mathematics Curriculum Framework (0096) constitutes the long-term plan for schools to follow at each stage across the school year. By closely reflecting the Curriculum Framework and the Cambridge Primary Mathematics Scheme of Work, the Collins International Primary Maths course embodies this long-term plan.

Medium-term plans

The Collins International Primary Maths Units and Recommended Teaching and Learning Sequence in sections ⑦ and ⑧ (see pages 37–47) show termly/semester outlines of units of work with Cambridge Primary Mathematics Curriculum Framework (0096) references (including the Curriculum Framework codes). By using Collins International Primary Maths Extended Teacher's Guide, these plans including curriculum coverage, delivery and timing, can be easily adapted to meet the specific needs of individual schools and teachers as well as learners' needs.

Short-term plans

Individual lesson plans and accompanying Additional practice activities represent the majority of each Teacher's Guide. The lessons provide short-term plans that can easily be followed closely, or used as a 'springboard' and varied to suit specific needs of particular classes. An editable 'Weekly Planning Grid' is also provided as Digital content, which individual teachers can fully adapt.

This includes modifying short-term planning in order to build on learners' responses to previous lessons, thereby enabling them to make greater progress in their learning.

The most important role of teaching is to promote learning and to raise learners' attainment. To best achieve these goals, Collins International Primary Maths believes in the importance of teachers:

- promoting a 'can do' attitude, where all learners can achieve success in, and enjoy, mathematics
- having high, and ambitious, expectations for *all* learners
- adopting a philosophy of equal opportunity that means *all* learners have full access to the same curriculum content
- generating high levels of engagement (*Active learning*) and commitment to learning
- offering sharply focused and timely support and intervention that matches learners' individual needs
- being *language aware* in order to understand the possible challenges and opportunities that language presents to learning
- systematically and effectively checking learners' understanding throughout lessons, anticipating where they may need to intervene, and doing so with notable impact on the quality of learning
- consistently providing high-quality marking and constructive feedback to ensure that learners make rapid gains.

To help teachers achieve these goals, Collins International Primary Maths provides:

- highly focused and clearly defined learning objectives
- examples of targeted questioning, using appropriate mathematical vocabulary, that is aimed at both encouraging and checking learner progress
- a proven lesson structure that provides clear and accurate directions, instructions and explanations
- meaningful and well-matched activities for learners, at all levels of understanding, to practise and consolidate their learning
- a balance of individual, pair, group and whole-class activities to develop both independence and collaboration and to enable learners to develop their own thinking and learn from one another
- highly effective models and images (representations) that clearly illustrate mathematical concepts, including interactive digital resources.

The lesson sequence in Collins International Primary Maths focuses on supporting learners' understanding of the learning objectives of the Cambridge Primary Mathematics Curriculum Framework (0096), as well as building their mathematical proficiency and confidence.

Based on a highly effective and proven teaching and learning sequence, each lesson is divided into six key teaching strategies that take learners on a journey of discovery. This approach is shown on the outer ring of the diagram below.

The inner ring shows the link between the six key teaching strategies and the five phases of a Collins International Primary Maths lesson plan, as well as when to use the Student's Book and Workbook.

Pedagogical cycle

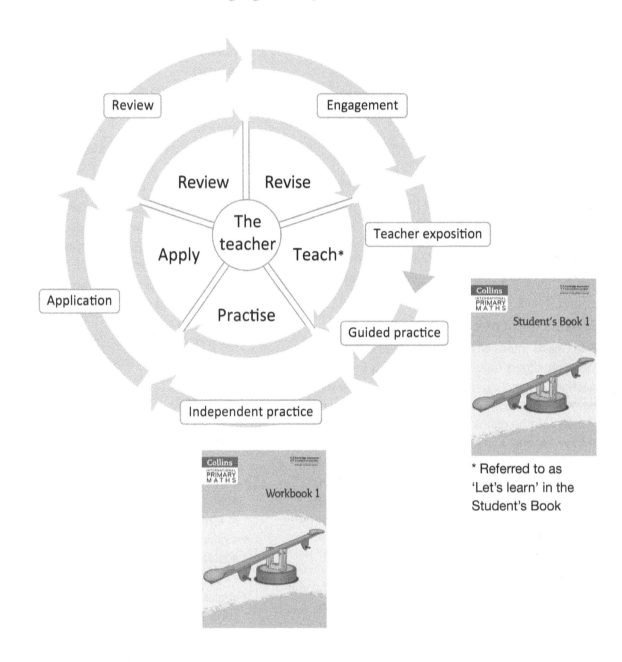

* Referred to as 'Let's learn' in the Student's Book

The chart below outlines the purposes of each phase in the Collins International Primary Maths lesson sequence, as well as the learner groupings and approximate recommended timings.

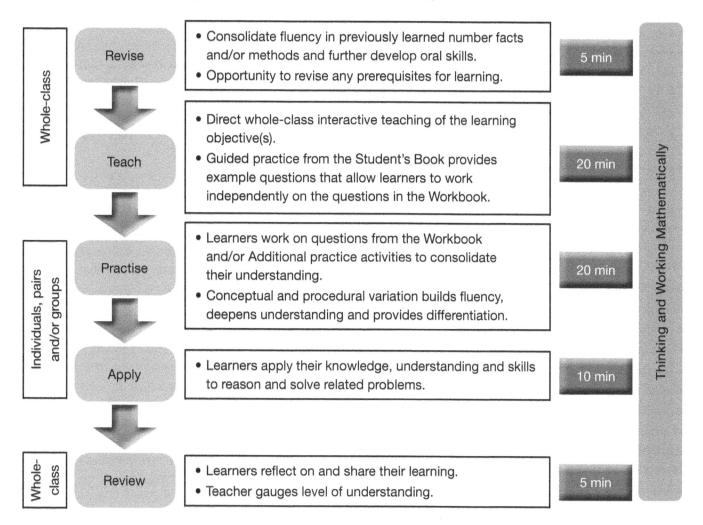

Monitoring, evaluation and feedback continue the teaching and learning cycle and are used to form the basis for adjustments to the teaching programme. Collins International Primary Maths offers meaningful, manageable and useful assessment on two of the following three levels:

Short-term 'on going' assessment

Short-term assessments are an informal part of every lesson. A combination of carefully crafted recall, observation and thought questions is provided in each lesson of Collins International Primary Maths and these are linked to specific learning objectives.

They are designed to monitor learning and provide immediate feedback to learners and to gauge learners' progress in order to enable teachers to adapt their teaching.

Success Criteria are also provided in each unit to assist learners in identifying the steps required to achieve the unit's learning objectives.

Each unit in Collins International Primary Maths begins with a Unit introduction. One of the features of the Unit introduction is 'Common difficulties and remediation'. This feature can be used to help identify why learners do not understand, or have difficulty with, a topic or concepts and to use this information to take appropriate action to correct mistakes or misconceptions.

Medium-term 'formative' assessment

Medium-term assessments are used to review and record the progress learners make, over time, in relation to the learning objectives of the Cambridge Primary Mathematics Curriculum Framework (0096). They are used to establish whether learners have met the learning objectives or are on track to do so.

'Assessment *for* learning' is the term generally used to describe the conceptual approach to both short-term 'ongoing' assessment and medium-term 'formative' assessment.

Assessment *for* learning involves both learners and teachers finding out about the specific strengths and weaknesses of individual learners, and the class as a whole, and using this to inform future teaching and learning.

Assessment *for* learning:
- is part of the planning process
- is informed by learning objectives
- engages learners in the assessment process
- recognises the achievements of all learners
- takes account of how learners learn
- motivates learners.

In order to assist teachers with monitoring both short- and medium-term assessments, and to ensure that evidence collected is meaningful, manageable and useful, Collins International Primary Maths includes a class record-keeping document on pages 299–304.

The document helps teachers:
- identify whether learners are on track to meet end-of-stage expectations
- identify those learners working *above* and *below* end-of-stage expectations
- make long-term 'summative' assessments
- report to parents and guardians
- inform the next year's teacher about which sub-strands of the Cambridge Primary Mathematics Curriculum Framework (0096) individual learners, and the class as a whole, are exceeding, meeting or are below in expectations.

For further details on how to use the class record-keeping document, please refer to pages 299–304.

Long-term 'summative' assessment

Long-term assessment is the third level of assessment. It is used at the end of the school year in order to track progress and attainment against school and external targets, and to report to other establishments and to parents on the actual attainments of learners. By ensuring complete and thorough coverage of the Cambridge Primary Mathematics Curriculum Framework (0096), Collins International Primary Maths provides an excellent foundation for the Cambridge Primary end-of-stage tests (Cambridge Primary Progression Tests) as well as the end of primary Cambridge Primary Checkpoint.

Mental strategies and Cambridge Primary Mathematics

Mental strategies learning objectives are not included in the Cambridge Primary Mathematics Curriculum Framework (0096). However, working mentally is an important feature in the curriculum framework and is embedded not just within the Number strand but throughout all strands in the curriculum framework.

Mental strategies should be applied across all of the Cambridge Primary stages (1 to 6) and to all mathematical strands. The Cambridge Primary Mathematics Curriculum Framework (0096) is, however, less prescriptive about the specific strategies that should be learned and practised by learners at each stage. Allowing teachers greater flexibility in teaching mental strategies, and allowing learners to view mental strategies as a more personal and less formal choice, means that learners will have greater ownership over the mental strategies that they choose to use, thereby developing a deeper conceptual understanding of the number system.

In keeping with the changes that have been introduced in the Cambridge Primary Mathematics Curriculum Framework (0096), mental strategies are embedded throughout Collins International Primary Maths. Learners are given opportunities to develop and practise mental strategies, using carefully chosen numbers, and are continually encouraged to articulate their strategies verbally. This is of particular importance for the updated curriculum framework, where there are no specific learning objectives relating to mental strategies.

It is not possible to exhaustively list all of the mental strategies that can be used, and there will not be one correct strategy for any particular calculation. The most appropriate mental strategy will depend on individual learners' knowledge of mathematical facts, their working memory and their conceptual understanding of different parts of the number system. Mental strategies can be explicitly learned and practised, and doing so will enable a learner to add that strategy to their 'repertoire'. It is important therefore that learners are exposed to a wide range of strategies.

Below are some of the different mental strategies that learners may employ and that are featured throughout Collins International Primary Maths.

Addition and subtraction

– counting on and back in steps

– using known addition and subtraction number facts/number bonds/complements

– applying knowledge of place value and partitioning (i.e. compose, decompose and regroup numbers)

– compensation

– putting the larger number first and counting on (addition)

– counting back from the larger number (take away)

– counting on from the smaller number (find the difference)

– recognising that when two numbers are close together it's easier to find the difference by counting on, not counting back

– using the commutative and associative properties

– using the inverse relationship between addition and subtraction

Multiplication and division

– counting on and back in steps of constant size

– using known multiplication and division facts and related facts involving multiples of 10 and 100

– applying knowledge of place value and partitioning (i.e. compose, decompose and regroup numbers), including multiplying and dividing whole numbers and decimals by 10, 100 and 1000

– using doubling

– recognising and using factor pairs

– using the commutative, distributive and associative properties

– using the inverse relationship between multiplication and division

The use of calculators and Cambridge Primary Mathematics

When used well, calculators can assist learners in their understanding of numbers and the number system.

Calculators should be used as a teaching aid to promote mental calculation and mental strategies and to explore mathematical patterns. Learners should understand when it is best to use calculators to assist calculations and when to calculate mentally or use written methods.

As Cambridge International includes calculator-based assessments at Stages 5 to 9, it is recommended that learners begin to use calculators for checking calculations from the end of Stage 3, and for performing and checking calculations from Stage 4. At Stages 5 and 6, learners should be developing effective use of calculators so that they are familiar with the functionality of a basic calculator in readiness for Stage 7 onwards.

④ Thinking and Working Mathematically

In the Cambridge Primary (0096) and Lower Secondary (0862) Mathematics Curriculum Framework, the problem – solving strand and associated learning objectives have been replaced with four pairs of Thinking and Working Mathematically (TWM) characteristics.

The TWM characteristics represent one of the most significant changes to the Cambridge Primary Mathematics Curriculum Framework (0096). In response to this, this new edition of Collins International Primary Maths has incorporated and interwoven TWM throughout all of the components; this reflects the course's most substantial change to the teaching and learning of mathematics.

Thinking and Working Mathematically is based on work by Mason, Burton and Stacey*; it places an emphasis on learners:

– actively engaging with their learning of mathematics

– talking with others, challenging ideas and providing evidence that validates conjectures and solutions

– seeking to make sense of ideas

– building connections between different facts, procedures and concepts

– developing higher-order thinking skills that assist them in viewing the world in a mathematical way.

This contrasts with learners simply following instructions and carrying out processes that they have been shown how to do, without appreciating

* Mason, J., Burton, L. and Stacey, K. (2010) *Thinking Mathematically*, 2nd edition, Harlow: Pearson

why such processes work or what the results mean. Through the development of each of the TWM characteristics, learners are able to see the application of mathematics in the real world more clearly and also, crucially, to develop the skills necessary to function as citizens who are autonomous problem solvers.

If learners at any of the Cambridge International stages are to gain meaning and satisfaction from their study of mathematics, then it is vital that TWM underpins their experience of learning the subject.

The four pairs of TWM characteristics that Cambridge International identifies as fundamental to a meaningful experience of learning mathematics are represented diagrammatically and referred to as 'The Thinking and Working Mathematically Star'.

The Thinking and Working Mathematically Star

Conjecturing and Convincing relate to forming, questioning and justifying own or others' mathematical ideas.

Conjecturing and Convincing

Specialising and Generalising

Specialising and Generalising relate to finding and considering examples that meet specific mathematical criteria.

Critiquing and Improving

Characterising and Classifying

Critiquing and Improving relate to reflecting on and refining mathematical strategies, representations or solutions.

Critiquing and improving run through the six other characteristics, demonstrating the importance of considering how to refine your own methods or strategies when Thinking and Working Mathematically.

Characterising and Classifying relate to identification and grouping of objects (shapes, numbers, graphical representations, etc.) according to their mathematical properties.

The Thinking and Working Mathematically Star, © Cambridge International, 2018.

The eight characteristics of Thinking and Working Mathematically

Characteristic	Definition
TWM.01 Specialising	Choosing *an example* and checking to see if it satisfies or does not satisfy specific mathematical criteria.
TWM.02 Generalising	Recognising an underlying pattern by identifying *many* examples that satisfy the same mathematical criteria.
TWM.03 Conjecturing	Forming mathematical questions or ideas.
TWM.04 Convincing	Presenting evidence to *justify or challenge* a mathematical idea or solution.
TWM.05 Characterising	Identifying and describing the mathematical properties of an object.
TW.06 Classifying	Organising objects into groups according to their mathematical properties.
TWM.07 Critiquing	Comparing and evaluating mathematical ideas, representations or solutions to identify advantages and disadvantages.
TWM.08 Improving	Refining mathematical ideas or representations to develop a more effective approach or solution.

All eight TWM characteristics can be applied across all of the Cambridge Primary and Lower Secondary stages (1 to 9) and across all mathematical strands and sub-strands, although the prominence of different characteristics may change as learners move through the stages.

Any characteristic can be combined with any other characteristic; characteristics should be taught alongside content learning objectives and should **not** stand alone.

The four pairs of characteristics intertwine and are interdependent, and a high-quality mathematics task may draw on one or more of them.

Thinking and Working Mathematically should **not** consist of a separate end-of-lesson or unit activity, but should be embedded throughout lessons in every unit of work. All of the characteristics identified above can be combined with most teaching topics so, when planning a unit of work, teachers should begin with one or more learning objectives and seek to draw on one or more TWM characteristics.

Thinking and Working Mathematically also enables learners' thinking to become visible, which is a crucial aspect of formative assessment.

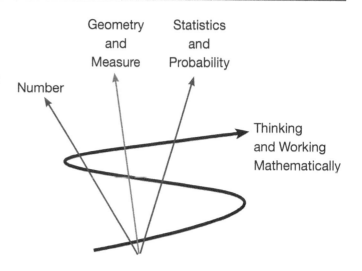

Just as TWM is at the very heart of Cambridge Primary Mathematics, so too is this approach to the teaching and learning of mathematics a core feature of Collins International Primary Maths. Opportunities are provided in each of the 27 units for learners to develop the TWM characteristics.

Specific guidance is provided at the start of each unit as part of the Unit introduction, which highlights teaching and learning opportunities, particularly in relation to the unit's learning objectives, that promote the TWM characteristics.

Unit introduction

Promoting Thinking and Working Mathematically

TWM.03 Conjecturing
Learners ask: **How many objects are there in these two sets?** They discover that by counting the objects in both sets together, they find the sum of those objects.

At each phase of the lesson (Revise, Teach, Practise, Apply and/or Review), guidance is given in the lesson plan on how to promote the TWM characteristics.

Whenever any of the eight TWM characteristics is being promoted in a lesson plan this is shown using the initials 'TWM', followed by the Cambridge Primary Mathematics Curriculum Framework (0096) code that identifies exactly which of the eight characteristics is being developed.

Teach [SB] [TWM.03/07]
• Look at the picture in the Student's book. [TWM.03] [T&T] Ask: **What would you like to know about the shells in the jar? Are you wondering how many there are? How many do you think? How do you think could you make a better guess?** Tell learners that sometimes when we want to know how many of something there are, we can have a 'best guess' instead of counting them. This is called 'estimating'.

> The **Teach** phase of this lesson aims to promote the Conjecturing [TWM.03] and Critiquing [TWM.07] characteristics.

Apply 👥 🖥 [TWM.06]
• Display **Slide 2**.
• Give pairs of learners a small selection of local currency coins and notes in a dish. Also add some coins and notes from other countries. Learners examine the coins and notes to work out whether they are local currency or not. They separate the money into local currency and not local currency.

> The **Apply** phase of this lesson aims to promote the Classifying [TWM.06] characteristic.

Practise [WB] [TWM.03]
• Workbook
Title: Pictograms

> The **Practise** phase of this lesson aims to promote the Conjecturing [TWM.03] characteristic.

In addition to the support provided in the Unit introductions and individual lesson plans, The Thinking and Working Mathematically Star below provides a list of prompting questions that teachers may find helpful when asking learners questions specifically aimed at developing each of the TWM characteristics.

The Thinking and Working Mathematically Star
Teacher prompting questions

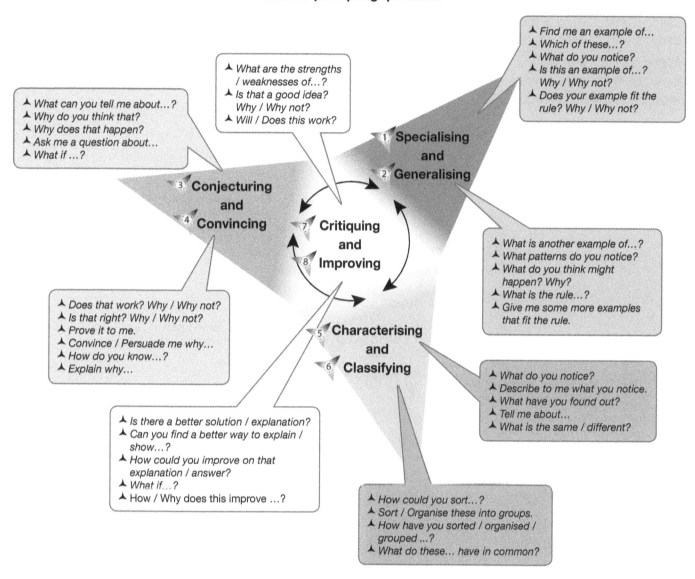

Similar to 'The Thinking and Working Mathematically Star – Teacher prompting questions' on page 17, the star below, which is located at the back of the Stages 1 and 2 Student's Books, defines in learner-friendly language, each of the eight TWM characteristics and numbers them 1 to 8 accordingly.

The star is aimed at helping learners think specifically about what is required when they are undertaking an activity designed to develop a specific TWM characteristic.

If used in conjunction with the Teacher's Guide which uses the initials 'TWM', followed by the Framework code to identify exactly which of the eight characteristics an activity, discussion prompt or practise question is developing, this star will help learners better understand the meaning and purpose of each of the eight TWM characteristics.

In Stages 3 to 6 a similar star is provided at the back of the Student's Book. However, as well as defining each of the eight TWM characteristics, it also includes some sentence stems to help learners to talk with others, challenge ideas and explain their reasoning.

The Thinking and Working Mathematically Star

Think about an idea or solution and decide what's good and bad about it

Find an example and see if it fits a rule

Ask questions and form ideas

1. Specialising
and
2. Generalising

Recognise patterns and find more examples

3. Conjecturing
and
4. Convincing

7. Critiquing
and
8. Improving

Prove a mathematical idea or solution to others

Improve a mathematical idea to come up with a better explanation or solution

5. Characterising
and
6. Classifying

Identify and describe a mathematical object or idea

Sort and organise mathematical objects or ideas into groups

⑤ Cambridge Global Perspectives™

Cambridge Global Perspectives is a unique programme that helps learners develop outstanding transferable skills, including critical thinking, research and collaboration. The programme is available for learners aged 5–19, from Cambridge Primary through to Cambridge Advanced. For Cambridge Primary and Lower Secondary learners, the programme is made up of a series of Challenges covering a wide range of topics, using a personal, local and global perspective. The programme is available to Cambridge schools but participation in the programme is voluntary. However, whether or not your school is involved with the programme, the six skills it focuses on are relevant to **all** students in the modern world. These skills are: research, analysis, evaluation, reflection, collaboration and communication.

More information about the Cambridge Global Perspectives programme can be found on the Cambridge Assessment International Education website: www.cambridgeinternational.org/programmes-and-qualifications/cambridge-global-perspectives.

Collins supports Cambridge Global Perspectives by including activities, tasks and projects in our Cambridge Primary and Lower Secondary courses which develop and apply these skills. Note that the content of the activities is not intended to correlate with the specific topics in the Cambridge Challenges; rather, they encourage practice and development of the Cambridge Global Perspectives to support teachers in integrating and embedding them into students' learning across all school subjects.

Activities in this book that link to the Cambridge Global Perspectives are listed at the back of this book on page 304.

⑥ The components of Collins International Primary Maths

Each of the six stages in Collins International Primary Maths consists of these four components:

Teacher's Guide 1

Student's Book 1

Workbook 1

Extended Teacher's Guide 1

Teacher's Guide

The Teacher's Guide comprises:

- a bank of **Revise activities** for the first 'warm up' phase of the lesson
- **Teaching and learning units**, which consist of Unit introductions, Lesson plans and Additional practice activities
- **Resource sheets** for use with particular lessons and activities
- **Answers** to the questions in the Workbook
- **Record-keeping document** to assist teachers with both short-term 'on-going' and medium-term 'formative' assessments.

A key aim of Collins International Primary Maths is to support teachers in planning, teaching and assessing a successful mathematics programme of work, in line with the Cambridge Primary Mathematics Curriculum Framework (0096).

To ensure complete curriculum coverage and adequate revision of the learning objectives, for each stage the learning objectives from the Cambridge Primary Mathematics Curriculum Framework (0096) have been grouped into 27 topic areas or 'units'. For a more detailed explanation of the 27 units in Collins International Primary Maths Stage 1, including a recommended teaching sequence, refer to Section ⑦ (pages 37–40).

The charts in Section ⑧ (pages 41–47) provide a medium-term plan, showing each of the 27 units in Collins International Primary Maths Stage 1 and which Stage 1 Cambridge Primary Mathematics Curriculum Framework (0096) strand, sub-strand and learning objectives (and codes) each of the units is teaching.

Similarly, the charts in Section ⑨ (pages 48–50) show when each of the Stage 1 learning objectives in the Cambridge Primary Mathematics Curriculum Framework (0096) are taught in the 27 units in Collins International Primary Maths Stage 1.

Icons used in Collins International Primary Maths

  work individually

 work in pairs

work in groups

work as a whole class

[T&T] turn and talk (*Talk Partners*)

  progress check question

 SB refer to the Student's Book

 WB refer to the Workbook

  interactive digital resource

slide

 activity that promotes Thinking and Working Mathematically

 question/activity number typeset on a circle indicates that the question/activity is suitable for learners who require additional support with either easier questions/activities or revising pre-requisite knowledge

 question/activity number typeset on a triangle indicates that the question/activity is suitable for the majority of learners to practise and consolidate the lesson's learning objective(s)

 question/activity number typeset on a square indicates that the question/activity is suitable for learners who require enrichment and/or extension

A note on the use of dice

Although some activities in Collins International Primary Maths suggest the use of dice, these are not always readily available in some countries. Where dice are unavailable or the use of dice is not appropriate, we suggest using a spinner, and have provided spinners on several Resource sheets along with instructions on how to use them using a pencil and paper clip.

Revise activities

A bank of 5-minute 'warm-up' or 'starter' activities is provided for teachers to use at this first phase of the mathematics lesson. Reference is given in each lesson plan to appropriate Revise activities.

The majority of activities are for whole-class work. However, some activities may involve individual learners demonstrating something to the rest of the class, or pairs or groups working together on an activity or a game.

Strand and sub-strand

The relevant Cambridge Primary Mathematics Curriculum Framework (0096) strand and sub-strand covered is stated in the sidebar.

Title

Each activity is given a title. This is designed to help both teacher and learners identify a particular activity.

Unit number and title

Learning objective(s)

Each activity has clearly defined learning objective(s) to assist teachers in choosing the most appropriate activity for the concept they want the learners to practise and consolidate.

2 The majority of Revise activities are suitable for the majority of learners to practise and consolidate the activity's learning objective(s).

Resources

To aid preparation, any resources required are listed, along with whether they are for the whole class, per group, per pair or per learner. Most of these resources are readily available in classrooms.

What to do

The activity is broken down into clear steps to support teachers in achieving the objective(s) of the activity and facilitate interactive whole-class teaching.

Classroom organisation

Icons are used to indicate whether the activity is designed to be used by the whole class working together, or for some activities for learners working in groups, pairs or individually.

Variations

Where appropriate, variations are included. Variations may be designed to make the activity easier **1** or more difficult **3**, or change the focus of the activity completely. Where the variation affects the challenge level of the activity, the new challenge level is given.

Collins International Primary Maths units

There are 27 units in Collins International Primary Maths, each consisting of:

- Unit introduction
- Lesson plans
- Additional practice activities.

Unit introduction

The one-page introduction to each unit in the Teacher's Guide is designed to provide background information to help teachers plan, teach and assess that unit.

Unit overview

General description of the knowledge, understanding and skills taught in the unit.

Prerequisites for learning

A list of knowledge, understanding and skills that are prerequisites for learning in the unit. This list is particularly useful for diagnostic assessment.

Vocabulary

A summary is provided of key mathematical terms particularly relevant to the unit.

Common difficulties and remediation

Common errors and misconceptions, along with useful remediation hints are offered where appropriate.

Supporting language awareness

Key strategy or idea to help learners access the mathematics of the unit and overcome any barriers that the language of mathematics may present.

Learning objectives

The Cambridge Primary Mathematics Curriculum Framework (0096) learning objectives covered in the unit.

Collins International Primary Maths unit number and title

Collins International Primary Maths Recommended Teaching and Learning Sequence

Cambridge Primary Mathematics Curriculum Framework (0096) strand and sub-strand

Promoting Thinking and Working Mathematically (TWM)

Specific guidance on how the unit promotes the TWM characteristics.

Success criteria

Success criteria are provided to help both teachers and learners identify what learners are required to know, understand and do in order to achieve the unit's learning objective(s).

Unit introduction

Unit 1: Counting and sequences to 10

Collins International Primary Maths Recommended Teaching and Learning Sequence: Term 1, Week 2

Learning objectives

Code	Learning objective
1Nc.01	Count objects from 1 [0] to 10 [20], recognising conservation of number and one-to-one correspondence.
1Nc.02	Recognise the number of objects presented in familiar patterns up to 10, without counting.
1Nc.03	Estimate the number of objects or people (up to 10 [20]), and check by counting.
1Nc.04	Count on in ones[, twos or tens,] and count back in ones [and tens], starting from any number (from 1 [0] to 10 [20]).

Unit overview

In this unit, learners acquire the skills they need to estimate up to ten objects and then check to see if they are correct. They learn to count objects with one-to-one correspondence, recognising conservation of number, and to count on and back in ones, noting what happens when they add one more object to the set or take one away. They learn what numbers of objects up to 10 look like when they are arranged in familiar patterns, which helps them to visualise amounts when estimating, and discover that estimation is about guessing the number as closely as possible, based on what you already know.

Prerequisites for learning

Learners need to:
- be able to count to 10
- be able to recognise, read and write numbers to 10.

Vocabulary

number, count, ones, estimate, guess, count on, forwards, count back, backwards, pattern, more than, less than

Common difficulties and remediation

Learners may not understand that, when counting objects, each number said is a label for one object, and that each object must be counted only once. Practise organising objects so that they are easier to count accurately (into lines or pairs) and counting slowly and deliberately together as a class, using one number name for each object.

To help learners with the concept of estimation, show them a group of ten objects, a group of five objects and one object so that they have a basis for comparison. Ask questions such as: **Has this group got more or less than five in it?**

Spend time counting and re-counting sets of objects after moving the objects into new positions to build the understanding of conservation of numbers and that the number of objects is always the same if no objects are added or taken away from the set.

Supporting language awareness

Use every possible opportunity to count with learners, for example counting them as they line up to leave the classroom or counting out resources as you hand them out to learners.

Promoting Thinking and Working Mathematically

TWM.03 Conjecturing
Learners ask: **How many objects are in that set?** and question how they can discover this information by counting.

TWM.07 Critiquing
Learners use the skills needed to estimate and check. They choose from a variety of ways to estimate a number of objects.

TWM.08 Improving
Learners organise objects to make them easier to count or estimate.

Success criteria

Learners can:
- accurately count up to ten objects in a set
- recognise what comes next in a sequence of numbers appearing on familiar objects
- recognise numbers of objects presented in familiar patterns
- count on in ones and back in ones from any number from 1 to 10
- give a reasonable estimate for how many are in a set of up to ten objects.

Note

In order to conform to the terminology of the Cambridge Primary Mathematics Framework (0096), and also current common usage, the term 'less than' is used predominantly at this stage to compare and order numbers. However, it is important to note that, if being grammatically correct, the word 'less' should be used when quantities cannot be individually counted, for example: 'It should take less time', whereas 'fewer' should be used when referring to items that can be counted individually, for example: 'Fewer than ten people attended'.

For those teachers/schools following the CIPM Recommended Teaching and Learning Sequence, zero is not introduced until Term 1, Week 3 (Unit 16, LO1Np.01: Understand that zero represents none of something). Therefore, in this unit learners are working with numbers 1 to 10 and not zero. In subsequent units, however, learners are reciting, counting, reading and writing from zero.

Number – Counting and sequences

93

There are two different types of teaching and learning opportunities provided for each of the 27 units in Collins International Primary Maths:

• four lesson plans
• two Additional practice activities.

The lesson plans provide a clear, structured, step-by-step approach to teaching mathematics according to the learning objective(s) being covered throughout a unit. Each of the lessons has been written in a comprehensive way in order to give teachers maximum support for mixed-ability whole-class interactive teaching. It is intended, however, that the lessons will act as a model to be adapted to the particular needs of each class.

The Additional practice activities provide teachers with a bank of practical, hands-on activities that give learners opportunities for independent practice of the learning objective(s) being taught throughout a unit.

In most instances, the Additional practice activities are designed to be undertaken by pairs or small groups of learners as part of the 'Practise' phase of a Collins International Primary Maths lesson. Teachers choose which of the two Additional practice activities provided in the unit is most appropriate for the lesson's learning objective(s) and the needs of individual learners. Guidance as to which Additional practice activity consolidates a lesson's learning objective(s) is stated in each lesson plan.

The 'Practise' phase of a Collins International Primary Maths lesson also consists of written exercises found in the accompanying Workbook. Teachers need to decide how they wish to use these two different types of independent practice for individual learners or groups of learners. For example, depending on the lesson's learning objective(s), and the needs of individual learners, learners may:

- only complete the Additional practice activity

- start with the Additional practice activity and then move onto exercises in the Workbook

- start with exercises in the Workbook and then move onto the Additional practice activity

- only complete exercises in the Workbook.

It is important that the Additional practice activities are used at any time throughout a unit and therefore incorporated into each lesson as and when necessary in order to supplement, or provide an alternative to, the exercises in the Workbook. They should not be seen solely as providing teaching and learning content for the 'fifth' lesson of the week.

These two different types of teaching and learning opportunities form the weekly structure for each of the 27 units and are aimed at supporting flexibility so that the course can be tailored to meet the needs of individual classes.

Experience gained from other courses similar to Collins International Primary Maths shows that individual classes take different lengths of time to learn the content of a lesson. In light of this, rather than providing five lesson plans for each week, the decision was made to provide four core lessons which cover the unit's learning objective(s).

The intention is that as part of their weekly short-term planning, teachers make decisions as to how they will spread out the four core lessons, over the course of five days. Alternatively, as teachers progress through the week, and as part of their ongoing monitoring and evaluation, they may decide to alter their short-term planning and spend more time on teaching a particular lesson, or provide additional teaching and learning opportunities (including the Additional practice activities) in order to ensure that the class are developing a secure understanding of the unit's learning objective(s).

Lesson plan

Lesson number and title

Cambridge Primary Mathematics Curriculum Framework (0096) strand and sub-strand

Lesson objective(s)

The Cambridge Primary Mathematics Curriculum Framework (0096) learning objective(s) covered in the lesson.

Resources

To aid preparation, all the resources necessary to teach the lesson are listed. Each resource clearly states whether it is for the whole class, per group, per pair or per learner. Icons are displayed within the lesson plan to indicate any digital resources used in the lesson.

Collins International Primary Maths unit number and title

Reference to accompanying Student's Book page and Workbook page

Student's Book page 6

Workbook page 6

Unit **1** Counting and sequences to 10

Lesson 1: **Counting objects**

Learning objectives

Code	Learning objective
1Nc.01	Count objects from 1 [0] to 10 [20], recognising conservation of number and one-to-one correspondence.

Resources

ten interlocking cubes (per learner)

(Side tab: Number – Counting and sequences)

Revise

Use the activity *Counting people* from Unit 1: *Counting and sequences to 10* in the Revise activities.

Teach 📖 📊 **[TWM.08]**

• Direct learners to the picture in the Student's Book. Ask: **Which fish tank do you think contains more fish?** Take suggestions, then count the fish in each tank, with learners, and establish that both tanks contain the same number of fish. Explain to learners that sometimes the same number can look different if it is arranged differently.

• Show the **Tree tool** and put five birds in the tree. **[T&T]** Say: **When we use new or different skills to make a method more effective, we are** *improving*. Elicit that it would help if you lined up the birds so you don't accidentally count any of them twice or miss any out, and that you must use one number name for each bird and only say the next number name when you touch the next bird.

• Count the birds with learners, following all of their counting advice. Establish that there are five birds.

• Move the birds so that they are further apart and ask: **How many birds are there now?** Take suggestions and then count the birds together to demonstrate that the number of birds is the same. Explain that when objects are spread out, our brain tries to trick us because it thinks that there are more objects as the group looks bigger. Experiment with this, moving the birds closer together and further apart, counting them each time and getting the same result. Say: **We haven't put any more birds on the tree or taken any off the tree, so the**

• Set out three sets of objects (such as interlocking cubes) on each group's table: a set of three cubes lined up but spaced far apart, a set of eight cubes lined up and pushed close together and a set of six cubes in a random group (not lined up).

• Learners count the cubes in each set and write down how many are in the set.

• 🗣 Ask: **Which set of cubes was the easiest/ hardest to count? Why?**

Review

• Ask two learners to come to the front and stand behind a table. Give each learner seven cubes.

• Ask: **Who do you think has more cubes?** Take suggestions.

• Count the cubes in each set with learners and establish that there are seven cubes in each set.

• Move the cubes in one set so that they are widely spaced and push those in the other set closer together. 🗣 Ask: **Who do you think has more cubes now?** Praise any answers mentioning that they still have the same number of cubes as you didn't give them any more or take any away, or saying that the numbers are still the same but that they look different because you've moved the cubes.

• Count the cubes again, reinforcing that both learners still have the same number of cubes.

Assessment for learning

• How could you make counting easier?

• How many objects have you got?

• I have moved these objects – how many are

Revise

Recommended teaching time: 5 min

A bank of Revise activities can be found on pages 51–92. Revise activities are designed to consolidate fluency in number facts and/or provide an opportunity to revise any prerequisites for learning.

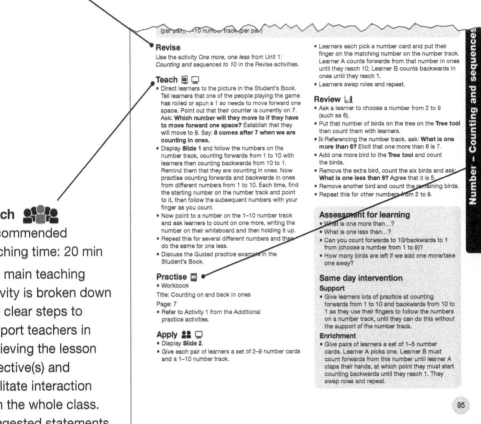

Teach

Recommended teaching time: 20 min

The main teaching activity is broken down into clear steps to support teachers in achieving the lesson objective(s) and facilitate interaction with the whole class. Suggested statements and questions are provided to support the teacher. During this phase of the lesson teachers also draw learners' attention to the 'Let's learn' and 'Guided practice' features in the Student's Book.

Other features of Teach include:

T&T: Turn and talk – Using *Talk Partners* helps to create a positive learning environment. Many learners feel more confident discussing with a partner before giving an answer to the whole class, and learners get opportunities to work with different students.

ᄆ: Progress check questions – These questions are designed to obtain an overview of learners' prior experiences before introducing a new concept or topic, to provide immediate feedback to learners and to gauge learner progress in order to adapt teaching.

TWM: Teaching and learning opportunities aimed at learners developing specific Thinking and Working Mathematically characteristics.

NOTE: Timings are approximate recommendations only.

Practise and/or and/or

Recommended teaching time: 20 min

Teach is followed by independent practice and consolidation, which provides an opportunity for all learners to focus on their newly acquired knowledge. Practice and consolidation consists of both written exercises and practical hands-on activities, with reference to the relevant Workbook page and bank of Additional practice activities.

All of the tasks are differentiated into three different ability levels:

1 question/activity number typeset on a circle indicates that the question/activity is suitable for learners who require additional support with either easier questions/activities or revising pre-requisite knowledge

2 question/activity number typeset on a triangle indicates that the question/activity is suitable for the majority of learners to practise and consolidate the lesson's learning objective(s)

3 question/activity number typeset on a square indicates that the question/activity is suitable for learners who require enrichment and/or extension.

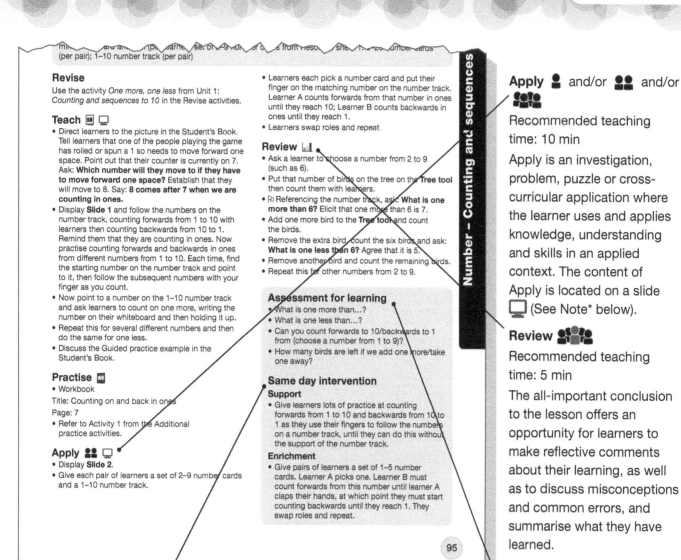

Same day intervention

Support and/or Enrichment

Offers same day intervention suggestions so that teachers can effectively provide either support or Enrichment where appropriate.

The aim is to provide guidance to teachers to ensure that:

• all learners reach a certain level of understanding by the end of the day, preventing an achievement gap from forming

• the needs of all learners are being met with respect to the lesson's learning objective(s).

Assessment for learning

Specific questions designed to assist teachers in checking learners' understanding of the lesson objective(s). These questions can be used at any time throughout the lesson.

NOTE*:

As learners will inevitably complete the 'Practise' activities at different times, it is recommended that teachers introduce the 'Apply' activity to the whole class at the end of 'Teach' and before learners work independently on the 'Practise' activities (Workbook page and/or Additional practice activities).

There should also be no expectation that *every* learner in *every* lesson should complete the 'Apply' activity. However, opportunities should be given, either during or outside the Maths lesson, for learners to work on 'Apply'. This could include (but not always) part of the teaching and learning content for the 'fifth' lesson of the week.

Finally, it is also recommended that as part of the whole class 'Review' phase of the lesson, learners are given the opportunity to share with the rest of the class the work they carried out as part of the 'Apply' activity. This includes not just providing the solution, answer or result (if there is one), but also the different methods and strategies they used, and what they learned from the activity.

Additional practice activities

As well as four lesson plans, each unit in Collins International Primary Maths provides two Additional practice activities.

Strand and Sub-strand

The relevant Cambridge Primary Mathematics Curriculum Framework (0096) strand and sub-strand covered is stated in the sidebar.

Collins International Primary Maths unit number and title

Learning objective(s)

Each activity has clearly defined learning objective(s) to assist teachers in choosing the most appropriate activity for the concept they want the learners to practise and consolidate.

Resources

To aid preparation, any resources required are listed, along with whether they are per group, per pair or per learner. Most of these resources are readily available in classrooms

What to do

The activity is broken down into clear steps to support teachers in explaining the activity to the learners.

Unit **1** Counting and sequences to 10

Additional practice activities
Activity 1

Learning objectives
• Count up to ten objects, recognising conservation of number
• Find the numbers one more and one less than a given number to 10
• Estimate whether there are more, less or the same amount of objects

Resources
10 countable objects (e.g. building blocks); mini whiteboard and pen (per group) teddy bear or other soft toy (per group); 1 to 10 number track (per group)

What to do
• Ask a learner to count out six blocks. Check that there are six blocks by counting them as a group and write the number 6 on the mini whiteboard.
• Tell learners that Teddy likes to move the blocks to trick us. Sometimes he moves the objects further apart or closer together and sometimes he adds an extra object or takes one away.
• Ask: **If Teddy moves the blocks but doesn't add any or take any away, will the amount of blocks change?**
• Learners close their eyes while Teddy moves the

• When learners open their eyes, ask them to guess (estimate) whether there are more or fewer objects or whether the number has stayed the same. Then count the objects together.
• If there is a different number of blocks now, check whether there is one more or one less by finding 6 on a number track and looking at the number before and after it.
• Repeat this with different starting numbers.

Variations
1 Ask learners to count out five or fewer blocks to allow them to secure these skills to five before they attempt to apply them to up to ten objects.
2 Choose which part of this lesson you want most to focus on as the unit progresses: estimation, counting and number conservation or counting

Number – Counting and sequ

Classroom organisation

Icons are used to indicate whether the activity is designed to be used by learners working in groups, pairs or individually.

Challenge level

The challenge level for each activity is given:

1 suitable for learners who require additional support

2 suitable for the majority of learners to practise and consolidate the lesson's learning objective(s)

3 suitable for learners who require enrichment and/ or extension.

Variations

Where appropriate, variations are included. Variations may be designed to make the activity easier **1** or more difficult , or change the focus of the activity completely. Where the variation affects the challenge level of the activity, the new challenge level is given.

Resource sheets

Where specific paper-based resources are needed for individual lesson plans or Additional practice activities, these are provided as Resource sheets in the Teacher's Guide. Use of Resource sheets is indicated in the resources list of the relevant lesson plan or Additional practice activity.

Some Resource sheets use a spinner to generate numbers or other forms of data.

To use a spinner, hold a paper clip in the centre of the spinner using a pencil and gently flick the paper clip with your finger to make it spin.

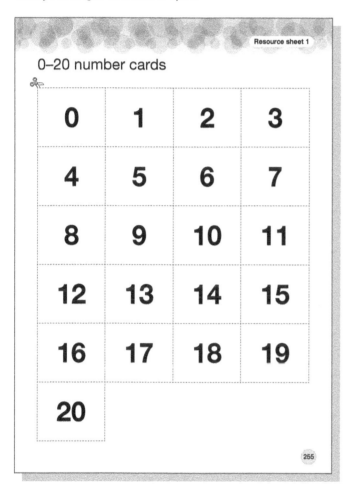

Answers

Answers are provided for all the Workbook pages.

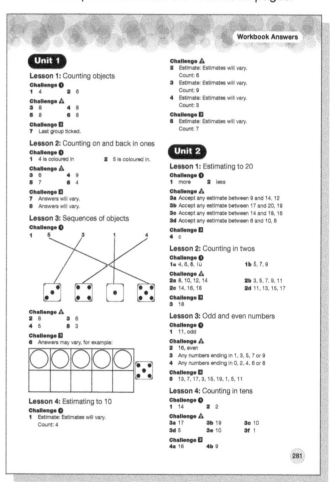

Class record-keeping document

In order to assist teachers with making both short-term 'ongoing' and medium-term 'formative' assessments manageable, meaningful and useful, Collins International Primary Maths includes a class record-keeping document on pages 299–304.

It is intended to be a working document that teachers start at the beginning of the academic year and continually update and amend throughout the course of the year.

Teachers use their own professional judgement of each learner's level of mastery in each of the sub-strands, taking into account:

- mastery of the learning objectives associated with each particular sub-strand
- performance in whole-class discussions
- participation in group work
- work presented in the Workbook
- any other evidence.

Once a decision has been made regarding the degree of mastery achieved by a learner in the particular sub-strand, teachers then write the learner's name (or initials) in the appropriate column:

A: Exceeding expectations in this sub-strand

B: Meeting expectations in this sub-strand

C: Below expectations in this sub-strand.

Given that this is a working document intended to be used throughout the entire academic year, teachers may decide to write (T1), (T2) or (T3) after the learner's name to indicate in which term/semester the judgement was made. This will also help to show the progress (or regress) that learners make during the course of a year.

Schools and/or individual teachers may decide to use a photocopy of the document at the back of this Teacher's Guide or printout the Word version from the Digital download (either enlarged to A3 if deemed appropriate). It is recommended that whichever option is taken, throughout the year teachers use pencil to fill out the document, and then at the end of the academic year complete the document in pen.

As an alternative to using a photocopy or printout, schools and/or teachers may decide to complete the document electronically using the Digital download version.

Cambridge Primary Mathematics Curriculum Framework (0096) strand and sub-strand

Cambridge Primary Mathematics Curriculum Framework (0096) learning objectives (and codes)

Overall level of mastery in the sub-strand

The degree of mastery achieved by a learner in each sub-strand is shown by writing the learner's name (or initials) in the appropriate column:

A: Exceeding expectations in this sub-strand

B: Meeting expectations in this sub-strand

C: Below expectations in this sub-strand.

An additional bonus of enlarging the document to A3 or completing it electronically, is that it will provide additional space in each of the three columns linked to each sub-strand for teachers to provide more qualitative data should they wish to do so. Teachers can write specific comments that they feel are appropriate for individual learners related not only to the entire sub-strand, but also for specific learning objectives within the sub-strand.

Finally, the class record-keeping document should be seen as an extremely useful document as it can be used to:

- identify those learners who are working *above* and *below* expectations, thereby helping teachers to better plan for the needs of individual learners
- report to parents and guardians
- inform the next year's teacher about which sub-strands of the Cambridge Primary Mathematics Curriculum Framework (0096) individual learners, and the class as a whole, are *exceeding* (A), *meeting* (B) or are *below* (C) in expectations
- assist senior managers within the school in determining whether individual learners, and the class as a whole, are on track to meet end-of-stage expectations.

Year group   **Class and academic year reference**

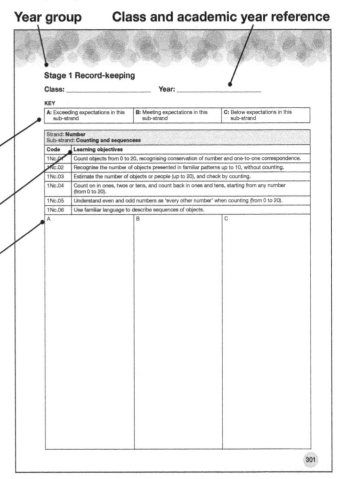

Student's Book

There is one Student's Book for each stage in Collins International Primary Maths with one page provided for each lesson plan.

The content provided in the Student's Book is designed to be used during the 'Teach' phase of a typical Collins International Primary Maths lesson.

However, it is recommended that during the 'Practise' phase of a lesson, if appropriate, learners also use the page in the Student's Book to help them answer the questions on the accompanying Workbook page.

On page 5 of the Student's Book is a guidance page referred to as 'How to use this book', which explains to learners the features of the book.

The back of the Student's Book includes the TWM Star which defines, in learner-friendly language, the eight Thinking and Working Mathematically characteristics.

Reference to accompanying Workbook page

Collins International Primary Maths unit number and title

Cambridge Primary Mathematics Curriculum Framework (0096) strand

Lesson number and title

Lesson objective(s)

The Cambridge Primary Mathematics Curriculum Framework (0096) learning objective(s) covered in the lesson, written in language appropriate for learners.

Key words

A list of key mathematical terms particularly relevant to the lesson.

Let's learn

Content that presents the key mathematical idea of the lesson being taught by the teacher in the 'Teach' phase.

Guided practice

Worked example(s) designed to prepare learners to work independently on the questions in the Workbook.

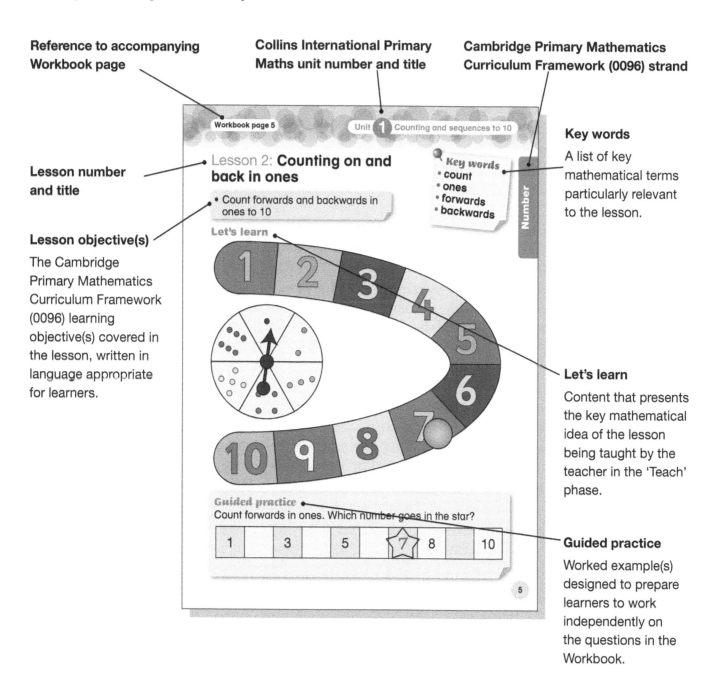

Workbook page 5

Unit 1 Counting and sequences to 10

Lesson 2: **Counting on and back in ones**

- Count forwards and backwards in ones to 10

Let's learn

Key words
• count
• ones
• forwards
• backwards

Number

Guided practice
Count forwards in ones. Which number goes in the star?

| 1 | | 3 | | 5 | | 7 | 8 | | 10 |

5

Workbook

All Workbook page exercises reinforce and build upon the main teaching points and learning objective(s) of a particular lesson in the Teacher's Guide. The work is intended to allow all learners in the class to practise and consolidate their newly acquired knowledge, understanding and skills.

The content provided in the Workbook is designed to be used during the 'Practise' phase of a typical Collins International Primary Maths lesson.

On pages 4 and 5 of the Workbook is a guidance page referred to as 'How to use this book', which explains to learners the features of the book.

In Stage 1, one Workbook page is provided for each lesson plan. There is no Workbook page for the Additional practice activities.

Each Workbook page has three levels of challenge designed to cater, not only for the different abilities that occur in a mixed-ability or mixed-aged class, but also to assist those schools who 'set' or 'stream' their learners into ability groups. The three different levels of challenge are identified as follows:

 Question number typeset on a circle indicates that the question is suitable for learners who require additional support with either easier questions or revising pre-requisite knowledge.

 Question number typeset on a triangle indicates that the question is suitable for the majority of learners to practise and consolidate the lesson's learning objective(s).

 Question number typeset on a square indicates that the question is suitable for learners who require enrichment and/or extension.

Teachers should think carefully about which of the three different levels of challenge individual learners are asked to complete. There should be no expectation that *every* learner must always answer the questions in *all* three challenge levels.

It is therefore good practice to look carefully at the questions in the Workbook and assign specific questions to specific groups of learners. An effective way of doing this is to tell learners which questions you would like them to answer, and for them to circle those question numbers on their Workbook page.

When appropriate, learners should also be encouraged to work in pairs to answer some or all of the questions. Not only does this help learners learn from each other, thereby reinforcing learners' knowledge, understanding and skills, but it also encourages discussion and mathematical talk, helps create positive self-esteem, and removes the frustrations and feelings of intellectual isolation which can so often be associated with learners working alone.

Teachers may also on occasion decide to work with a group of learners to complete some or all of the questions in the Workbook.

During the 'Review' phase of the lesson, when the whole class is back working together, teachers may decide to complete and/or mark some or all of the questions – perhaps using a different-coloured pencil to differentiate those questions that learners answered independently from those they answered with some assistance.

Finally, it is important to be aware that the Workbook is *not* designed for assessment purposes, nor as a record of what learners can or can't do, nor as proof of what has been taught / learned. Its purpose is for learners to practise and consolidate the mathematical ideas that have been taught during the lesson.

Cambridge Primary Mathematics Curriculum Framework (0096) strand

Collins International Primary Maths unit number and title

Reference to accompanying Student's Book page

Lesson number and title

Lesson objective(s)

The Cambridge Primary Mathematics Curriculum Framework (0096) learning objective(s) covered in the lesson, written in language appropriate for learners.

You will need:

Lists any resources learners will need when using the Workbook.

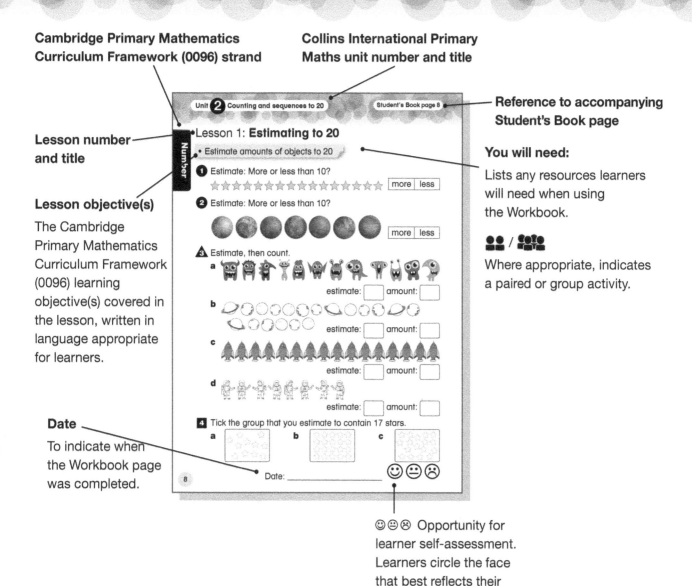

Where appropriate, indicates a paired or group activity.

Date

To indicate when the Workbook page was completed.

☺☺☹ Opportunity for learner self-assessment. Learners circle the face that best reflects their level of understanding of the lesson's learning objectives.

Digital content

Collins International Primary Maths also includes a comprehensive set of digital tools and resources, designed to support teachers and learners. The digital content is organised into three sections: Teach, Interact and Support, and is available as the Extended Teacher's Guide.

Teach

The Teach section contains all of the teaching content from the Teacher's Guide, organised into units. This includes:

- Unit introductions
- Lesson plans
- Additional practice activities
- Resource sheets
- Answers

In addition to the above, the Teach section also contains:

- Weekly planning grids

  Editable short-term planning grids provide a synopsis of the teaching and learning opportunities in each of the 27 weekly units. Each Weekly planning grid highlights the content of the five phases for each of the four lessons in the unit (the 5th lesson of the week is left empty for teachers to complete) as well as providing background information, assessment opportunities and teacher and learner evaluation. The intention is for teachers to adapt the grid in order to create a bespoke weekly planning overview for the specific needs of their class.

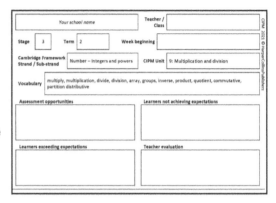

- Slideshows

  Slideshows are provided as visual aids to be shown to the whole class at various phases of a lesson (Revise, Teach, Apply and/or Review), as directed in the lesson plan.

- Interactive whiteboard mathematical tools

  Flexible interactive whiteboard (IWB) teaching tools provide additional visual representations to display to the whole class at various phases of a lesson (Revise, Teach, and/or Review), as directed in the lesson plan.

  These 41 highly adaptable teaching tools are particularly useful in generating specific examples and questions. By doing this, it enables teachers to individualise the content displayed to the class, thereby creating teaching and learning opportunities that better meet the needs of their class.

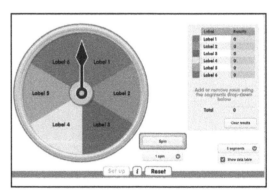

A brief description of the functionality of each of the 40 interactive whiteboard mathematical tools is provided below.

Strand	Tool name	Description of functionality
Number	Counting tool	Use the counting tool to assist with counting from 1–20.
	Place value	Explore how multiple-digit numbers are made up of millions, hundreds of thousands, tens of thousands, thousands, hundreds, tens, ones, tenths, hundredths and thousandths.
	Base 10	Demonstrate the relationship between ones, tens, hundreds and thousands.
	Place value counters	Similar in functionality to the Base ten tool above, i.e. demonstrate the relationship between ones, tens, hundreds, thousands (and millions and decimals – tenths, hundredths and thousandths). However this tool uses place value counters and not Dienes, as is used in the Base ten tool.
	Number line	Use the number line to assist with counting, calculations and exploring decimals.
	Fractions	Demonstrate fractions visually and display the accompanying numerical fraction, decimal, ratio or percentage alongside the pictorial fraction.
	Fraction wall	Demonstrate fractions visually with the fully customisable fraction wall.
	Snake fraction tool	Demonstrate fractions visually with the fully customisable fraction snake.
	Spinner	Demonstrate the concept of probability and making a calculated estimate.
	Bead sticks	Assist with counting, and show the relationship between thousands, hundreds, tens, ones, tenths, hundredths and thousandths.
	Number cards	Display sets of numbers and calculations on movable number cards.
	Number square	Demonstrate and explore counting and number patterns.
	Multiplication square	Demonstrate and explore multiplication and number patterns.
	Function machine	Demonstrate one- and two-step calculations with this animated tool.
	Tree tool	Use real-life objects to practise addition and subtraction.
	Dice tool	Demonstrate the concept of probability or use the tool in conjunction with activities which require dice.
	Ten frame	Use a ten frame template to demonstrate calculations.
Geometry and Measure	Co-ordinates	Create and interpret labelled co-ordinate grids, in one, two or four quadrants.
	Geoboard	Join the dots to depict shapes, routes between two points, to draw nets or to make patterns.
	Geometry set	Use the ruler, protractor or set square to measure and draw lines and angles.
	Rotate and reflect	Demonstrate the rotation and reflection across a horizontal, vertical and diagonal mirror line of a range of 2D shapes.
	Symmetry	Demonstrate lines of horizontal, vertical and diagonal symmetry on a range of 2D shapes.
	Pattern tool	Create patterns and sequences of shapes and design a jumper.

Strand	Tool name	Description of functionality
Geometry and Measure	Beads and laces	Create repeating patterns with the beads and laces tool.
	Shape set	Use the shape set to compare 2D and 3D shapes.
	Nets	Explore the nets of 3D shapes such as prisms, pyramids, tetrahedrons, cubes, and cuboids.
	Money	Practise counting money, or solving calculations involving money.
	Clock	Demonstrate the features of analogue and digital clocks, and explore time.
	Thermometer	Demonstrate how to measure and record temperature, using either Fahrenheit and Celsius scales.
	Capacity	Demonstrate how to measure capacity with this range of animated water containers.
	Weighing	Demonstrate mass by weighing objects of different masses on a range of animated weighing scales.
Statistics and Probability	Bar charter	Demonstrate the creation and interpretation of bar graphs.
	Pie charter	Demonstrate how to create and interpret pie charts with this dynamic data tool.
	Pictogram	Demonstrate the creation and interpretation of pictograms.
	Line grapher	Demonstrate the plotting and interpretation of line graphs with this dynamic data tool.
	Carroll diagram	Demonstrate how to classify and group data with this sorting tool.
	Venn diagram	Demonstrate how to classify and group data with this sorting tool. Import images to give a real-life context.
	Waffle diagram	Demonstrate the creation and interpretation of waffle diagrams.
	Frequency diagram	Demonstrate the creation and interpretation of frequency diagrams for continuous data.
	Scatter graph	Demonstrate the creation and interpretation of scatter graphs.
	Dot plot	Demonstrate the creation and interpretation of dot plot diagrams.

Within Teach, the planning tool allows schools and individual teachers to customise the sequence of units in Collins International Primary Maths within and across all stages. This allows schools and individual teachers to develop their own unique scheme of work.

Interact

The Interact section contains 16 interactive mathematical games. The audio glossary of terms for all stages is also located here.

Support

The Support section contains useful documents for the teacher, such as the medium-term plan, Record-keeping documents described on pages 29 and 30 and the Collins International Primary Maths Training Package.

Ebook

Ebooks are available for all of the components: Teacher's Guide, Student's Book and Workbook. These enable greater teacher-learner interaction during the whole-class 'Teach' phase of the lesson and also assist teachers in explaining activities and questions to learners as well as in discussing results, solutions and answers once learners have completed an activity or set of questions.

The ebooks can be used in a reader view on computer screens and are also designed to be used with interactive whiteboards (IWB) and if available, iPads and tablets.

Each ebook has standard functionality such as scrolling, zooming, an interactive Contents page and the ability to make notes and highlight sections digitally.

⑦ Collins International Primary Maths Stage 1 Units and Recommended Teaching and Learning Sequence

The Stage 1 learning objectives from the Cambridge Primary Mathematics Curriculum Framework (0096) have been grouped into the following 27 topic areas or 'units'.

The Thinking and Working Mathematically characteristics are developed throughout each unit.

Cambridge Primary Mathematics Curriculum Framework (0096)		Collins International Primary Maths	
Strand	Sub-strand	Unit number	Topic
Number	Counting and sequences	1	Counting and sequences to 10
		2	Counting and sequences to 20
	Integers and powers	3	Reading and writing numbers to 10
		4	Reading and writing numbers to 20
		5	Addition as combining two sets
		6	Addition as counting on
		7	Subtraction as take away
		8	Subtraction as counting back
		9	Subtraction as difference
		10	Addition and subtraction to 10 (A)
		11	Addition and subtraction to 10 (B)
		12	Addition and subtraction to 20 (A)
		13	Addition and subtraction to 20 (B)
		14	Doubling
	Money	15	Money
	Place value, ordering and rounding	16	Place value and ordering to 10
		17	Place value and ordering to 20
	Fractions, decimals, percentages, ratio and proportion	18	Half (A)
		19	Half (B)
Geometry and Measure	Time	20	Time
	Geometrical reasoning, shapes and measurements	21	2D shapes
		22	3D shapes
		23	Length and mass
		24	Capacity and temperature
	Position and transformation	25	Position and direction
Statistics and Probability	Statistics	26	Statistics (A)
		27	Statistics (B)

The Cambridge Primary Mathematics Scheme of Work offers an approach to organising the learning objectives of the Stage 1 curriculum. An overview of this can be seen below.

Unit 1.1   Numbers to 20
Unit 1.2   Time
Unit 1.3   Shapes, direction and movement
Unit 1.4   Addition, subtraction, doubles
Unit 1.5   Money
Unit 1.6   Measurement
Unit 1.7   Fractions
Unit 1.8   Statistical methods

The table below shows how the 27 units of Collins International Primary Maths Stage 1 link to the Cambridge Primary Mathematics Stage 1 Scheme of Work units. Please note that while the units in the Cambridge Primary Mathematics Scheme of Work may differ from that of Collins International Primary Maths, guidance from Cambridge International states that there is no requirement for endorsed resources to follow the teaching order suggested in the Cambridge scheme of work. If a resource is endorsed, schools can be confident that all the learning objectives are covered.

Cambridge Primary Mathematics Stage 1 Scheme of Work units		Collins International Primary Maths Stage 1 units	
Unit number	Topic	Unit number	Topic
1.1	Numbers to 20	1	Counting and sequences to 10
		2	Counting and sequences to 20
		3	Reading and writing numbers to 10
		4	Reading and writing numbers to 20
		16	Place value and ordering to 10
		17	Place value and ordering to 20
1.2	Time	20	Time
1.3	Shapes, direction and movement	21	2D shapes
		22	3D shapes
		25	Position and direction
1.4	Addition, subtraction, doubles	5	Addition as combining two sets
		6	Addition as counting on
		7	Subtraction as take away
		8	Subtraction as counting back
		9	Subtraction as difference
		10	Addition and subtraction to 10 (A)
		11	Addition and subtraction to 10 (B)
		12	Addition and subtraction to 20 (A)
		13	Addition and subtraction to 20 (B)
		14	Doubling
1.5	Money	15	Money
1.6	Measurement	23	Length and mass
		24	Capacity and temperature
1.7	Fractions	18	Half (A)
		19	Half (B)
1.8	Statistical methods	26	Statistics (A)
		27	Statistics (B)

STRAND: | Number | Geometry and Measure | Statistics and Probability |

The table on page 40 shows a recommended teaching and learning sequence (often referred to as a 'medium-term plan') for the 27 units in Collins International Primary Maths Stage 1.

However, as with the Cambridge Primary Mathematics Scheme of Work, schools and individual teachers are free to teach the learning objectives in any order to best meet the needs of individual schools, teachers and learners.

It is important to note that in order to allow for greater flexibility, the 27 units in each stage in Collins International Primary Maths are **not** ordered according to the recommended teaching sequence. Instead, they are in numerical order: Units 1 to 27, according to how the Strands and Sub-strands are arranged in the Cambridge Primary Mathematics Curriculum Framework (0096).

In other words, progression through the components in Collins International Primary Maths does not start at the beginning of the Teacher's Guide, Student's Book and Workbook and end at the back of the guide/book. Rather, units are covered as and when is appropriate, according to the Recommended Teaching and Learning Sequence provided by Collins International Primary Maths, or your school's specific scheme of work.

As a note of caution, the Collins International Primary Maths Recommended Teaching and Learning Sequence has been carefully written to ensure continuity and progression both *within* the units at a particular stage and also *across* Stages 1 to 6 and onwards into Lower Secondary.

This is extremely important in ensuring that learners have the pre-requisite knowledge, understanding and skills they require in order to successfully engage with new mathematical ideas at a deeper level and in different contexts.

Learners need to develop mastery of the learning objectives of the Strands and Sub-strands *within* a stage before they are able to apply and transfer their newly acquired knowledge, understanding and skills *across* other Strands and Sub-strands and into later stages. It is for this reason that in the Collins International Primary Maths Recommended Teaching and Learning Sequence, terms/semesters begin with units from the Number Strand, so that learners develop knowledge and skills in number that they can then apply to the other strands of mathematics (Geometry and Measure, and Statistics and Probability) and to their own lives.

Therefore, schools need to think extremely carefully when altering the Collins International Primary Maths Recommended Teaching and Learning Sequence in order to ensure that new learning builds on learners' prior knowledge, understanding and skills. In order to be confident with making such amendments, it is important that teachers are extremely familiar with the lines of progression both *within* and *across* stages of the Cambridge Primary and Lower Secondary Mathematics Curriculum Frameworks.

As with the Cambridge Primary Mathematics Scheme of Work, Collins International Primary Maths has assumed an academic year of three terms/semesters, each of 10 weeks duration. This is the minimum length of a school year and thereby allows flexibility for schools to add in more teaching time as necessary to meet the needs of the learners, and also to comfortably cover the content of the curriculum into an individual schools specific term/semester times.

Collins International Primary Maths Stage 1 Recommended Teaching and Learning Sequence

	Term 1	Term 2	Term 3
Week 1	Unit 3: Reading and writing numbers to 10	Unit 7: Subtraction as take away	Unit 14: Doubling
Week 2	Unit 1: Counting and sequences to 10	Unit 8: Subtraction as counting back	Unit 12: Addition and subtraction to 20 (A)
Week 3	Unit 16: Place value and ordering to 10	Unit 9: Subtraction as difference	Unit 13: Addition and subtraction to 20 (B)
Week 4	Unit 5: Addition as combining two sets	Unit 10: Addition and subtraction to 10 (A)	Unit 18: Half (A)
Week 5	Unit 6: Addition as counting on	Unit 11: Addition and subtraction to 10 (B)	Unit 19: Half (B)
Week 6	Unit 20: Time	Unit 4: Reading and writing numbers to 20	Unit 15: Money
Week 7	Unit 21: 2D shapes	Unit 2: Counting and sequences to 20	Unit 27: Statistics (B)
Week 8	Unit 22: 3D shapes	Unit 17: Place value and ordering to 20	Unit 23: Length and mass
Week 9	Unit 25: Position and direction	Unit 26: Statistics (A)	Unit 24: Capacity and temperature
Week 10	Revision	Revision	Revision

STRAND:	Number	Geometry and Measure	Statistics and Probability

No material is provided in Collins International Primary Maths for the three Revision weeks each term/semester. Individual teachers will decide the content to cover during these weeks, based on monitoring and evaluation made over the course of the term/semester, and learners' levels of achievement on the topics covered throughout the term/semester.

Teachers may decide to revisit certain topics and provide further practice of various concepts that have been taught during the term/semester, or they may use this week to catch up if lessons or units have taken longer than expected.

There is also no expectation that a Revision week will only take place at the end of each term/semester. If individual teachers feel that better use can be made of this week at another time throughout the term/semester, then they should feel free to do so.

⑧ Collins International Primary Maths Stage 1 units match to Cambridge Primary Mathematics Curriculum Framework (0096) Stage 1

The recommended teaching time for each unit is 1 week.

The Thinking and Working Mathematically characteristics are developed throughout each unit. Square brackets within objectives indicate parts of the objective that are not covered in this unit, but are covered elsewhere.

These learning objectives are reproduced from the Cambridge Primary Mathematics curriculum framework (0096) from 2020. This Cambridge International copyright material is reproduced under licence and remains the intellectual property of Cambridge Assessment International Education.

Unit 1 – Counting and sequences to 10			
Cambridge Primary Mathematics Curriculum Framework (0096)			
Strand	**Sub-strand**	**Code**	**Learning objectives**
Number	Counting and sequences	1Nc.01	Count objects from 1 [0] to 10 [20], recognising conservation of number and one-to-one correspondence.
		1Nc.02	Recognise the number of objects presented in familiar patterns up to 10, without counting.
		1Nc.03	Estimate the number of objects or people (up to 10 [20]), and check by counting.
		1Nc.04	Count on in ones[, twos or tens,] and count back in ones [and tens], starting from any number (from 1 [0] to 10 [20]).
		Collins International Primary Maths Recommended Teaching and Learning Sequence:	**Term 1 Week 2**

Unit 2 – Counting and sequences to 20			
Cambridge Primary Mathematics Curriculum Framework (0096)			
Strand	**Sub-strand**	**Code**	**Learning objectives**
Number	Counting and sequences	1Nc.01	Count objects from 0 to 20, recognising conservation of number and one-to-one correspondence.
		1Nc.02	Recognise the number of objects presented in familiar patterns up to 10, without counting.
		1Nc.03	Estimate the number of objects or people (up to 20), and check by counting.
		1Nc.04	Count on in ones, twos or tens, and count back in ones and tens, starting from any number (from 0 to 20).
		1Nc.05	Understand even and odd numbers as 'every other number' when counting (from 0 to 20).
		1Nc.06	Use familiar language to describe sequences of objects.
		Collins International Primary Maths Recommended Teaching and Learning Sequence:	**Term 2 Week 7**

Unit 3 – Reading and writing numbers to 10			
Cambridge Primary Mathematics Curriculum Framework (0096)			
Strand	**Sub-strand**	**Code**	**Learning objectives**
Number	Integers and powers	1Ni.01	Recite, read and write number names and whole numbers (from 1 [0] to 10 [20]).
	Counting and sequences	1Nc.01	Count objects from 1 [0] to 10 [20], recognising [conservation of number and] one-to-one correspondence.
		Collins International Primary Maths Recommended Teaching and Learning Sequence:	**Term 1 Week 1**

Unit 4 – Reading and writing numbers to 20			
Cambridge Primary Mathematics Curriculum Framework (0096)			
Strand	**Sub-strand**	**Code**	**Learning objectives**
Number	Integers and powers	1Ni.01	Recite, read and write number names and whole numbers (from 0 to 20).
	Counting and sequences	1Nc.01	Count objects from 0 to 20, recognising conservation of number and one-to-one correspondence.
		Collins International Primary Maths Recommended Teaching and Learning Sequence:	**Term 2 Week 6**

Unit 5 – Addition as combining two sets			
Cambridge Primary Mathematics Curriculum Framework (0096)			
Strand	**Sub-strand**	**Code**	**Learning objective**
Number	Integers and powers	1Ni.02	Understand addition as: [– counting on] – combining two sets.
		Collins International Primary Maths Recommended Teaching and Learning Sequence:	**Term 1 Week 4**

Unit 6 – Addition as counting on			
Cambridge Primary Mathematics Curriculum Framework (0096)			
Strand	**Sub-strand**	**Code**	**Learning objectives**
Number	Integers and powers	1Ni.02	Understand addition as: – counting on [– combining two sets.]
	Counting and sequences	1Nc.04	Count on in ones [, twos or tens, and count back in ones and tens,] starting from any number (from 0 to 10 [20]).
		Collins International Primary Maths Recommended Teaching and Learning Sequence:	**Term 1 Week 5**

Unit 7 – Subtraction as take away			
Cambridge Primary Mathematics Curriculum Framework (0096)			
Strand	**Sub-strand**	**Code**	**Learning objective**
Number	Integers and powers	1Ni.03	Understand subtraction as: [– counting back] – take away [– difference.]
		Collins International Primary Maths **Recommended Teaching and Learning Sequence:**		**Term 2** **Week 1**

Unit 8 – Subtraction as counting back			
Cambridge Primary Mathematics Curriculum Framework (0096)			
Strand	**Sub-strand**	**Code**	**Learning objectives**
Number	Integers and powers	1Ni.03	Understand subtraction as: – counting back [– take away – difference.]
	Counting and sequences	1Nc.04	Count [on in ones, twos or tens, and count] back in ones [and tens,] starting from any number (from 0 to 10 [20]).
		Collins International Primary Maths **Recommended Teaching and Learning Sequence:**		**Term 2** **Week 2**

Unit 9 – Subtraction as difference			
Cambridge Primary Mathematics Curriculum Framework (0096)			
Strand	**Sub-strand**	**Code**	**Learning objective**
Number	Integers and powers	1Ni.03	Understand subtraction as: [– counting back – take away] – difference.
		Collins International Primary Maths **Recommended Teaching and Learning Sequence:**		**Term 2** **Week 3**

Unit 10 – Addition and subtraction to 10 (A)			
Cambridge Primary Mathematics Curriculum Framework (0096)			
Strand	**Sub-strand**	**Code**	**Learning objectives**
Number	Integers and powers	1Ni.04	Recognise complements of 10.
		1Ni.05	[Estimate,] add and subtract whole numbers (where the answer is from 0 to 10 [20]).
	Counting and sequences	1Nc.04	Count on in ones, [two or tens,] and count back in ones [and tens], starting from any number (from 0 to 10 [20]).
		Collins International Primary Maths **Recommended Teaching and Learning Sequence:**		**Term 2** **Week 4**

Unit 11 – Addition and subtraction to 10 (B)			
Cambridge Primary Mathematics Curriculum Framework (0096)			
Strand	Sub-strand	Code	Learning objectives
Number	Integers and powers	1Ni.05	Estimate, add and subtract whole numbers (where the answer is from 0 to 10 [20]).
	Counting and sequences	1Nc.04	Count on in ones, [two or tens,] and count back in ones [and tens], starting from any number (from 0 to 10 [20]).
		Collins International Primary Maths Recommended Teaching and Learning Sequence:	**Term 2** **Week 5**

Unit 12 – Addition and subtraction to 20 (A)			
Cambridge Primary Mathematics Curriculum Framework (0096)			
Strand	Sub-strand	Code	Learning objectives
Number	Integers and powers	1Ni.04	Recognise complements of 10.
		1Ni.05	Estimate, add [and subtract] whole numbers (where the answer is from 0 to 20).
		1Ni.06	Know doubles up to double 10.
	Counting and sequences	1Nc.04	Count on in ones, [twos or tens, and count back in ones and tens,] starting from any number (from 0 to 20).
		Collins International Primary Maths Recommended Teaching and Learning Sequence:	**Term 3** **Week 2**

Unit 13 – Addition and subtraction to 20 (B)			
Cambridge Primary Mathematics Curriculum Framework (0096)			
Strand	Sub-strand	Code	Learning objectives
Number	Integers and powers	1Ni.05	Estimate, add and subtract whole numbers (where the answer is from 0 to 20).
	Counting and sequences	1Nc.04	Count on in ones, twos, or tens, and count back in ones and tens, starting from any number (from 0 to 20).
		Collins International Primary Maths Recommended Teaching and Learning Sequence:	**Term 3** **Week 3**

Unit 14 – Doubling			
Cambridge Primary Mathematics Curriculum Framework (0096)			
Strand	Sub-strand	Code	Learning objective
Number	Integers and powers	1Ni.06	Know doubles up to double 10.
		Collins International Primary Maths Recommended Teaching and Learning Sequence:	**Term 3** **Week 1**

Unit 15 – Money			
Cambridge Primary Mathematics Curriculum Framework (0096)			
Strand	Sub-strand	Code	Learning objective
Number	Money	1Nm.01	Recognise money used in local currency.
		Collins International Primary Maths Recommended Teaching and Learning Sequence:	**Term 3** **Week 6**

Unit 16 – Place value and ordering to 10			
Cambridge Primary Mathematics Curriculum Framework (0096)			
Strand	**Sub-strand**	**Code**	**Learning objectives**
Number	Place value, ordering and rounding	1Np.01	Understand that zero represents none of something.
		1Np.03	Understand the relative size of quantities to compare and order numbers from 0 to 10 [20].
		1Np.04	Recognise and use ordinal numbers from 1st to 10th.
		Collins International Primary Maths Recommended Teaching and Learning Sequence:	**Term 1 Week 3**

Unit 17 – Place value and ordering to 20			
Cambridge Primary Mathematics Curriculum Framework (0096)			
Strand	**Sub-strand**	**Code**	**Learning objectives**
Number	Place value, ordering and rounding	1Np.02	Compose, decompose and regroup numbers from 10 to 20.
		1Np.03	Understand the relative size of quantities to compare and order numbers from 0 to 20.
		Collins International Primary Maths Recommended Teaching and Learning Sequence:	**Term 2 Week 8**

Unit 18 – Half (A)			
Cambridge Primary Mathematics Curriculum Framework (0096)			
Strand	**Sub-strand**	**Code**	**Learning objectives**
Number	Fractions, decimals, percentages, ratio and proportion	1Nf.01	Understand that an object or shape can be split into two equal parts or two unequal parts.
		1Nf.02	Understand that a half can describe one of two equal parts of a quantity or set of objects.
		Collins International Primary Maths Recommended Teaching and Learning Sequence:	**Term 3 Week 4**

Unit 19 – Half (B)			
Cambridge Primary Mathematics Curriculum Framework (0096)			
Strand	**Sub-strand**	**Code**	**Learning objectives**
Number	Fractions, decimals, percentages, ratio and proportion	1Nf.03	Understand that a half can act as an operator (whole number answers).
		1Nf.04	Understand and visualise that halves can be combined to make wholes.
		Collins International Primary Maths Recommended Teaching and Learning Sequence:	**Term 3 Week 5**

Unit 20 – Time			
Cambridge Primary Mathematics Curriculum Framework (0096)			
Strand	**Sub-strand**	**Code**	**Learning objectives**
Geometry and Measure	Time	1Gt.01	Use familiar language to describe units of time.
		1Gt.02	Know the days of the week and the months of the year.
		1Gt.03	Recognise time to the hour and half hour.
		Collins International Primary Maths Recommended Teaching and Learning Sequence:	**Term 1 Week 6**

Unit 21 – 2D shapes			
Cambridge Primary Mathematics Curriculum Framework (0096)			
Strand	**Sub-strand**	**Code**	**Learning objectives**
Geometry and Measure	Geometrical reasoning, shapes and measurements	1Gg.01	Identify, describe and sort 2D shapes by their characteristics or properties, including reference to number of sides and whether the sides are curved or straight.
		1Gg.07	Identify when a shape looks identical as it rotates.
Number	Counting and sequences	1Nc.06	Use familiar language to describe sequences of objects

Collins International Primary Maths
Recommended Teaching and Learning Sequence: | **Term 1**
Week 7

Unit 22 – 3D shapes			
Cambridge Primary Mathematics Curriculum Framework (0096)			
Strand	**Sub-strand**	**Code**	**Learning objectives**
Geometry and Measure	Geometrical reasoning, shapes and measurements	1Gg.03	Identify, describe and sort 3D shapes by their properties, including reference to the number of faces, edges and whether faces are flat or curved.
		1Gg.06	Differentiate between 2D and 3D shapes.
		1Gg.07	Identify when a shape looks identical as it rotates.

Collins International Primary Maths
Recommended Teaching and Learning Sequence: | **Term 1**
Week 8

Unit 23 – Length and mass			
Cambridge Primary Mathematics Curriculum Framework (0096)			
Strand	**Sub-strand**	**Code**	**Learning objectives**
Geometry and Measure	Geometrical reasoning, shapes and measurements	1Gg.02	Use familiar language to describe length such as long, longer, longest, thin, thinner, thinnest, short, shorter, shortest, tall, taller and tallest.
		1Gg.04	Use familiar language to describe mass, including heavy, light, less and more.
		1Gg.08	Explore instruments that have numbered scales, and select the most appropriate instrument to measure length, mass, [capacity and temperature].

Collins International Primary Maths
Recommended Teaching and Learning Sequence: | **Term 3**
Week 8

Unit 24 – Capacity and temperature			
Cambridge Primary Mathematics Curriculum Framework (0096)			
Strand	**Sub-strand**	**Code**	**Learning objectives**
Geometry and Measure	Geometrical reasoning, shapes and measurements	1Gg.05	Use familiar language to describe capacity, including full, empty, less and more.
		1Gg.08	Explore instruments that have numbered scales, and select the most appropriate instrument to measure [length, mass,] capacity and temperature.

Collins International Primary Maths
Recommended Teaching and Learning Sequence: | **Term 3**
Week 9

Unit 25 – Position and direction			
Cambridge Primary Mathematics Curriculum Framework (0096)			
Strand	**Sub-strand**	**Code**	**Learning objective**
Geometry and Measure	Position and transformation	1Gp.01	Use familiar language to describe position and direction.
		Collins International Primary Maths Recommended Teaching and Learning Sequence:	**Term 1 Week 9**

Unit 26 – Statistics (A)			
Cambridge Primary Mathematics Curriculum Framework (0096)			
Strand	**Sub-strand**	**Code**	**Learning objectives**
Statistics and Probability	Statistics	1Ss.01	Answer non-statistical questions (categorical data).
		1Ss.02	Record, organise and represent categorical data using: – practical resources and drawings – lists and tables [– Venn and Carroll diagrams – block graphs and pictograms.]
		1Ss.03	Describe data, using familiar language including reference to more, less, most or least to answer non-statistical questions and discuss conclusions.
		Collins International Primary Maths Recommended Teaching and Learning Sequence:	**Term 2 Week 9**

Unit 27 – Statistics (B)			
Cambridge Primary Mathematics Curriculum Framework (0096)			
Strand	**Sub-strand**	**Code**	**Learning objectives**
Statistics and Probability	Statistics	1Ss.01	Answer non-statistical questions (categorical data).
		1Ss.02	Record, organise and represent categorical data using: [– practical resources and drawings] – lists and tables – Venn and Carroll diagrams – block graphs and pictograms.
		1Ss.03	Describe data, using familiar language including reference to more, less, most or least to answer non-statistical questions and discuss conclusions.
		Collins International Primary Maths Recommended Teaching and Learning Sequence:	**Term 3 Week 7**

⑨ Cambridge Primary Mathematics Curriculum Framework (0096) Stage 1 match to Collins International Primary Maths units

The charts below show when each of the Stage 1 learning objectives in the Cambridge Primary Mathematics Curriculum Framework (0096) are taught in the 27 units in Collins International Primary Maths Stage 1.

Cambridge Primary Mathematics Curriculum Framework (0096)				Collins International Primary Maths unit(s)
Strand	Sub-strand	Code	Learning objective	
Number	Counting and sequences	1Nc.01	Count objects from 0 to 20, recognising conservation of number and one-to-one correspondence.	1, 2, 3, 4
		1Nc.02	Recognise the number of objects presented in familiar patterns up to 10, without counting.	1, 2
		1Nc.03	Estimate the number of objects or people (up to 20), and check by counting.	1, 2
		1Nc.04	Count on in ones, twos or tens, and count back in ones and tens, starting from any number (from 0 to 20).	1, 2, 6, 8, 10, 11, 12, 13
		1Nc.05	Understand even and odd numbers as 'every other number' when counting (from 0 to 20).	2
		1Nc.06	Use familiar language to describe sequences of objects.	2, 21
	Integers and powers	1Ni.01	Recite, read and write number names and whole numbers (from 0 to 20).	3, 4
		1Ni.02	Understand addition as: – counting on – combining two sets.	5, 6
		1Ni.03	Understand subtraction as: – counting back – take away – difference.	7, 8, 9
		1Ni.04	Recognise complements of 10.	10, 12
		1Ni.05	Estimate, add and subtract whole numbers (where the answer is from 0 to 20).	10, 11, 12, 13
		1Ni.06	Know doubles up to double 10.	12, 14
	Money	1Nm.01	Recognise money used in local currency.	15

Cambridge Primary Mathematics Curriculum Framework (0096)				Collins International Primary Maths unit(s)
Strand	**Sub-strand**	**Code**	**Learning objective**	
Number	Place value, ordering and rounding	1Np.01	Understand that zero represents none of something.	16
		1Np.02	Compose, decompose and regroup numbers from 10 to 20.	17
		1Np.03	Understand the relative size of quantities to compare and order numbers from 0 to 20.	16, 17
		1Np.04	Recognise and use ordinal numbers from 1st to 10th.	16
	Fractions, decimals, percentages, ratio and proportion	1Nf.01	Understand that an object or shape can be split into two equal parts or two unequal parts.	18
		1Nf.02	Understand that a half can describe one of two equal parts of a quantity or set of objects.	18
		1Nf.03	Understand that a half can act as an operator (whole number answers).	19
		1Nf.04	Understand and visualise that halves can be combined to make wholes.	19
Geometry and Measure	Time	1Gt.01	Use familiar language to describe units of time.	20
		1Gt.02	Know the days of the week and the months of the year.	20
		1Gt.03	Recognise time to the hour and half hour.	20
	Geometrical reasoning, shapes and measurements	1Gg.01	Identify, describe and sort 2D shapes by their characteristics or properties, including reference to number of sides and whether the sides are curved or straight.	21
		1Gg.02	Use familiar language to describe length such as long, longer, longest, thin, thinner, thinnest, short, shorter, shortest, tall, taller and tallest.	23
		1Gg.03	Identify, describe and sort 3D shapes by their properties, including reference to the number of faces, edges and whether faces are flat or curved.	22
		1Gg.04	Use familiar language to describe mass, including heavy, light, less and more.	23

Cambridge Primary Mathematics Curriculum Framework (0096)				Collins International Primary Maths unit(s)
Strand	**Sub-strand**	**Code**	**Learning objective**	
Geometry and Measure	Geometrical reasoning, shapes and measurements	1Gg.05	Use familiar language to describe capacity, including full, empty, less and more.	24
		1Gg.06	Differentiate between 2D and 3D shapes.	22
		1Gg.07	Identify when a shape looks identical as it rotates.	21, 22
		1Gg.08	Explore instruments that have numbered scales, and select the most appropriate instrument to measure length, mass, capacity and temperature.	23, 24
	Position and transformation	1Gp.01	Use familiar language to describe position and direction.	25
Statistics and Probability	Statistics	1Ss.01	Answer non-statistical questions (categorical data).	26, 27
		1Ss.02	Record, organise and represent categorical data using: – practical resources and drawings – lists and tables – Venn and Carroll diagrams – block graphs and pictograms.	26, 27
		1Ss.03	Describe data, using familiar language including reference to more, less, most or least to answer non-statistical questions and discuss conclusions.	26, 27

Revise

Counting people

Resources

cubes (e.g. building blocks) in red, blue, green and yellow (per class); chalk (per class)

What to do

- Ask up to ten learners to come to the front of the class and line up.
- Starting at one end of the line, the first learner starts to count from **one**, the next learner says **two**, and so on. When they reach the end of the line, ask the class how many learners there are in total.
- Divide the learners into two unequal groups and have them stand close together, i.e. not in a line.

- Ask: **How many learners are there now?** Take feedback and let another learner check. As they touch each learner in turn, that learner sits down so they are not counted twice. You could vary this by asking learners to move around as they are counted. Ask: **Is it easier or more difficult to count the learners now?**

Variation

2 Ask the seated learners to estimate how many learners are standing at the front before you count them.

One more, one less

Resources

1–10 number track (per pair); red counter (per pair); two yellow counters (per pair)

What to do

- Learners work in pairs.
- Learner A drops the red counter onto the number track, then places it on the number closest to where it falls.
- Learner B says the number that the red counter is on (e.g. 4), then says the number that is one more than that number (i.e. 5) and places a yellow counter on it.

- Learner B then says the red counter number again (4), followed by the number that is one less than that number (i.e. 3), and places the other yellow counter on it.
- Learners swap roles and repeat.

Variation

3 Learners do the same activity without the resources. Learner A chooses a number from 2 to 9 and learner B must say the number that is one more and one less than the given number. They check the answers on a number track, then swap roles and repeat.

Number – Counting and sequences

Revise

Roll the dice

Learning objectives

Code	Learning objective
1Nc.02	Recognise the number of objects presented in familiar patterns up to 10, without counting.

Resources

1–6 dice (per pair); 10 counters or similar (per pair); mini whiteboard and pen (for variation) (per learner)

What to do

- Learners work in pairs and take turns to roll the dice.
- The first learner in the pair to say the correct number of dots on the top of the dice takes a counter.
- If there is any doubt, learners count the number of dots.

- The winner is the learner with more counters after ten rounds.

Variation

Each learner chooses a different number from 1 to 6 and writes it on their mini whiteboard.

Learners take turns to roll the dice.

If the number rolled matches the number chosen by either learner, that learner takes a counter.

The winner is the learner with more counters after all ten counters have been won.

Revise

Estimation contest

Learning objectives

Code	Learning objective
1Nc.01	Count objects from 0 to 20, recognising conservation of number and one-to-one correspondence.
1Nc.02	Recognise the number of objects presented in familiar patterns up to 10, without counting.
1Nc.03	Estimate the number of objects or people (up to 20), and check by counting.
1Nc.04	Count on in ones, twos [or tens, and count back in ones and tens], starting from any number (from 0 to 20).

Resources

20 building blocks (per class); mini whiteboard and pen (per learner)

What to do

- Ask the learners to sit in a circle. Give each learner a mini whiteboard and pen.
- Put an amount of building blocks (up to 20) in the middle of the circle.
- Learners estimate how many blocks are in the circle and write their estimate on their mini whiteboards. Give learners a short time limit (e.g. 20 seconds) to deter them from counting the blocks rather than estimating.

- On the count of three, everyone holds up their estimate. Compare the estimates and discuss as a class why they chose their numbers.
- Count the blocks as a class, moving each block as you count it and congratulate learners whose estimates were the closest to the real amount.

Variation

2 Count the blocks in twos after estimating.

What is the pattern?

Learning objectives

Code	Learning objective
1Nc.05	Understand even and odd numbers as 'every other number' when counting (from 0 to 20).
1Nc.06	Use familiar language to describe sequences of objects.

Resources

coloured cubes in two or three colours (or sizes) (one per learner); sets of 1–20 number cards from Resource sheet 1: 0–20 number cards (per class) (for variation)

What to do

- Learners sit in a circle on the floor.
- Give each learner a coloured cube in a sequence around the circle (e.g. red, blue, red, blue or red, blue, yellow, red, blue, yellow).
- All learners hold up their cubes. Choose learners to describe the sequence that they see and predict what would come next. You could vary this by using blocks of different sizes (e.g. big, small, big, small) or by giving each learner a 0–20 number card (in order) and asking a learner to describe the sequence of numbers (odd, even, odd, even), instead of different colours.

Variation

2 Learners sit in a circle as above.

Hold up a number card (e.g. 11 or 12). Pick a learner and tell them to move around the circle by that number of places. If the number is odd, the learner jumps around the circle (11 jumps). If the number is even, the learner must hop (12 hops).

When they reach their destination, they swap places with the learner that they are standing behind. This learner is the next to have a turn.

Number – Counting and sequences

Revise

Ten more

Learning objectives

Code	Learning objective
1Nc.04	Count on in [ones, twos or] tens, [and count back in ones and tens,] starting from any number (from 0 to 10 [20]).

Resources

sets of 1–9 number cards from Resource sheet 1: 0–20 number cards (per pair); 20 interlocking cubes (per pair)

What to do

- Learners work in pairs.
- Learner A picks a 1–9 number card and makes a tower of that many interlocking cubes. Learner B makes a tower of 10 interlocking cubes.
- When you call out: **10 more** learners combine their towers and count the cubes to find out how many they have after adding 10 more.

Variation

Do a quickfire version of this game with learner A counting out between one and nine fingers and learner B holding up ten fingers next to theirs, then counting all the fingers together.

Revise

Counting rhymes

Learning objectives

Code	Learning objective
1Ni.01	Recite[, read and write] number names and whole numbers (from 1 [0] to 10 [20]).

What to do

- Ask learners to sit in a circle.
- Start at 'one' and go around the circle, each learner saying the next number in the sequence.
- Stop at 'ten' and start at 'one' again from the next learner.
- Note any learners who struggle to recite numbers to 10 and offer them extra assistance.

Variations

3 Ask learners to choose a different number from 1 to 9 to start counting from, to practise counting from any number to 10.

1 Teach learners the words and actions to the chant 'One, two, three, four, five, once I caught a fish alive'.

2 Alternatively, use any song or chant (up to 10) that learners are familiar with.

Recognising numbers

Learning objectives

Code	Learning objective
1Ni.01	Recite, read and write number names and whole numbers (from 1 [0] to 10 [20]).
1Nc.01	Count objects from 1 [0] to 10 [20], recognising [conservation of number and] one-to-one correspondence.

Resources

1–10 number cards from Resource sheet 1: 0–20 number cards (per class); jar of ten counters (per class); book with numbered pages (for variation) (per pair); old magazines or newspapers (for variation) (per group)

What to do

- Display the numbers 1–10 around the classroom (for example, on the wall, on desks or on chairs).
- Place a random number of counters onto the desk (ten or fewer).
- As a whole class, count them to establish how many there are.
- Ask learners to find the matching number in the classroom and stand next to it.
- Repeat for different numbers.

Variations

2 Give pairs of learners a book of at least ten pages in length.

Place counters from the jar onto a desk and count how many.

Ask learners to find that number page in their book.

2 Give each group some old magazines or newspapers.

Ask learners to find examples of numbers from 1 to 10 in their magazines or newspapers.

Number – Integers and powers/Counting and sequences

Revise

Counting circle

Learning objectives

Code	Learning objective
1Ni.01	Recite[, read and write] number names and whole numbers (from 0 to 20).

What to do

- Learners stand in a circle.
- They count from 1 to 20, with each learner saying a number name in turn. When a learner has said their number name they turn and point to the learner to the left of them to indicate that it is now their turn to say the next number.
- When the learners reach 20, they count backwards from 20 to 1, continuing around the circle.

Variation

2 Give learners different starting numbers, from 1 to 19, and ask them to count forwards or backwards.

Number race

Learning objectives

Code	Learning objective
1Ni.01	Recite, read and write number names and whole numbers (from 0 to 20).

Resources

shuffled set of 0–20 number cards from Resource sheet 1: 0–20 number cards (per group); mini whiteboard and pen (per group) (for variation)

What to do

- Learners work in groups of three.
- Learner A calls out a number from 0 to 20.

- Learners B and C look through their set of number cards to find the number. The winner is the first person to find the number.
- Learners swap roles and repeat.

Variation

2 Learner A calls out a number from 0 to 20. Learner B looks through a set of number cards for the number. Learner C writes the number on a mini whiteboard. Then they swap roles and repeat.

Counting out 20

Learning objectives

Code	Learning objective
1Ni.01	Recite, read and write number names and whole numbers (from 0 to 20).
1Nc.01	Count objects from 0 to 20, recognising conservation of number and one-to-one correspondence.

Resources

set of 0–20 number cards from Resource sheet 1: 0–20 number cards (per class); set of countable objects (per group); set of 0–20 number word cards from Resource sheet 6: 0–20 number word cards (for variation)

What to do

- Hold up a 0–20 number card.
- Learners count out that many countable objects.

- They then swap seats with another learner and count each other's objects to check that they have counted correctly.
- Repeat with a different number card.

Variation

 Hold up a 0–20 number word card instead.

Revise

Add the apples

Learning objectives

Code	Learning objective
1Ni.02	Understand addition as: [- counting on] - combining two sets.

What to do

- Display the **Tree tool**, showing two trees.
- Put four apples on one tree and two apples on the other tree. Ask: **How many apples are there altogether?**
- Count each set of apples separately, then count them together, removing the apples from the trees and putting them in one group before you count them.

- Ask learners to choose the numbers of apples for the trees and to come to the front to count the apples.

Variation

 Write an addition on the board and put apples on the tree to match the numbers. Count the apples together to find the total, then fill in the total for the number sentence.

How many fingers?

Learning objectives

Code	Learning objective
1Ni.02	Understand addition as: [- counting on] - combining two sets.

Resources

chalk (per group) (variation only); mini whiteboard and pen (per group) (variation only)

What to do

- Pairs of learners face each other with their hands behind their backs. On the count of three, they show each other one hand with any number of fingers held up.
- They put their hands next to each other and work together to count how many fingers they are holding up altogether.

Variation

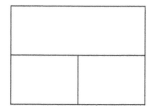 Learners work in groups of three. Draw a rectangular part–whole diagram on the floor with chalk.

They start the activity as above, then two learners, holding up fingers, each stand in one of the bottom 'part' sections of the diagram. The remaining learner counts how many fingers there are altogether and writes the total on a mini whiteboard, then places the whiteboard with the total on it in the 'whole' part of the diagram.

Revise

Adding more birds

Learning objectives

Code	Learning objective
1Ni.02	Understand addition as: - counting on [- combining two sets].
1Nc.04	Count on in ones, [twos or tens, and count back in ones and tens,] starting from any number (from 0 to 20).

What to do

- Display the **Tree tool**, showing one tree.
- Put five birds on the tree and two in the sky. Say: **There are five birds already in the tree and two more want to join them.**
- Ask a learner to come to the front to move the birds from the sky to the tree, counting on from five in ones as they do so. Check that the learners understand that the last number they said was how many birds are now on the tree altogether. Count the birds to check that the answer is correct.
- Repeat the activity using different starting amounts and volunteers.

Variation

 Write an addition number sentence on the board and put birds on the tree and in the sky to match it. Fill in the total when learners have moved and counted the birds.

Giant number line

Learning objectives

Code	Learning objective
1Ni.02	Understand addition as: - counting on [- combining two sets].
1Nc.04	Count on in ones, [twos or tens, and count back in ones and tens,] starting from any number (from 0 to 20).

Resources

skipping rope (per class); two sets of 0–10 number cards from Resource sheet 1: 0–20 number cards (per class); chalk (optional) (variation only); set of large 0–10 number cards (variation only)

What to do

- Use the skipping rope and one of the sets of 0–10 number cards to make a number line on the floor.
- Pick two 0–10 number cards from the remaining set and write them as an addition on the board (e.g. 4 + 3 =).
- Ask a learner to start on 4 and count on three more by making three jumps to the next number. Note the number that they land on and write the total.
- Repeat with another set of starting numbers and a different volunteer learner.

Variation

 This activity could be used without the number sentences, at the start of the unit, to give learners practice at counting on in ones from a given number. As the unit progresses, use the activity as above but focus on the skills required, for example using the larger number as the starting number. You could use chalk to record the jumps if necessary. You could use large number cards instead of the skipping rope and smaller number cards to make a number track instead of a number line.

Revise

Taking away beads

Learning objectives

Code	Learning objective
1Ni.03	Understand subtraction as: [- counting back] - take away [- difference].

Resources

string of ten beads (per pair)

What to do

- Learners work in pairs.
- Give each pair of learners a string of ten beads.
- Learner A 'takes away' an amount of beads specified by you by moving them to the other side of the string. Learner B counts how many are left.

- They swap roles and repeat, taking away another amount of beads until there are no beads left in the original set.

Variation

 Write a number sentence on the board and ask learners to thread the minuend amount of beads onto the string, then remove the subtrahend amount.

Part–whole subtraction race

Learning objectives

Code	Learning objective
1Ni.03	Understand subtraction as: [- counting back] - take away [- difference].

Resources

Resource sheet 10: Part–whole diagram (per learner); ten counters (per learner); set of 0–10 number cards from Resource sheet 1: 0–20 number cards (per learner)

What to do

- Give each learner a part–whole diagram resource sheet, a set of 0–10 number cards and ten counters.
- Give the class or each group a starting number (such as 6). All learners put that amount of counters in the 'whole' part of their diagram.
- Now call out an amount to 'take away' (such as 2).
- Learners race to take away that amount by placing them in the bottom left part of their diagram. They move the leftover counters in the 'whole' section to the bottom right section, count them and hold up the number card that matches the amount that they have left.

- The winners are the first learners in each group to hold up the correct number card.
- As the learners are competing in ability groups rather than as a whole class, you can scaffold the starting amounts and amounts to take away according to ability.

Variation

Write a number sentence on the board (such as 6 – 2 =). Learners do the activity as above to find the answer.

Revise

Rocket launch

Learning objectives

Code	Learning objective
1Ni.03	Understand subtraction as: - counting back [- take away - difference].
1Nc.04	Count [on in ones, twos or tens, and count] back in ones [and tens], starting from any number (from 0 to 10 [20]).

What to do

- Ask learners to crouch down.
- Count back from 10 to 0 with learners. When they reach 0, they launch themselves up into the air like rockets blasting off.
- Repeat, starting from a different number between 0 and 10.

Variation

 Write a subtraction number sentence on the board (such as 9 – 3 =). Learners crouch down and start counting backwards to find the answer (8, 7, 6). Hold up fingers as a visual guide to how many they have counted back so far. When they reach the number that is the answer, they blast off like rockets.

Teddy number track subtraction

Learning objectives

Code	Learning objective
1Ni.03	Understand subtraction as: - counting back [- take away - difference].
1Nc.04	Count [on in ones, twos or tens, and count] back in ones [and tens], starting from any number (from 0 to 10 [20]).

Resources

ten stuffed toys (per class); ten stickers numbered 0–10 (per class)

What to do

- Stick a 0–10 number sticker on each stuffed toy and line them up from 0 to 10 like a number track.
- Write a subtraction on the board (e.g. 8 – 5 =).
- Choose a learner to count back along the line of teddies by starting at toy number 8 and counting back five toys.
- The learner holds up the toy that they land on after counting back the amount shown.
- Check their answer as a class by repeating the activity together.
- Repeat with a different subtraction and learner volunteer.

Variation

Do the activity as above but instead of counting five jumps, count backwards (for example, 7, 6, 5, 4, 3), holding up one finger for each number you say until you are holding up five fingers.

Revise

Counting back race

Learning objectives

Code	Learning objective
1Ni.03	Understand subtraction as: - counting back [- take away - difference].
1Nc.04	Count [on in ones, twos or tens, and count] back in ones [and tens], starting from any number (from 0 to 10 [20]).

Resources

ten interlocking cubes for half of the class (per learner); Resource sheet 11: 0–10 number line for half of the class (per learner); Resource sheet 4: 0–10 number track for half of the class (per learner); 0–10 number cards from Resource sheet 1: 0–20 number cards (per learner)

What to do

- Divide learners into two groups, down the middle of the room.
- Give half of the learners a 0–10 number line each (from Lesson 3 onwards) or a 0–10 number track each (from Lesson 2 onwards) and the other half of the learners ten interlocking cubes each. Give every learner a set of 0–10 number cards.
- Tell learners that each side of the room is going to race to see which method is quicker at finding the answer to a subtraction.

- Write a subtraction on the board (such as 7 – 4 =).
- Learners use their equipment to work out the answer. When they know the answer they hold up the corresponding number card.
- The winning side is the side in which more learners hold the correct number card up first.

Variations

 Depending on where you are in the unit, you could just give learners cubes to solve the subtraction or later just give them number lines or tracks.

1 Learners work in pairs for support during this activity.

Revise

Hands up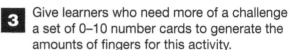

Learning objectives

Code	Learning objective
1Ni.03	Understand subtraction as: [- counting back - take away] - difference.

Resources

set of 0–5 number cards from Resource sheet 1: 0–20 number cards (per pair)

What to do

- Learners work in pairs.
- Each learner chooses a number card. They hold up the amount of fingers that matches the number on their card.
- They face their partner and put their hands together, still holding up the amount of fingers shown on their card.

- Learners work out the difference between the amounts of fingers that they are holding up by counting the fingers that don't have a matching finger held up by the other learner in the pair.

Variation

3 Give learners who need more of a challenge a set of 0–10 number cards to generate the amounts of fingers for this activity.

Find the difference towers

Learning objectives

Code	Learning objective
1Ni.03	Understand subtraction as: [- counting back - take away] - difference.

Resources

20 building blocks (per class); Resource sheet 11: 0–10 number line (optional) (per pair)

What to do

- Write a subtraction on the board. Keep the minuend to 10 or less.
- Invite two learners to the front of the class to make towers from building blocks to match the two numbers. Put the towers next to each other.
- Ask: **What is the difference between the amounts of blocks in the towers?** The rest of the learners count the extra blocks in the taller tower to find the answer.

- Learners then check their answer by finding the difference between the two numbers on a number line.

Variation

2 Omit the use of the number line to check the answer if you have not yet covered finding the difference on a number line in this unit.

Revise

Bird combinations

Learning objectives

Code	Learning objective
1Ni.04	Recognise complements of 10.
1Ni.05	[Estimate,] add and subtract whole numbers (where the answer is from 0 to 10 [20]).
1Nc.04	Count on in ones, [two or tens,] and count back in ones [and tens], starting from any number (from 0 to 10 [20]).

What to do

- Display the **Tree tool** with two trees and ten birds in the sky.
- Ask individual learners to find different ways of sharing the birds between the two trees, counting how many birds are in each tree, then counting all the birds together every time.

Variations

 As the unit progresses, write an addition and/ or subtraction number sentence to match the birds in the trees (such as 6 + 4 = 10, 10 – 4 = 6). Encourage learners to use the addition and subtraction strategies they have learned in previous units to check the answers.

 For Lesson 3 or 4, start with any amount of birds from 3 to 10 in the sky and share the birds between the trees in different ways to find complements of other numbers to 10.

Number pairs ➊

Learning objectives

Code	Learning objective
1Ni.04	Recognise complements of 10.
1Ni.05	[Estimate,] add and subtract whole numbers (where the answer is from 0 to 10 [20]).
1Nc.04	Count on in ones, [two or tens,] and count back in ones [and tens], starting from any number (from 0 to 10 [20]).

Resources

mini whiteboard and pen (per learner)

What to do

- Ask each learner to choose a number from 0 to 10. They write their number on a mini whiteboard.
- Learners walk around the classroom holding up their whiteboards and looking for a number that makes 10 when added to their number.
- If there are any learners left without a partner at the end, ask the class which number each leftover learner would need to make 10.

Variations

 As the unit progresses, write an addition and/ or subtraction number sentence to match the pairs of learners (such as 3 + 7 = 10, 10 – 7 = 3). Encourage learners to use the addition and subtraction strategies they have learned in previous units to check the answers.

 For Lesson 3 or 4, give learners a different target number from 3 to 9.

Number – Integers and powers/Counting and sequences

Revise

Higher or lower

Learning objectives

Code	Learning objective
1Ni.05	Estimate, add and subtract whole numbers (where the answer is from 0 to 10 [20]).
1Nc.04	Count on in ones, [twos or tens,] and count back in ones [and tens], starting from any number (from 0 to 10 [20]).

Resources

set of large 0–5 number cards (per class); ten large countable objects (optional) (per class)

What to do

- Ask two volunteer learners each to pick a large 0–5 number card. They hold the cards up at the front of the class.
- The other learners estimate whether the total would be more or less than 5 when the numbers are added together. Those who think the answer would be more than 5 stand up and those who think it would be less than 5 sit down. Those who think the answer would be exactly 5 stay sitting down but put their hands on their head.
- Add the numbers together as a class by counting on, either with objects or mentally.

Variation

2 Adapt this activity for subtraction by learners picking two 0–10 number cards. Show the rest of the class the card with the larger number, then ask them to estimate what the answer would be if the smaller number was subtracted.

Addition and subtraction machines

Learning objectives

Code	Learning objective
1Ni.05	Estimate, add and subtract whole numbers (where the answer is from 0 to 10 [20]).
1Nc.04	Count on in ones, [twos or tens,] and count back in ones [and tens], starting from any number (from 0 to 10 [20]).

Resources

Resource sheet 10: Part–whole diagram and ten counters (optional) (per pair); ten interlocking cubes (optional) (per pair); Resource sheet 11: 0–10 number line (optional) (per pair)

What to do

- Display the **Function machine tool**.
- Write an addition or subtraction to 10 on the board (such as 6 + 3 = or 8 − 5 =).
- Ask pairs of learners to solve it by using one of the strategies that they have been taught in previous units (counting on or back/finding the difference/taking away/combining sets). Provide any equipment that will support them with this, such as a part–whole diagram sheet and counters/ten interlocking cubes/a number line.
- Ask a learner to check the answer to the addition or subtraction by inputting the numbers and the operation into the function machine and pressing GO.

Variations

2 As the unit progresses, write an addition and/or subtraction number sentence to match (such as 3 + 7 = 10, 10 − 7 = 3). Encourage learners to use the addition and subtraction strategies they have learned in previous units to check the answers.

2 For Lesson 3 or 4, give learners a different target number between 3 and 9.

Number – Integers and powers/Counting and sequences

Revise

Counting on race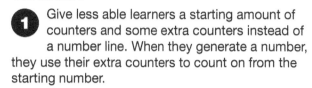

Learning objectives

Code	Learning objective
1Ni.05	Estimate, add [and subtract] whole numbers (where the answer is from 0 to 20).
1Nc.04	Count on in ones, [twos or tens, and count back in ones and tens,] starting from any number (from 0 to 20).

Resources

Resource sheet 12: 0–20 number line (per pair); counter (per pair); dice or Resource sheet 9: 1–6 spinner (per pair); paper clip and pencil, for the spinner (per pair)

What to do

- Give pairs of learners a 0–20 number line, a counter and a dice or 1–6 spinner (and a paper clip and pencil).
- Tell the learners a starting number. All learners put their counter on the starting number.

- Working together, pairs then roll their dice/spin their spinner to generate a number and count on that many with their counter on the number line. They repeat this until they reach or pass 20.

Variation

1 Give less able learners a starting amount of counters and some extra counters instead of a number line. When they generate a number, they use their extra counters to count on from the starting number.

Ten and some left over

Learning objectives

Code	Learning objective
1Ni.04	Recognise complements of 10.
1Ni.05	Estimate, add [and subtract] whole numbers (where the answer is from 0 to 20).
1Nc.04	Count on in ones, [twos or tens, and count back in ones and tens,] starting from any number (from 0 to 20).

Resources

mini whiteboard and pen (per learner); 7–9 interlocking cubes (per learner)

What to do

- Give each learner 7–9 interlocking cubes. Ask learners to sit in circles in groups.
- Give the first learner six interlocking cubes. They add cubes from their original pile to make a tower of 10, then pass on all remaining cubes to the next learner in the circle.

- The next learner adds cubes to make 10 and passes the left-over cubes on as before. Continue this until all learners have made a tower of ten cubes.
- Ask: **How many cubes were left over at the end? How many cubes were left over each time? Why do you think that was?**

Variation

2 Give each learner a different amount of cubes as their starting amount. They add enough cubes from their pile to make 10, all doing this at the same time. Discuss how many cubes each learner has left after making 10.

Revise

Double or near double?

Learning objectives

Code	Learning objective
1Ni.05	Estimate, add [and subtract] whole numbers (where the answer is from 0 to 20).
1Ni.06	Know doubles up to double 10.

Resources

5–20 interlocking cubes (per learner)

What to do

- Give each learner between 5 and 20 cubes.
- Learners make two equal towers from the cubes. If a cube is left over, they add it to one of the towers at the end.

- Discuss with the class who has 'double' towers and who has 'near double' towers, referencing how many cubes are in each tower, for example: **You have double 6** or **You have a near double – double 6 add 1.**

Variation

2 Build towers to answer a near doubles addition such as 5 + 6 = ?

Revise

Subtracting the ones

Learning objectives

Code	Learning objective
1Ni.05	Estimate, add and subtract whole numbers (where the answer is from 0 to 20).
1Nc.04	Count on in ones, [twos or tens,] and count back in ones [and tens,] starting from any number (from 0 to 20).

Resources

20 interlocking cubes (per pair); set of 0–5 number cards from Resource sheet 1: 0–20 number cards (per pair)

What to do

- Give pairs of learners 20 interlocking cubes and a set of 0–5 number cards.
- Write a number from 15 to 19 on the board (e.g. 18).
- Learners make a tower of ten cubes and another tower of cubes to match the ones digit in the number (e.g. 8).

- They pick a number card and subtract their number from the ones tower by removing that many cubes.
- They count how many cubes they have left by starting on 10 and counting on the extra ones.

Variation

1 Ask learners to use the cubes to make an addition for the number on the board after they have completed the subtraction. They write the addition and subtraction as an equal statement, for example: 18 – 5 = 10 + 3.

Number facts

Learning objectives

Code	Learning objective
1Ni.05	Estimate, add and subtract whole numbers (where the answer is from 0 to 20).
1Nc.04	Count on in ones, [twos or tens,] and count back in ones [and tens,] starting from any number (from 0 to 20).

Resources

Resource sheet 1: 0–20 number cards (one card per learner)

What to do

- Give each learner a 0–20 number card.
- Ask the class questions related to the addition and subtraction number facts to 20, such as: **12 add something equals 18. Who has the missing number?** Or **If I take 9 away from something I'll have 11. Who has the starting number?** The learner(s) with the missing number in the statement holds up the card.

Variation

2 Challenge learners to look around the class for someone whose number will make a given number to 20 when added to or subtracted from their number. You could extend this to make equal statements.

Number – Integers and powers/Counting and sequences

Number – Integers and powers

Revise

Double the monkeys

Learning objectives

Code	Learning objective
1Ni.06	Know doubles up to double 10.

What to do

- Display the **Tree tool** with two trees. Put from one to five monkeys in one of the trees.
- Ask learners to come to the front of the class to make double the monkeys by putting the same number of monkeys on the other tree.

Variation

 Add up to ten monkeys to the first tree as the unit progresses.

Number track doubles

Learning objectives

Code	Learning objective
1Ni.06	Know doubles up to double 10.

Resources

tape, safety pins or similar (per class); set of 1–10 number cards from Resource sheet 1: 0–20 number cards (per class); ball or beanbag (per class)

What to do

- Ask ten learners to come to the front. They stand in a line facing the rest of the class.
- Fasten a 1–10 number card to each learner's chest. The numbers must be in order, as for a number track.
- Ask another learner to choose a number from 1 to 5 (e.g. 3). Give the learner with that number card the ball to hold.
- The learners in the number track pass the ball up the track the number of spaces that matches the starting number (such as 3 more). When they have stopped passing the ball, the number assigned to the learner left holding the ball is double the starting number.
- Repeat with other numbers to 10.

Variation

Make a longer number track for numbers from 1 to 20, or practise doubling numbers from 5 to 10 with a 5–20 number track.

Revise

I went shopping... or ![icon] ▲2

Learning objectives

Code	Learning objective
1Nm.01	Recognise money used in local currency.

Resources

dish of coins and notes in your local currency (per class or group); bag of small objects that could be mistaken for coins and notes (for variation) (per class or group)

What to do

- Seat the whole class (or smaller groups) in a circle.
- One learner holds a dish of the local currency and says, 'I went shopping to spend my [insert name of currency] and I bought [insert item].' They then pass the dish onto the next learner in the circle.
- The next learner says, 'I went shopping to spend my [insert name of currency] and I bought [the previous learner's item and insert new item].'

- This continues around the circle with each consecutive learner trying to remember to list as many of the previous items as possible.
- Afterwards, ask: **As you buy more items, will the total price of your shopping list go up or down?** Elicit that the price will go up as you add more items to your shopping list.

Variation

▲2 Pass the coins and notes around in a bag that learners cannot see through. Put into the bag other small objects that could be mistaken for coins and notes, for example small items of jewellery, buttons, bottle tops and pieces of paper. On their turn, each learner must try to identify a coin or note by touch only and take it out of the bag.

Describing coins and notes ▲2

Learning objectives

Code	Learning objective
1Nm.01	Recognise money used in local currency.

Resources

coins and notes from local currency (per class); large book (per class)

What to do

- Choose one learner to come to the front of the class.
- The learner chooses a coin or note and holds it behind a large book so that the rest of the class can't see it. The learner looks at it carefully and describes it in detail. The rest of the learners must identify the coin or note from the description.

Variation

▲2 The learner holds a coin behind their back and describes the way it feels, for example it is round/it has five straight sides/it is a large coin.

Revise

Zero!

Learning objectives

Code	Learning objective
1Np.01	Understand that zero represents none of something.

What to do

- Clear a space in the room and ask the learners to stand up.
- Give learners an action to do (such as jump/hop/clap/turn around) followed by the amount of times you want them to do it (for example, **jump... four times**).
- The learners follow your instructions. They must listen carefully because if the number of times you ask them to do something is zero, they must stand as still as they can and not do the action

at all. Remind learners that zero means none of something, so you are asking them to do no hops/jumps/claps when you tell them to do the action zero times.

Variation

3 Learners take turns to be the leader and call out instructions. Another learner is their 'assistant' and looks out for people accidentally doing the action when they are asked to do it zero times.

More or less?

Learning objectives

Code	Learning objective
1Np.03	Understand the relative size of quantities to compare [and order] numbers from 0 to 10 [20].

Resources

set of large 0–10 number cards from Resource sheet 1: 0–20 number cards (per class)

What to do

- Display the **Number track tool**, shuffle a set of large 0–10 number cards and invite two learners to come to the front of the class.
- Learner A picks a number card and holds it up. Highlight this number on the number track and ask the other learners to vote for whether they think learner B's number will be more or less than this.

- Learner B picks a card and holds it up. Highlight their number on the number track and discuss with learners which number is more/less and how we can tell by looking at their places on the number track.

Variation

1 Give Challenge 1 learners counters to support them when working out which number is more or less than the other.

Revise

Order the numbers **2**

Learning objectives

Code	Learning objective
1Np.03	Understand the relative size of quantities to [compare and] order numbers from 0 to 10 [20].
1Np.04	Recognise and use ordinal numbers from 1st to 10th. (Variation only)

Resources

sets of 0–10 number cards from Resource sheet 1: 0–20 number cards (enough for every learner to have a card each); ordinal number cards from Resource sheet 14: Ordinal number cards (enough for every learner to have a card each) (variation only)

What to do

• Give each learner a 0–10 number card, selected at random.

• Learners walk around the classroom looking for other learners who have numbers to complete a set from 0 to 10. This will take careful thinking as there will probably be two or three full sets of 0–10 numbers in the whole class.

• When learners think they have found all the numbers, they line up in number order from 0 to 10.

Variation

2 Do the same activity but with ordinal number cards (1st–10th).

Revise

Tens and ones

Learning objectives

Code	Learning objective
1Np.02	Compose, decompose and regroup numbers from 10 to 20.

Resources

large + and = signs (per group) (for variation)

What to do

• Ask three or four learners in the class to be 'tens' and the rest of the class to be 'ones'. 'Tens' take big slow strides around the room or playground, with their arms in the air to make them look tall. 'Ones' take small fast steps and bend over to make them look small.

• Call out a number from 10 to 20 (such as 14). Each of the 'tens' must find enough 'ones' to represent that number. They then stand together as a group with the ten first, followed by the ones.

• Swap the roles of the learners and repeat with a different number.

Variation

 Give learners large + and = signs to hold between themselves as they line up in their number groups to show how their number is composed.

Number race

Learning objectives

Code	Learning objective
1Np.02	[Compose, decompose and] regroup numbers from 10 to 20.

Resources

mini whiteboard and pen (per learner); ten counters (per pair) (for variation) **[TWM.05]**

What to do

• Give each learner a mini whiteboard and pen.

• Write a number on the board from 10 to 20 (such as 17).

• Learners write a way of making that number on their whiteboards (for example, 17 = 10 + 7 or 17 = 10 + 6 + 1).

• Learners hold up their whiteboards. Ask learners who chose the same way of making the number to stand together. Which calculation was the most popular? Were there any other ways of making the number that nobody used? Make a list on the board.

Variation

 Learners do this activity in pairs, choosing a number and each writing a number sentence to make that number. If they write two different calculations, they win a counter. If both of their calculations are the same, they do not win a counter. The winning pair is the pair with the most counters at the end of the given time.

Number – Place value, ordering and rounding

Revise

Order the numbers

Learning objectives

Code	Learning objective
1Np.03	Understand the relative size of quantities to compare and order numbers from 0 to 20.

Resources

any six cards from a set of 0–20 number cards from Resource sheet 1: 0–20 number cards (per pair)

What to do

- Learners work in pairs.
- Give each pair six 0–20 number cards at random. They put the cards face down on the table.
- When you clap your hands, the pairs can turn over their cards. They work as quickly as they can to put them in ascending order. As soon as they have finished they must stand up.

- The winners are the first pair standing who have correctly ordered their numbers.
- Each pair swaps their cards with another pair and the activity is repeated.

Variation

1 **3** This activity could be done individually for Challenge 3 learners or in an adult-led group for Challenge 1 learners.

Number – Place value, ordering and rounding

Revise

Match the halves

Learning objectives

Code	Learning objective
1Nf.01	Understand that an object or shape can be split into two equal parts or two unequal parts.

Resources

paper cut-out shapes (triangles, squares, circles, rectangles), some of which have been cut exactly into halves, others that have been cut into two unequal parts (per group)

What to do

- Give each group some cut-up paper shapes.
- Learners work together to match halves to make whole shapes. They put any unequal parts aside.

- At the end of the activity, show some of the shapes from each group to the whole class, demonstrating how the halves are the same size and shape and what they look like when they are put together and taken apart again.

Variation

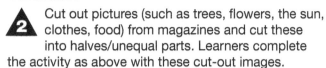 Cut out pictures (such as trees, flowers, the sun, clothes, food) from magazines and cut these into halves/unequal parts. Learners complete the activity as above with these cut-out images.

Tree tool halving

Learning objectives

Code	Learning objective
1Nf.02	Understand that a half can describe one of two equal parts of a quantity or set of objects.

What to do

- Display the **Tree tool** with two trees. Ask a learner to put ten apples on one tree.
- Now choose a learner to move half of the apples to the other tree. Remind the learner that when the number of apples has been halved, each tree will contain the same number of apples.

- Repeat this with a different starting number of apples.

Variation

Keep the number of apples on the tree to 6 or fewer.

Number – Fractions, decimals, percentages, ratio and proportion

Revise

Halve the towers

Learning objectives

Code	Learning objective
1Nf.03	Understand that a half can act as an operator (whole number answers).

Resources

20 interlocking cubes (per learner)

What to do

- Choose a number from 2 to 20 and ask learners to make towers of that many interlocking cubes.
- When you call out: **Halve it!** learners must halve their towers.
- If half of the starting amount is an even number, ask the learners to halve it again – and so on until the amount of cubes can't be halved any more.

- Discuss what happens to the towers (they get smaller every time they are halved) and the starting number (it decreases every time it is halved).

Variation

1 Do this activity with learners at the front of the class instead of cubes for a larger visual representation of what halving does to an amount/number.

How many wholes?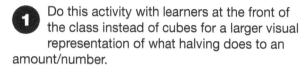

Learning objectives

Code	Learning objective
1Nf.04	Understand and visualise that halves can be combined to make wholes.

Resources

at least six paper cut-out objects from magazines or newspapers, e.g. food, trees, flowers, toys, cut into halves (per group)

What to do

- Give each group about 12 halves of paper cut-out objects. Ensure that some of the halves do not match (are not from the same object).
- Learners work together to match halves to make whole objects.
- At the end of the activity, the winning group is the one with the most whole objects.
- Count each group's whole objects and then count the halves that made the whole objects.

Variation

2 Do the same activity but with circles cut into halves. Each half of a circle should have a number of dots drawn on it (e.g. three dots) and a target amount written on it as a number (e.g. 6). Learners must find the other half of their circles that makes the full amount when combined with the first half.

Number – Fractions, decimals, percentages, ratio and proportion

Revise

On Monday...

Learning objectives

Code	Learning objective
1Gt.01	Use familiar language to describe units of time.
1Gt.02	Know the days of the week and the months of the year.

What to do

- Ask all learners to stand up.
- Go around the class one learner at a time. The first learner says 'On Monday...' then mimes an activity (for example, playing football, baking a cake, wrapping a present).
- The whole class must guess what they are doing.
- The next learner starts with 'On Tuesday...' and so on.

Variations

1 Support these learners by displaying a calendar with the days of the week in order for them to refer to.

2 **3** Use months instead of days of the week.

Stop the clock!

Learning objectives

Code	Learning objective
1Gt.01	Use familiar language to describe units of time.
1Gt.03	Recognise time to the hour and half hour.

What to do

- Display the **Clock tool**.
- Say a time to the hour and choose a volunteer to come up and make the time on the tool.
- Ask: **Are they going the right way around the clock? Is the minute/hour hand in the right place?**
- Play a few rounds with different volunteers.

Variation

2 Do the same activity, replacing o'clock times with half past times.

Geometry and Measure – Time

Revise

Make the set

Learning objectives

Code	Learning objective
1Gt.01	Use familiar language to describe units of time.
1Gt.02	Know the days of the week and the months of the year.
1Gt.03	Recognise time to the hour [and half hour].

Resources

Resource sheet 23: Days of the week cards (one per learner); Resource sheet 16: Months of the year cards (one per learner) (variation only); Resource sheet 24: O'clock times from 1 o'clock to 12 o'clock (one per learner) (variation only)

What to do

• Give each learner a days of the week card.
• Learners group themselves into sets, making sure that their group has one of each of the days of the week in it.
• If they are able to, learners can order their days of the week.

Variations

Replace days of the week with months of the year.

Replace days of the week with o'clock times. Either each group must contain the o'clock times from 1 o'clock to 12 o'clock or learners can make smaller, random groups and arrange their o'clock times in order (e.g. 2 o'clock, 5 o'clock, 6 o'clock, 11 o'clock).

Geometry and Measure – Time

Revise

Hunt the shapes

Learning objectives

Code	Learning objective
1Gg.01	Identify[, describe and sort] 2D shapes by their characteristics or properties, including reference to number of sides and whether the sides are curved or straight.

Resources

box containing a selection of 2D shapes: circles, triangles, squares and rectangles – about 12 shapes altogether (per group)

What to do

- One learner has the group's box of shapes in front of them.
- Say the name of a common 2D shape, for example: **triangle.**
- Each learner with a box quickly finds and holds up one of the shapes that you have named – one triangle.

- The learner with the box then passes it to the next learner in the group.
- Repeat this several times, quickening the pace.

Variations

 Adapt as the unit progresses, for example: **Find a shape with a curved side. Find a shape with three sides.**

 Hide shapes in a classroom sandbox; name a shape and invite individual learners to come up to find it.

Shape reveal

Learning objectives

Code	Learning objective
1Gg.01	Identify, describe and sort 2D shapes by their characteristics or properties, including reference to number of sides and whether the sides are curved or straight.
1Gg.07	Identify when a shape looks identical as it rotates.
1Nc.06	Use familiar language to describe sequences of objects

Resources

various circles, squares, rectangles and triangles of different sizes (per class); a large book or a scarf to hide the shapes behind (per class)

What to do

- Hide the shapes behind the book or scarf.
- Choose a shape and slowly reveal it to the learners. How quickly can they work out which shape it is?
- Occasionally use the same shape twice in succession, but rotate it through half a turn the second time to check if learners still recognise it as the same shape.

Variation

 Make a repeating pattern with one variable (including rotation if appropriate) with the shapes. Choose whether to use two different shapes or the same shape in different colours or sizes. Hide the pattern under the scarf then reveal it and ask learners to discuss what they see in the pattern with a partner. Ask partners to give their observations to the class. Repeat with a different pattern.

Geometry and Measure – Geometrical reasoning, shapes and measurements

Revise

The shape shop

Learning objectives

Code	Learning objective
1Gg.01	Identify, describe [and sort] 2D shapes by their characteristics or properties, including reference to number of sides and whether the sides are curved or straight.

Resources

box containing at least one circle, triangle, square and rectangle (per pair)

What to do

- Learners work in pairs. One learner is the 'shopper' and the other is the 'shopkeeper'.
- The shopper describes to the shopkeeper the shape(s) that they want to buy, for example the number of sides, curved or straight.
- Pairs repeat the activity several times, alternating roles.

Variation

1 Make this a teacher-led activity for a small group.

Revise

Shape feely bags [TWM.05]

Learning objectives

Code	Learning objective
1Gg.03	Identify, describe and sort 3D shapes by their properties, including reference to the number of faces, edges and whether faces are flat or curved.
1Gg.06	Differentiate between 2D and 3D shapes. (see variation)

Resources

small feely bag (per class); collection of spheres, cones, cylinders, cubes and cuboids (per class); collection of circles, squares, triangles and rectangles (per class) (variation only)

What to do

- Ask volunteers to come forward and feel inside the bag.
- Ask: **Can you describe the shape using words such as curved or flat?**
- Encourage the learner to name the shape.
- Reveal the shape and clear up any misconceptions or confirm the description.

- As the unit progresses, ask for a more detailed description of the shape, for example: 'A cylinder rolls when it is on its side and stands still when it is upright; a cuboid can be as big as a building or as small as an eraser.'

Variation

 Hide one 2D and one 3D shape in the bag and ask the volunteers to pull out the 2D shape first.

The same or different?

Learning objectives

Code	Learning objective
1Gg.03	Identify, describe and sort 3D shapes by their properties, including reference to the number of faces, edges and whether faces are flat or curved.
1Gg.06	Differentiate between 2D and 3D shapes. (see variation)
1Gg.07	Identify when a shape looks identical as it rotates.

Resources

collection of spheres, cones, cylinders, cubes and cuboids (per class); collection of circles, squares, triangles and rectangles (per class) (variation only)

What to do

- Pick a shape from the class 3D shapes collection and show it to the learners. Put the shape on a table where they can see it.

- Now choose either a different shape or a shape that is the same as the one on the table, but rotated. Hold up the shape to show the learners.
- Learners must now decide with a partner whether they think the shapes are the same or different.

Variation

 Include 2D shapes to address misconceptions, for example a square and a cube are the same.

Geometry and Measure – Geometrical reasoning, shapes and measurements

Revise

3D shape sequences

Learning objectives

Code	Learning objective
1Gg.03	Identify, describe and sort 3D shapes by their properties, including reference to the number of faces, edges and whether faces are flat or curved.
1Nc.06	Use familiar language to describe sequences of objects.

Resources

set of classroom 3D shapes

What to do

- Make a repeating pattern with the classroom 3D shapes (e.g. pyramid, cuboid, pyramid, cuboid, pyramid). Ask: **What shape will come next in this sequence?** Invite a learner to the front of the class to add the next shape.

- Discuss the sequence of shapes with the learners, then challenge one learner to come to the front of the class to change the repeating pattern by swapping one of the types of shape (e.g. the cuboids) for another. They could choose a different shape entirely (e.g. a cylinder) or the same shape as the existing shape in the pattern but in a different size or colour. The class must then describe the new sequence and what has changed.

Name and describe 3D shapes

Learning objectives

Code	Learning objective
1Gg.03	Identify, describe and sort 3D shapes by their properties, including reference to the number of faces, edges and whether faces are flat or curved.

What to do

- Introduce the **Falling shapes game** (3D shapes only) to learners, demonstrating how to catch the shape specified and avoid the other shapes. Stop to ask individual learners questions at various points throughout, for example: **What 3D shape just fell? Are we catching cuboids? How will we know if the shape that's falling is a sphere?**
- At the end of the five-minute session, revise some of the shape names and describe the shape together.

Variation

Display the **Shape set tool** set to 3D shapes. Ask pairs of learners to say three things about a shape so that the whole class can then identify it. Drag and drop the shape onto the screen.

Geometry and Measure – Geometrical reasoning, shapes and measurements

Revise

Mystery box

Learning objectives

Code	Learning objective
1Gg.02	Use familiar language to describe length including long, longer, longest, thin, thinner, thinnest, short, shorter, shortest, tall, taller and tallest.
1Gg.04	Use familiar language to describe mass, including heavy, light, less and more. (variation)

Resources

box that can be closed so that the contents are not visible, with at least six items inside, e.g. pencil, ribbon, crayon, shoe, notebook and string – anything from the role-play area that can be measured (per class)

What to do

- This is a fast-paced teacher-led activity to develop using key words for comparison.
- Walk around the classroom with the box and invite volunteers to remove two objects (without looking) and compare their lengths, using the words 'long', 'short', 'longer' and 'shorter'.

Variations

Ask learners to find things in the classroom that could go into the box and lead a discussion as to whether the object should be measured in terms of length, height or both.

Put objects in the box that can be compared by mass rather than length (such as an eraser, a shoe, a balloon, a book, a soft toy and a bottle of water).

Comparing length and height

Learning objectives

Code	Learning objective
1Gg.02	Use familiar language to describe length including long, longer, longest, thin, thinner, thinnest, short, shorter, shortest, tall, taller and tallest.

Resources

20 interlocking cubes (per group)

What to do

- This is a whole-class activity but learners work at their group table. Put a pile of interlocking cubes into the middle of each table and give learners two minutes to join as many as they can individually.
- Note: Make sure that if learners hold up their *rods* horizontally, they use the terms 'longer' and 'longest', and if they hold up *towers* vertically they use the terms 'taller' and 'tallest'. Point this out to learners before starting.

- Stop and choose two learners at a time to hold up and compare their cube *rods/towers*, using the terms 'longer' and 'shorter' or 'taller' and 'shorter'. Ensure that learners use the key words in their comparisons.
- As the unit develops, choose three or more learners at a time to compare and order their towers.

Variation

Choose one learner's tower and challenge the class to make a tower that is longer or shorter.

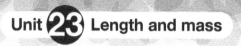

Revise

Being balance scales

Learning objectives

Code	Learning objective
1Gg.04	Use familiar language to describe mass, including heavy, light, less and more.

Resources

objects of different mass (per class) (for variation)

What to do

- Ask the whole class to stand up. Learners will be balance scales, modelling 'the same' with their arms outstretched at equal heights; 'heavy and light' with one arm higher and the other lower.
- Model before the first session and play the game at a swift pace. Before starting, make sure learners can spread their arms out without touching anybody.
- Say: **Show me 'the same'. Show me 'not balanced'. Wave your heavier hand. Wave your lighter hand. Show me 'not balanced'. Show me 'the same'**.

Variation

Ask a learner to come to the front to be a set of balance scales.

Put objects of different mass in their hands. They adjust the position of their hands to show which object feels heavier and which feels lighter.

Repeat with different learners and objects.

Revise

Show me

Learning objectives

Code	Learning objective
1Gg.05	Use familiar language to describe capacity, including full, empty, less and more.

What to do

- Teach learners the following instructions in response to these gestures: full – hands on top of head; half full – hands on hips; holds the least – arms to the floor and shake; holds the most – arms up and shake.
- Call out an instruction (e.g. half full). Learners must show you the correct gesture for that instruction.
- Learners respond to quickfire instructions that should be mixed up and changed constantly.

Variation

 Mix up instructing the whole class or saying a learner's name. If a class member follows the instruction when it was to a different individual learner, they are out of the game.

Capacity comparisons

Learning objectives

Code	Learning objective
1Gg.05	Use familiar language to describe capacity, including full, empty, less and more.

Resources

six pieces of card with these statements written on them: 'Could a hippopotamus fit in a cooking pot?' 'Does an elephant fit in a bath?' 'Would [teacher's name] drink out of a bucket?' 'Could you fit the ocean in a bucket?', 'What sort of container is bigger than your water bottle?', 'If you wanted to fill a paddling pool with water, would you use a bucket or a cup?' (per class); mini whiteboard and pen (per learner)

What to do

- This is a quickfire starter game.
- Shuffle the cards, choose one, ask the question and count to 10.

- Learners write 'yes' or 'no' or draw the answer on their mini whiteboards.
- Say: **Show me** and discuss the answers with the class.

Variation

 Adapt for temperature lessons by substituting the card questions and statements with the following: 'Is an ice cube hot or cold?'; 'Would you wear a woolly hat and scarf when it's hot?'; 'Does food get hot or cold in an oven?'; 'Do we use a fridge to cool things down?'; 'Is a fire hot or cold?'; 'Do you wear a sun hat in cold weather?'

Revise

Which scale?

Learning objectives

Code	Learning objective
1Gg.08	Explore instruments that have numbered scales, and select the most appropriate instrument to measure length, mass, capacity and temperature.

Resources

selection of scales for measuring such as different thermometers (Lesson 4 only), measuring jugs and beakers, measuring scales, rulers and metre sticks (per group); mini whiteboard and pen (per group) (for variation)

What to do

- Give each group a selection of measurement scales and allow them a few minutes to look at these and discuss what they are for, as a group.
- Ask the following questions, pausing after each one for a member of each group to hold up an appropriate scale in answer to your questions: **What would you use to find out how hot it is today? What would you use to measure the length of a pencil? What would you use to measure out drinks? What would you use to find out how heavy you are? What would you use to measure out nuts? What would you use to measure the length of the classroom? What would you use to set the oven to hot?**
- There will sometimes be more than one appropriate answer to a question. Discuss this with learners where necessary.

Variation

2 If you don't have enough measuring scales for each group, give the groups a mini whiteboard and pen each and hold up one scale at a time and ask the groups to discuss what they think it would be used for and write or draw the answer on the whiteboards.

Geometry and Measure – Geometrical reasoning, shapes and measurements

Geometry and Measure – Position and transformation

Revise

Teacher says...

Learning objectives

Code	Learning objective
1Gp.01	Use familiar language to describe position and direction.

Resources

key words from Resource sheet 17: Direction cards (per class/group)

What to do

- This game is similar to the traditional 'Simon says' activity. Move quickly on, from instructing the whole class, to rows/groups or individual learners, bearing in mind space, safety issues and time constraints.
- Say: **Stand UP; sit DOWN; TURN AROUND and smile at the person BEHIND you; sit IN an imaginary paddling pool...**

- Add extra positional language as the unit progresses (for example, **TURN AROUND; go ALL THE WAY AROUND the table; stand NEAR the door; step INSIDE the reading corner.**

Variation

2 Add left and right to your instructions after Lesson 2.

Ants crossing

Learning objectives

Code	Learning objective
1Gp.01	Use familiar language to describe position and direction.

Resources

squared paper (per class) (for variation)

What to do 📊

- Display the **Ant crossing game**.
- Take instructions from the whole class for which arrow to press to move the ant across the water.
- Count each step the ant takes to land on the other side.

Variation

2 Make class versions of the game, drawn up on squared paper.

Revise

Treasure hunt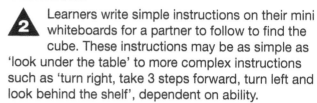

Learning objectives

Code	Learning objective
1Gp.01	Use familiar language to describe position and direction.

Resources

one interlocking cube or another similar-sized item (per learner); mini whiteboard and pen (per pair) (for variation)

What to do

- Working in pairs each learner hides their cube somewhere. Suggest that they hide it inside, underneath, on top of or behind something.
- They say simple instructions to their partner to follow to find the cube.

Variations

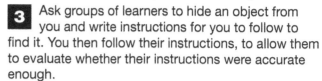 Learners write simple instructions on their mini whiteboards for a partner to follow to find the cube. These instructions may be as simple as 'look under the table' to more complex instructions such as 'turn right, take 3 steps forward, turn left and look behind the shelf', dependent on ability.

3 Ask groups of learners to hide an object from you and write instructions for you to follow to find it. You then follow their instructions, to allow them to evaluate whether their instructions were accurate enough.

Geometry and Measure – Position and transformation

Revise

Would you rather...

Learning objectives

Code	Learning objective
1Ss.01	Answer non-statistical questions (categorical data).
1Ss.02	[Record,] organise and represent categorical data using: - practical resources [and drawings - lists and tables - Venn and Carroll diagrams - block graphs and pictograms].
1Ss.03	Describe data, using familiar language including reference to more, less, most or least to answer non-statistical questions and discuss conclusions.

What to do

- Learners all stand together in the middle of the room. Ensure that - tables and chairs have been moved out of the way for safety.
- Ask a question with two possible answers (such as: **Would you rather play football or go swimming?**) and tell learners that one side of the room is for the first answer and the other side of the room is for the second answer.
- Learners move to the side of the room that matches the answer they have chosen.

- Count the learners on each side of the room and discuss which is the more popular answer and why that might be the case, for example: **18 people like football and 12 people like swimming. Maybe that is because a lot of people in this class go to a football club.**

Variation

3 Add a third possible answer to the question. Learners that choose the third answer stay in the middle of the room.

Revise

Sorting contest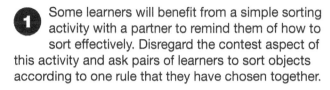

Learning objectives

Code	Learning objective
1Ss.01	Answer non-statistical questions (categorical data).
1Ss.02	[Record,] organise and represent categorical data using: - practical resources [and drawings - lists and tables - Venn and Carroll diagrams - block graphs and pictograms].
1Ss.03	Describe data, using familiar language including reference to more, less, most or least to answer non-statistical questions and discuss conclusions.

Resources

eight sortable objects (e.g. pencils of differing colour and length, different-coloured and sized buttons with two and four holes, 2D shapes of differing colour and size) (per group); two sorting hoops (per group)

What to do

- Learners work in groups.
- Give groups of learners a group of at least eight objects to sort. The objects must all fall under the same category but have variables (such as eight buttons with varying numbers of holes in different colours and sizes) and two sorting hoops.
- Give learners one minute to sort their objects according to their rule. If all members of the group choose the same sorting rule they will be able to do this easily. If different sorting rules are chosen, they will discover that it is impossible to sort all the objects as different members of the group will keep removing objects from one sorting hoop and putting it into another.
- When the minute is up, discuss the problems encountered. Ask: **Why can't we sort objects if we are both using different sorting rules?** Explain that we must agree what we are trying to learn about the objects or the information at the end won't be accurate.
- Try the activity again with learners aiming to use the same sorting rule.

Variation

1 Some learners will benefit from a simple sorting activity with a partner to remind them of how to sort effectively. Disregard the contest aspect of this activity and ask pairs of learners to sort objects according to one rule that they have chosen together.

Revise

Shopping lists

Learning objectives

Code	Learning objective
1Ss.01	Answer non-statistical questions (categorical data).
1Ss.02	[Record,] organise and represent categorical data using: - practical resources [and drawings - lists and tables - Venn and Carroll diagrams - block graphs and pictograms].
1Ss.03	Describe data, using familiar language including reference to more, less, most or least to answer non-statistical questions and discuss conclusions.

Resources

mini whiteboard and pen (per group)

What to do

- Learners work in groups.
- Tell learners that they need to make a shopping list for fruit and that they will need to know how many of each fruit to buy.
- Learners ask each person in the group which fruit they like best. They draw a picture of that fruit on their whiteboards and write the learner's name (or initials) next to it.
- If a learner specifies a fruit that has already been chosen by another learner, they write the second learner's name (or initials) alongside the first learner's name next to the picture of the fruit.

- When all learners have been asked, they write a number next to each fruit by counting the number of names in each fruit category. Names can be rubbed out at this point to make the list clearer.

Variation

Extend the activity by allowing learners to 'shop' for their fruit from the snack table or role play a fruit stall with plastic fruit. This allows learners to see that the data collection and list-making had a real purpose.

Revise

Sorting with a Venn diagram

Learning objectives

Code	Learning objective
1Ss.01	Answer non-statistical questions (categorical data).
1Ss.02	Record, organise and represent categorical data using: [- practical resources and drawings - lists and tables] - Venn and Carroll diagrams [- block graphs and pictograms].
1Ss.03	Describe data, using familiar language including reference to more, less, most or least to answer non-statistical questions and discuss conclusions.

What to do

- Write the numbers 1 to 10 on the board and draw a simple Venn diagram with the ring having the heading 'straight lines'.
- Learners direct sorting according to the following criteria: numbers with straight lines only in the ring, numbers with curved or curved and straight lines in the frame outside the ring.

Variations

1 Present this as a simple sorting activity without the Venn diagram to remind learners of how to sort before they commence work on Venn and Carroll diagrams: Numbers with straight lines only on one side of the board, numbers with curved or curved and straight lines on the other side of the board.

2 Use a Carroll diagram for sorting using the same criteria 'Straight lines only' / 'Not straight lines' (swap numbers from 1 to 10 for letters from a to j if you need to).

Human pictogram

Learning objectives

Code	Learning objective
1Ss.01	Answer non-statistical questions (categorical data).
1Ss.02	Record, organise and represent categorical data using: [- practical resources and drawings - lists and tables - Venn and Carroll diagrams] - block graphs and pictograms.
1Ss.03	Describe data, using familiar language including reference to more, less, most or least to answer non-statistical questions and discuss conclusions.

Resources

cubes (e.g. building blocks) in red, blue, green and yellow (per class); chalk (per class)

What to do

- Take learners outside.
- Ask them what their favourite colour is out of red, blue, green and yellow.
- Place a cube in each of the four colours on the ground, one below the other in a line. Ask learners to arrange themselves in rows going across from the cube of their favourite colour.
- Count the learners in each row and write the number on the ground in chalk next to each coloured cube.

Variations

2 Use the chalk to draw the axes of a simple block graph. Place the coloured cubes along the bottom of the 'graph' and ask learners to line up above their favourite colour.

3 Set learners the challenge of finding out which colour has the longest/shortest line.

Revise

Block graph

Learning objectives

Code	Learning objective
1Ss.01	Answer non-statistical questions (categorical data).
1Ss.02	Record, organise and represent categorical data using: [- practical resources and drawings - lists and tables - Venn and Carroll diagrams] - block graphs[and pictograms].
1Ss.03	Describe data, using familiar language including reference to more, less, most or least to answer non-statistical questions and discuss conclusions.

What to do

- Display the **Bar chart tool**. Work with the top four colours, blue, red, orange, green. Remind learners that they can only vote once. Say: **Stand up to vote for your favourite colour from blue, red, orange or green.** Learners stand up in turn. Count them and add to the value on the bar charter.
- Now make the graph. Total the numbers shown. Ask: **Why must the total number equal the number of children in class today?**

Variation

Use the colours as a key for any variation appropriate to the learners, such as favourite drink or snack.

Unit 1: Counting and sequences to 10

Collins International Primary Maths Recommended Teaching and Learning Sequence: Term 1, Week 2

Learning objectives

Code	Learning objective
1Nc.01	Count objects from 1 [0] to 10 [20], recognising conservation of number and one-to-one correspondence.
1Nc.02	Recognise the number of objects presented in familiar patterns up to 10, without counting.
1Nc.03	Estimate the number of objects or people (up to 10 [20]), and check by counting.
1Nc.04	Count on in ones[, twos or tens,] and count back in ones [and tens], starting from any number (from 1 [0] to 10 [20]).

Unit overview

In this unit, learners acquire the skills they need to estimate up to ten objects and then check to see if they are correct. They learn to count objects with one-to-one correspondence, recognising conservation of number, and to count on and back in ones, noting what happens when they add one more object to the set or take one away. They learn what numbers of objects to 10 look like when they are arranged in familiar patterns, which helps them to visualise amounts when estimating, and discover that estimation is about guessing the number as closely as possible, based on what you already know.

Prerequisites for learning

Learners need to:
• be able to count to 10
• be able to recognise, read and write numbers to 10.

Vocabulary

number, count, ones, estimate, guess, count on, forwards, count back, backwards, pattern, more than, less than

Common difficulties and remediation

Learners may not understand that, when counting objects, each number said is a label for one object, and that each object must be counted only once. Practise organising objects so that they are easier to count accurately (into lines or pairs) and counting slowly and deliberately together as a class, using one number name for each object.

To help learners with the concept of estimation, show them a group of ten objects, a group of five objects and one object so that they have a basis for comparison. Ask questions such as: **Has this group got more or less than five in it?**

Spend time counting and re-counting sets of objects after moving the objects into new positions to build the understanding of conservation of numbers and that the number of objects is always the same if no objects are added or taken away from the set.

Supporting language awareness

Use every possible opportunity to count with learners, for example counting them as they line up to leave the classroom or counting out resources as you hand them out to learners.

Promoting Thinking and Working Mathematically

TWM.03 Conjecturing
Learners ask: **How many objects are in that set?** and question how they can discover this information by counting.

TWM.07 Critiquing
Learners use the skills needed to estimate and check. They choose from a variety of ways to estimate a number of objects.

TWM.08 Improving
Learners organise objects to make them easier to count or estimate.

Success criteria

Learners can:
• accurately count up to ten objects in a set
• recognise what comes next in a sequence of numbers appearing on familiar objects
• recognise numbers of objects presented in familiar patterns
• count on in ones and back in ones from any number from 1 to 10
• give a reasonable estimate for how many are in a set of up to ten objects.

Note

In order to conform to the terminology of the Cambridge Primary Mathematics Framework (0096), and also current common usage, the term 'less than' is used predominantly at this stage to compare and order numbers. However, it is important to note that, if being grammatically correct, the word 'less' should be used when quantities cannot be individually counted, for example: 'It should take less time', whereas 'fewer' should be used when referring to items that can be counted individually, for example: 'Fewer than ten people attended'.

For those teachers/schools following the CIPM Recommended Teaching and Learning Sequence, zero is not introduced until Term 1, Week 3 (Unit 16, LO1Np.01: Understand that zero represents none of something). Therefore, in this unit learners are working with numbers 1 to 10 and not zero. In subsequent units, however, learners are reciting, counting, reading and writing from zero.

Lesson 1: **Counting objects**

Learning objectives

Code	Learning objective
1Nc.01	Count objects from 1 [0] to 10 [20], recognising conservation of number and one-to-one correspondence.

Resources

ten interlocking cubes (per learner)

Revise

Use the activity *Counting people* from Unit 1: *Counting and sequences to 10* in the Revise activities.

Teach [SB] [📊] [TWM.08]

- Direct learners to the picture in the Student's Book. Ask: **Which fish tank do you think contains more fish?** Take suggestions, then count the fish in each tank, with learners, and establish that both tanks contain the same number of fish. Explain to learners that sometimes the same number can look different if it is arranged differently.

- Show the **Tree tool** and put five birds in the tree. **[T&T]** Say: **When we use new or different skills to make a method more effective, we are *improving*.** Elicit that it would help if you lined up the birds so you don't accidentally count any of them twice or miss any out, and that you must use one number name for each bird and only say the next number name when you touch the next bird.

- Count the birds with learners, following all of their counting advice. Establish that there are five birds.

- Move the birds so that they are further apart and ask: **How many birds are there now?** Take suggestions and then count the birds together to demonstrate that the number of birds is the same. Explain that when objects are spread out, our brain tries to trick us because it thinks that there are more objects as the group looks bigger. Experiment with this, moving the birds closer together and further apart, counting them each time and getting the same result. Say: **We haven't put any more birds on the tree or taken any off the tree, so the number of birds has stayed the same.**

- Repeat the above for other numbers of objects to 10.

- Discuss the Guided practice example in the Student's Book.

Practise [WB]

- Workbook

Title: Counting objects

Page: 6

- Refer to Activity 1 from the Additional practice activities.

Apply [👤] [🖥]

- Display **Slide 1**.

- Set out three sets of objects (such as interlocking cubes) on each group's table: a set of three cubes lined up but spaced far apart, a set of eight cubes lined up and pushed close together and a set of six cubes in a random group (not lined up).

- Learners count the cubes in each set and write down how many are in the set.

- [🗣] Ask: **Which set of cubes was the easiest/ hardest to count? Why?**

Review

- Ask two learners to come to the front and stand behind a table. Give each learner seven cubes.

- Ask: **Who do you think has more cubes?** Take suggestions.

- Count the cubes in each set with learners and establish that there are seven cubes in each set.

- Move the cubes in one set so that they are widely spaced and push those in the other set closer together. [🗣] Ask: **Who do you think has more cubes now?** Praise any answers mentioning that they still have the same number of cubes as you didn't give them any more or take any away, or saying that the numbers are still the same but that they look different because you've moved the cubes.

- Count the cubes again, reinforcing that both learners still have the same number of cubes.

Assessment for learning

- How could you make counting easier?
- How many objects have you got?
- I have moved these objects – how many are there now?
- What would happen if I said the next number before I touched the next object while I was counting it?

Same day intervention

Support

- Practise counting up to three, four and five objects with learners until they are used to saying a number name for each object that they touch and are confident with conservation of number. When they are secure with this, add more objects.

Lesson 2: **Counting on and back in ones**

Learning objectives

Code	Learning objective
1Nc.04	Count on in ones[, twos or tens,] and count back in ones [and tens], starting from any number (from 1 [0] to 10 [20]).

Resources

mini whiteboard and pen (per learner); set of 2–9 number cards from Resource sheet 1: 0–20 number cards (per pair); 1–10 number track (per pair)

Revise

Use the activity *One more, one less* from Unit 1: *Counting and sequences to 10* in the Revise activities.

Teach 🔲 🖥

- Direct learners to the picture in the Student's Book. Tell learners that one of the people playing the game has rolled or spun a 1 so needs to move forward one space. Point out that their counter is currently on 7. Ask: **Which number will they move to if they have to move forward one space?** Establish that they will move to 8. Say: **8 comes after 7 when we are counting in ones.**

- Display **Slide 1** and follow the numbers on the number track, counting forwards from 1 to 10 with learners then counting backwards from 10 to 1. Remind them that they are counting in ones. Now practise counting forwards and backwards in ones from different numbers from 1 to 10. Each time, find the starting number on the number track and point to it, then follow the subsequent numbers with your finger as you count.

- Now point to a number on the 1–10 number track and ask learners to count on one more, writing the number on their whiteboard and then holding it up.

- Repeat this for several different numbers and then do the same for one less.

- Discuss the Guided practice example in the Student's Book.

Practise 🔲

- Workbook

Title: Counting on and back in ones

Page: 7

- Refer to Activity 1 from the Additional practice activities.

Apply 👥 🖥

- Display **Slide 2**.
- Give each pair of learners a set of 2–9 number cards and a 1–10 number track.

- Learners each pick a number card and put their finger on the matching number on the number track. Learner A counts forwards from that number in ones until they reach 10; Learner B counts backwards in ones until they reach 1.
- Learners swap roles and repeat.

Review 📊

- Ask a learner to choose a number from 2 to 9 (such as 6).
- Put that number of birds on the tree on the **Tree tool** then count them with learners.
- 🔲 Referencing the number track, ask: **What is one more than 6?** Elicit that one more than 6 is 7.
- Add one more bird to the **Tree tool** and count the birds.
- Remove the extra bird, count the six birds and ask: **What is one less than 6?** Agree that it is 5.
- Remove another bird and count the remaining birds.
- Repeat this for other numbers from 2 to 9.

Assessment for learning

- What is one more than...?
- What is one less than...?
- Can you count forwards to 10/backwards to 1 from (choose a number from 1 to 9)?
- How many birds are left if we add one more/take one away?

Same day intervention

Support

- Give learners lots of practice at counting forwards from 1 to 10 and backwards from 10 to 1 as they use their fingers to follow the numbers on a number track, until they can do this without the support of the number track.

Enrichment

- Give pairs of learners a set of 1–5 number cards. Learner A picks one. Learner B must count forwards from this number until learner A claps their hands, at which point they must start counting backwards until they reach 1. They swap roles and repeat.

Number – Counting and sequences

Lesson 3: **Sequences of objects**

Learning objectives

Code	Learning objective
1Nc.02	Recognise the number of objects presented in familiar patterns up to 10, without counting.

Resources

blank ten frames from Resource sheet 2: Ten frames (per learner); set of 1–10 number cards from Resource sheet 1: 0–20 number cards (per learner); ten counters (per learner); 1–10 ten frame card (filled in) from Resource Sheet 2: Ten frames or a 1–6 dice/domino pattern cards from Resource sheet 3: Number patterns (one frame or card per learner)

Revise

Use the activity *Roll the dice* from Unit 1: *Counting and sequences to 10* in the Revise activities.

Teach 🆂🅱 🖥 📊

- Ask learners if they know what the picture in the Student's Book shows. Elicit that it is a domino. Explain that the dots on each half of the domino represent numbers. Count the dots on each side and establish that there are five dots on one side and two on the other.
- [T&T] Ask: **What other things can you think of that have dots on to stand for a number?** Take suggestions and discuss how the objects are used.
- Explain to learners that it is useful to be able to recognise the number represented by the dots without having to count them (a concept known as 'subitising') because you can play games more quickly and also use this skill to work out how many objects there are in larger groups.
- Display **Slide 1** and ask learners to count / identify the number of dots on each dice face. Label each pattern, writing its number underneath.
- Now show the **Ten frame tool** or draw a ten frame on the board and show learners how to fill in dots to represent numbers from 1 to 10.
- Discuss the Guided practice example in the Student's Book.

Practise 🆆🅱

- Workbook

Title: Sequences of objects

Page: 8

- Refer to Activity 2 from the Additional practice activities.

Apply 👥 🖥 [TWM.07]

- Display **Slide 2**.
- Give each learner a blank ten frame template, a set of 1–10 number cards and ten counters.

- [TWM.07] Each learner secretly picks a number card and (without their partner knowing the number) puts counters on their ten frame to represent that number.
- Learners look at each other's ten frame and say what number their partner's ten frame represents. They check the number cards to see if they are correct.
- They discuss the results – were they correct? Did they mistake the number pattern for a different one?

Review

- Give each learner a 1–10 ten frame card (filled in) or a 1–6 dice/domino pattern card.
- Call out numbers from 1 to 10. Any learner who has a card that represents that number must hold up their card.
- Learners swap cards with another learner and the activity is repeated.

Assessment for learning

- What number does the domino/dice show?
- Show me how to make the pattern for 8 on the ten frame.
- What number is the ten frame showing?

Same day intervention

Support

- Focus on the number patterns for 1–5 until learners can recognise them immediately, then add more number patterns slowly.

Enrichment

- Each learner secretly picks a number card and puts counters on one ten frame to represent their number and on another ten frame to represent a different number to 10. They show both ten frames and the number card to their partner who must match the card to the correct ten frame as quickly as possible.

Lesson 4: **Estimating to 10**

Learning objectives

Code	Learning objective
1Nc.01	Count objects from 1 [0] to 10 [20], recognising conservation of number and one-to-one correspondence.
1Nc.02	Recognise the number of objects presented in familiar patterns up to 10, without counting.
1Nc.03	Estimate the number of objects or people (up to 10 [20]), and check by counting.

Resources

dishes containing 10 buttons, 2 buttons and 20 buttons (per class); 2–10 countable objects, e.g. small toys, counters, interlocking cubes (per group); sheet of paper on each table with the names of the different learner groups in the class (e.g. circle group, square group, triangle group, ... or red group, blue group, green group...) (per class)

Revise

Use the activity *Counting people* (variation) from Unit 1: *Counting and sequences to 10* in the Revise activities.

Teach [SB] [TWM.03/07]

- Look at the picture in the Student's book. **[TWM.03]** **[T&T]** Ask: **What would you like to know about the shells in the jar? Are you wondering how many there are? How many do you think? How do you think could you make a better guess?** Tell learners that sometimes when we want to know how many of something there are, we can have a 'best guess' instead of counting them. This is called 'estimating'.

- Show learners a dish containing about ten buttons. Explain that they are going to estimate how many buttons are in the dish.

- Show them an identical dish containing two buttons. Ask: **Are there more or fewer than two buttons in the first dish?** Elicit that there are definitely more than two buttons.

- **[T&T]** Ask: **Are there just one or two more buttons than two, or quite a lot more buttons than two? How can you tell?**

- Repeat this process, showing them a dish containing 20 buttons. Establish that you can tell that there are quite a lot fewer buttons than 20 but quite a lot more than two. Take estimates for how many buttons in the first dish and then count them together.

- Show learners another strategy by reminding them that they know what numbers from 1 to 10 look like in groups because they can recognise numbers as patterns.

- Draw the dice pattern for 5 on the board and next to it, draw the dice pattern for 4. Explain to learners that they know that there are five dots in one set and four dots in the other set without counting, so there must be more than five dots in total.

- Discuss the Guided practice example in the Student's Book.

Practise [WB]

- Workbook

Title: Estimating to 10

Page: 9

- Refer to Activity 1 from the Additional practice activities.

Apply 👥 🖥 [TWM.07]

- Display **Slide 1**.

- On each table place a set of 2–10 objects, along with the sheet of paper containing the names of the different learner groups in the class. Remind learners that they are estimating how many, not counting the objects.

- Each group must visit every table and leave an estimate there beside their group name on the sheet of paper.

- **[TWM.07]** At the end of the activity, discuss and compare the estimates. Which group got the closest at each table? How did each group decide on their estimate?

Review [TWM.08]

- Ask ten learners to stand at the front. From this group of ten, ask learners who prefer bananas to raise their hands and learners who prefer apples to fold their arms.

- The rest of the learners estimate how many learners prefer each fruit. **[TWM.08]** Ask: **How can we make a good estimate of how many learners prefer apples?** Invite discussion, eliciting that they must think about how many learners are at the front of the class in total, whether more or less prefer apples and ways to arrange the learners to inform their estimates.

Assessment for learning

- What information are we organising?
- How many... are there? How do you know?

Same day intervention

Support

- Give these learners quantities to estimate, with large differences such as two counters or nine counters, so that they can easily see the difference between an amount that is close to 10 and an amount that is close to 1.

Additional practice activities

Activity 1

Learning objectives
- Count up to ten objects, recognising conservation of number
- Find the numbers one more and one less than a given number to 10
- Estimate whether there are more, less or the same amount of objects

Resources
10 countable objects (e.g. building blocks); mini whiteboard and pen (per group) teddy bear or other soft toy (per group); 1 to 10 number track (per group)

What to do
- Ask a learner to count out six blocks. Check that there are six blocks by counting them as a group and write the number 6 on the mini whiteboard.
- Tell learners that Teddy likes to move the blocks to trick us. Sometimes he moves the objects further apart or closer together and sometimes he adds an extra object or takes one away.
- Ask: **If Teddy moves the blocks but doesn't add any or take any away, will the amount of blocks change?**
- Learners close their eyes while Teddy moves the blocks. You can choose to keep the number the same or add a block or take one away at the same time as moving them.

- When learners open their eyes, ask them to guess (estimate) whether there are more or fewer objects or whether the number has stayed the same. Then count the objects together.
- If there is a different number of blocks now, check whether there is one more or one less by finding 6 on a number track and looking at the number before and after it.
- Repeat this with different starting numbers.

Variations
1 Ask learners to count out five or fewer blocks to allow them to secure these skills to five before they attempt to apply them to up to ten objects.

2 Choose which part of this lesson you want most to focus on as the unit progresses: estimation, counting and number conservation or counting on and back in ones.

3 Ask pairs of learners to take it in turns to be Teddy moving/adding to/taking away the blocks so that this can be a paired activity without adult support.

Activity 2 or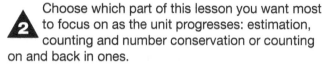

Learning objective
- Recognise the number of objects in familiar patterns

Resources
set of cards showing the dice/domino number patterns for numbers 1–6 from Resource sheet 3: Number patterns (per pair); set of 1–6 number cards from Resource sheet 1: 0–20 number cards (per pair)

What to do
- Give one learner from each pair a shuffled set of number pattern cards and the other learner from each pair a shuffled set of 1–6 number cards.
- Learners play 'Snap' with the number pattern cards and the number cards, taking turns to put down a card. If the number card put down matches the amount on the number pattern card put down, learners compete to shout 'Snap!' and put their hand on the cards.

- The first learner to do so wins one point and the game continues, reshuffling each set of cards if necessary.

Variations
1 Pairs of learners simply match the number patterns to the corresponding number cards.

2 Turn all the cards face down and play a matching pairs game with them, each learner taking turns to turn two cards over. If the two cards match, the learner gets to keep those cards. The winner is the learner with the most pairs of cards at the end.

3 Remove one number card and one number pattern card from each set of cards. Learners must work out which card is missing from each set.

Unit 2: Counting and sequences to 20

Collins International Primary Maths Recommended Teaching and Learning Sequence: Term 2, Week 7

Learning objectives

Code	Learning objective
1Nc.01	Count objects from 0 to 20, recognising conservation of number and one-to-one correspondence.
1Nc.02	Recognise the number of objects presented in familiar patterns up to 10, without counting.
1Nc.03	Estimate the number of objects or people (up to 20), and check by counting.
1Nc.04	Count on in ones, twos or tens, and count back in ones and tens, starting from any number (from 0 to 20).
1Nc.05	Understand even and odd numbers as 'every other number' when counting (from 0 to 20).
1Nc.06	Use familiar language to describe sequences of objects.

Unit overview

In this unit, learners extend the amount of objects they can count and estimate to 20. They also learn about different ways to count and organise numbers and amounts when counting as they begin to count in twos or tens. This progresses to learning to count on in twos or tens from any given number from 0 to 20. Learners also learn about odd and even numbers, recognising that these are 'every other number' when counting.

Prerequisites for learning

Learners need to:
- be able to recite, recognise, read and write numbers to 20
- have experience of estimating amounts
- be able to count reliably up to 20 objects.

Vocabulary

estimate, count, count on, odd, even, twos, tens, more, less, on, back

Common difficulties and remediation

Some learners may struggle to understand the concept of estimation and will choose a number at random as their estimate, rather than looking at the group of objects and thinking carefully about how many there may be. To help learners with this, show them a group of 20 objects, a group of ten objects, a group of five objects and one object so that they have a basis for comparison. Ask questions such as: **Has this group got more or less than ten in it?** while encouraging learners to compare it to the groups of five and 20 objects.

Some learners may readily take to counting on in twos from an even number but find counting on in twos from an odd number more difficult. Using a number line or track for support can help with this.

Supporting language awareness

Display the key words for each lesson and discuss them with the learners. Use and refer to these words throughout the lesson and encourage learners to use the words too, prompting them when necessary.

Display both the number names and the numerals that match them together and ensure that these are displayed in various places around your classroom, as number lines and also as labels for objects (e.g. 3/three windows).

Use every available opportunity to count with the learners, for example, counting them as they line up to leave the classroom or counting out resources as you hand them out to learners. Try to count learners or objects in twos when possible, to remind learners of this strategy, and refer to odd and even numbers in practical situations (e.g. **There are 15 learners in class today. That is an odd number, so if the learners work in pairs, there will be one learner left over**).

Promoting Thinking and Working Mathematically

TWM.03 Conjecturing
Learners ask: **Is this number odd or even?** They use number lines and pair objects to work out the answer.

TWM.06 Classifying
Learners sort numbers into odd and even numbers.

TWM.07 Critiquing
Learners gather the skills they need in order to estimate and check whether their estimate was correct. They choose from a variety of ways to estimate a number of objects.

TWM.08 Improving
Learners use different strategies to make objects easier to count or estimate.

Success criteria

Learners can:
- accurately count a set of up to 20 objects
- provide a reasonable estimate for an amount of objects to 20
- recognise odd and even numbers to 20
- count on in twos from any number to 20
- count on in tens from any number to 20
- describe a sequence using familiar language.

Number – Counting and sequences

Number – Counting and sequences

Lesson 1: Estimating to 20

Learning objectives

Code	Learning objective
1Nc.03	Estimate the number of objects or people (up to 20), and check by counting.
1Nc.01	Count objects from 0 to 20, recognising conservation of number and one-to-one correspondence.
1Nc.02	Recognise the number of objects presented in familiar patterns up to 10, without counting.

Resources

36 interlocking cubes (per group); mini whiteboard and pen (per learner); sheet of A4 paper (per group); pot of 18 pencils (per group); tray of 20 crayons (per group); dish of 15 paper clips (per group); tray of 11 paintbrushes (per group)

Revise

Use the activity *Estimation contest* from Unit 2: *Counting and sequences to 20* in the Revise activities.

Teach 👥 [SB] [TWM.07/08]

- Direct learners to the picture in the Student's Book. Ask: **How many eggs do you think there are in the basket?** Make a note of all of their estimates, then count the eggs together. [T&T] Ask: **Why might you need to know roughly how many eggs there are?** Elicit that you may need a certain amount for a recipe or meal and it is quicker to look at the eggs and estimate how many there are rather than counting them individually.
- [TWM.07] Give each group 36 interlocking cubes and ask them to make a group of 20 cubes, a group of 10, a group of 5 and 1 cube on its own. Now ask learners to look carefully at what each group looks like. Next, ask learners to arrange the cubes in each group into a familiar number pattern (e.g. four cubes with another cube in the middle to represent 5) and remind them that recognising patterns can help them to estimate amounts.
- Show learners a group of 17 cubes. Ask each learner to write, on their mini whiteboards, an estimate of how many cubes there are. Then ask questions such as: **Are there more or fewer than ten cubes here? Are there as many as 20? Do you think there are at least 5 more than 10? Are there any patterns of amounts that you recognise?** [TWM.08] Based on your discussion, ask learners to make a second estimate and write it on their whiteboards. Discuss their two estimates and whether they *improved* their estimates after looking more closely at the group of cubes.
- [T&T] Ask: **Why do you think your estimate was closer the second time? What helped you to improve your estimate?**
- Discuss the Guided practice example in the Student's Book.

Practise [WB]

- Workbook

Title: Estimating to 20

Page: 10

- Refer to Activity 1 from the Additional practice activities.

Apply 👥 🖥

- Display **Slide 1**.
- Put a piece of A4 paper, a pot of 18 pencils, a tray of 20 crayons, a dish of 15 paper clips and a tray of 11 paintbrushes on each table.
- Each learner estimates how many items are in each container and writes their estimates and initials on the sheet of paper.
- When all learners have estimated, they count how many objects are in each container.

Review

- Discuss the containers of objects from the **Apply** activity, stating how many objects were in each container and looking at the estimates on the paper to see which estimates were the closest to the actual amounts. Address any misconceptions (e.g. that there were fewer paper clips because paper clips are small).

Assessment for learning

- How many cubes do you think are in this group?
- Can you see any number patterns in the group of cubes?
- Are there more/less than 5/10 cubes?
- How can you make your estimate better?

Same day intervention

Support

- Revisit estimating 1–10 objects before moving on to larger amounts.

Enrichment

- Give learners three sets of objects. Two of the sets must contain amounts of objects to 20. One set must contain more objects than this. Learners must first estimate which is the group with 'too many' objects, then estimate the two other quantities.

Lesson 2: **Counting in twos**

Learning objectives

Code	Learning objective
1Nc.04	Count on in [ones], twos [or tens, and count back in ones and tens], starting from any number (from 0 to 20).
1Nc.06	Use familiar language to describe sequences of objects.

Resources

Resource sheet 5: 0–20 number track (per pair); counter (per pair); Resource sheet 1: 0–20 number cards (per pair)

Revise

Use the activity *Estimation contest* (variation) from Unit 2: *Counting and sequences to 20* in the Revise activities.

Teach SB

- Direct learners to the picture in the Student's Book and ask them to look at the number track. Ask: **Can you see a pattern?** Elicit that the child has jumped onto every other number, missing one number out every time. Write a list on the board of the numbers that have been jumped on.
- Display the **Number line tool** for 0–20. Ask: **If we carried on the pattern, what number would come next?** Ask learners to refer to the number line to check. Continue like this until you have reached 20.
- Now make a list of all the numbers that were not jumped on and ask learners to refer to the number line to work out which numbers would come next if that pattern were continued. Stop when you reach 19.
- Count in twos along the number line, starting on 0 and stopping on 20. Try starting from different even numbers to demonstrate that you can count on from any number.
- Now count on in twos from 1 to 19. Ask: **What was different this time?** Agree that you said a different set of numbers. Try counting on from various odd numbers with learners.
- Ask learners to pick a number to count on from at random and ask them to guess which numbers you will be saying as you count on in twos each time. Some learners may be able to work this out by looking at the number line before they start to count.
- Discuss the Guided practice example in the Student's Book.

Practise WB

- Workbook

Title: Counting in twos

Page: 11

- Refer to Activity 2 from the Additional practice activities.

Apply 👥 🖥

- Display **Slide 1**.
- Give pairs of learners a 1–20 number track and a counter.
- Learner A puts the counter on any number on the number track. Learner B counts on in twos until they reach 19 or 20.
- Learners swap roles and repeat.

Review

- Give pairs of learners a set of 0–20 number cards.
- Call out 'What number will come next?' questions, for example: **What number will come next if I start on 8 and count on in twos?**
- Learners find the number that will come next and hold it up.

Assessment for learning

- Count on in twos from 4/7.
- What number will come next if you start on 15/16 and count on in twos?
- What numbers will you say if you start counting on in twos from 2/3?
- Show me how you would use a number line to help you count on in twos.

Same day intervention

Support

- Give learners a 1–10 number track and let them count on in twos to 10 only until they are confident with this.

Enrichment

- Challenge learners to create a 'counting in twos number track'. They should draw this on a whiteboard, decide whether to start on 1 or 2 and then write only the numbers that they would say when counting in twos along the track.

Number – Counting and sequences

Unit 2 Counting and sequences to 20

Lesson 3: **Odd and even numbers**

Number – Counting and sequences

Learning objectives

Code	Learning objective
1Nc.04	Count on in [ones], twos [or tens, and count back in ones and tens], starting from any number (from 0 to 20).
1Nc.05	Understand even and odd numbers as 'every other number' when counting (from 0 to 20).

Resources

Resource sheet 2: Ten frames (per learner); set of 1–20 number cards from Resource sheet 1: 0–20 number cards (shuffled) (per pair); 20 counters (per pair); two sorting hoops labelled 'odd' and 'even' (per group)

Revise

Choose an activity from Unit 2: *Counting and sequences to 20* in the Revise activities.

Teach 〔SB〕 📊

- Direct learners to the picture in the Student's Book and ask: **What do you notice about the two teams?** Elicit that the player numbers are the same as the numbers we used when counting on in twos in Lesson 2.
- Explain that the numbers 2, 4, 6 and 8 are called even numbers and that any number ending in one of these digits or 0 is an even number. Explain that the numbers 1, 3, 5, 7 and 9 are odd numbers and that any number ending in one of these digits is an odd number. Use the **Number line tool** to demonstrate counting on in twos as in Lesson 2, pointing out the odd and even numbers as you count. 🖸 Pick a number at random and ask: **Will I land on odd or even numbers if I count on in twos from this number?**
- Ask a learner to pick an even number from 2 to 10. Ask that many learners to come to the front. Demonstrate putting these learners into pairs, explaining that you can check if a number is odd or even because even numbers can always be put into pairs with nothing left over. Do this again with two different even numbers of learners. Now ask a learner to pick an odd number of learners and demonstrate that putting this amount of learners into pairs always leaves one left over. Do this with two more different odd numbers of learners.
- Discuss the Guided practice example in the Student's Book.

Practise 〔WB〕

- Workbook

Title: Odd and even numbers

Page: 12

- Refer to Activity 2 from the Additional practice activities.

Apply 👥 🖥 [TWM.06]

- Display **Slide 1**.
- Give each learner a blank ten frames template, and each pair of learners a shuffled set of 1–20 number

cards and 20 counters. Put sorting hoops labelled 'odd' and 'even' on each table.
- Working in pairs, learners pick a number card and put counters on their ten frames to represent that number.
- **[TWM.06]** They note whether the ten frame shows pairs or pairs with one left over to identify whether their number was odd or even. They put their number card into the correct sorting hoops then pick another number and start again.

Review 📊

- Count in twos along the **Number line tool** showing 0–20, starting on 8. Ask learners to call out 'odd' or 'even' for every number you say. Ask: **What did you notice about all the numbers that I said?**
- Do the same, this time starting on 7.
- Discuss your findings, explaining to the learners that if you count on in twos from any even number, all the numbers that you say will be even and if you count on in twos from any odd number, all the numbers that you say will be odd.

Assessment for learning

- Is 17 odd or even?
- How could you use a number line to check if a number is odd or even?
- How could you use pairs/ten frames to check if a number is odd or even?
- There are three pairs and none left over for the number 6 – is it odd or even?

Same day intervention
Support

- Focus on odd and even numbers to 10 before progressing to the numbers 11 to 20.

Enrichment

- Give learners a set of 1–20 number cards and a dish of counters. Turn the number cards face down and spread them out. Learners take it in turns to pick a number card. If their number card is even, they get a counter. If it is odd, they don't get a counter. The first learner to get five counters is the winner.

Lesson 4: **Counting in tens**

Learning objectives

Code	Learning objective
1Nc.04	Count on in [ones, twos or] tens, and count back in [ones and] tens, starting from any number (from 0 to 20).

Resources

Resource sheet 1: 0–20 number cards (per pair); mini whiteboard and pen (per pair)

Revise

Use the activity *Ten more* from Unit 2: *Counting and sequences to 20* in the Revise activities.

Teach 📘 📊

- Look at the picture in the Student's Book. Ask: **How fingers is the first child holding up? How many fingers is the second child holding up?** Count how many with the learners, then say: **They have made 10 more fingers than 3. How many is that altogether?** Count all the fingers and agree that there are 13 fingers.
- Write the numbers 3 and 13 on the board and ask: **What do you notice about these two numbers?** Elicit that they both contain 3. Explain that when you count on 10 more from any one-digit number, the answer will always have 1 in the tens position and the starting number in the ones position.
- Use the **Number line tool** to test this theory by asking learners to call out one-digit numbers and counting on 10 more from these, using your fingers and the 0–20 number line. Write the pairs of numbers on the board.
- Now explain that when you count back 10 from any two digit number up to 19, the answer will always be the number in the ones position as a single-digit number. Test this by counting back 10 along a number line from 14, 19 and 16.
- 🖐 Pick a single-digit number and ask: **If I count on ten, what number will I land on?**
- Remember to investigate what happens when you count on in tens from 0 and 10 and back in tens from 20 and 10.
- Discuss the Guided practice example in the Student's Book.

Practise 📓

- Workbook

Title: Counting in tens

Page: 13

- Refer to Activity 1 (variation) from the Additional practice activities.

Apply 👥 💻

- Display **Slide 1**.
- Give each pair a set of 0–20 number cards and a mini whiteboard and pen.
- Learner A chooses a number card. If it is a single-digit number, learner B must count on in tens. If it is a two-digit number, Learner B must count back in tens.
- Learner B looks through the remaining number cards to find the number that is 10 more or 10 less. They write the pair of numbers on their whiteboard.
- Learners swap roles and repeat.

Review

- Ensure pairs of learners have a set of 0–20 number cards.
- Ask a variety of 10 more/less questions, for example: **I had 15 pencils on my desk but I gave 10 away. How many are left?** Learners work in pairs to hold up the number card that displays the answer. Keep the questions rapid and quickfire to encourage learners to count on or back the whole 10 at once rather than counting on or back 10 in ones.

Assessment for learning

- What is 10 more than 5?
- What is 10 less than 17?
- If my starting number is 2 and I count on 10 more, what will the ones/tens digit be?

Same day intervention
Support

- Give learners interlocking cubes so that they can use them to physically count on 10 in ones and see the answers to reinforce the concept before they attempt counting on in tens in one go.

Number – Counting and sequences

Additional practice activities

Activity 1

Learning objectives
- Estimate and count up to 20 objects accurately
- Count on and back in tens (variation)

Resources

20 countable objects (e.g. marbles, counters or interlocking cubes) (per group); small plastic or paper bags and dishes (per group); Resource sheet 12: 0–20 number line (variation) (per group)

What to

- Set up a sweet shop role-play area with jars of 'sweets' (countable objects). Label the jars with the names of different sweets and provide small plastic or paper bags and dishes to pour the sweets into.
- [TWM.07] Two learners are the shopkeepers. The other learners in the group are customers. Each customer asks for an amount of sweets between 1 and 20. The shopkeepers pour the sweets from the jar into the dish and what they estimate to be that amount into a bag.

- The customer then counts how many 'sweets' are really in the bag. Encourage learners to talk through the process of estimation, explaining how they have come up with their estimate, comparing this to other methods they could have used.

Variations

Give pairs of learners a bag of 'sweets' and ask them to tip them onto the table and estimate how many they think there were in the bag.

Customers ask for an amount of sweets that is 10 more or less than their given number (e.g. 'I would like 10 more sweets than 3/I would like 10 less sweets than 20'). The shopkeeper works out how many sweets that is and counts them into the bag. The customers check the answer on a number line.

Activity 2

Learning objective
- Count on in twos and recognise odd and even numbers

Resources

string or rope to use as a washing line (per group); 10 pairs of socks or paper sock cut-outs (per group); 10 pegs (per group); set of 1–20 number cards from Resource sheet 1: 0–20 number cards (per group)

What to do

- Set up the string or rope as a washing line by tying it between two chairs or similar.
- One learner picks a 1–20 number card.
- The other learners count out that many socks.
- [TWM.03] Learners predict whether the amount of socks will be odd or even based on the odd and even numbers that they can recall already, then check this by hanging the socks on the washing line in pairs. If all socks are in pairs the number is even. If there is one sock hanging up on its own at the end, the number is odd.
- Learners swap roles and repeat with a new number.

Variations

Learners choose a 1–10 number card for these activities until they are secure with these concepts for number to 10.

As above but learners count the pairs of socks in twos, then add on the one at the end if it is an odd number.

Use the pegs to peg a sequence of 0–20 number cards on the washing line (you could peg all odd or even numbers in order or all numbers from 0–10 or 0–20 on the washing line). Learners describe the sequence to you (e.g. It is going up in ones/twos/all of the numbers are odd/even) and predict which number will come next.

Before the learners add the socks to the washing line, they say whether they think the number is odd or even then use the sock activity to check their answer.

Unit 3: Reading and writing numbers to 10

Learning objectives

Code	Learning objective
1Ni.01	Recite, read and write number names and whole numbers (from 1 [0] to 10 [20]).
1Nc.01	Count objects from 1 [0] to 10 [20], recognising [conservation of number and] one-to-one correspondence.

Unit overview

In this unit, learners recite the numbers from 1 to 10 in order and recognise, read and write the numbers from 1 to 10. In this unit, learners are only reading and writing numbers to 10 as numerals. The reading and writing of numbers as words is taught in Unit 5.

Learners will learn the number names and the symbols by which they are represented (e.g. 'one' is represented by '1'). Learners will discover that numbers have a fixed order when counting. They will begin to count up to ten objects with one-to-one correspondence – a skill that will be refined in later units. Recognising conservation of numbers to 10 and 20 is taught in Units 1 and 2.

Prerequisites for learning

Learners need to:
- recognise some numbers in the environment
- be able to copy symbols with a pen or pencil
- be able to say some numbers in context.

Vocabulary

number, count, symbols, digit

Common difficulties and remediation

Some learners may confuse number symbols with letters. You can help learners to differentiate between them by pointing out numbers in the classroom (e.g. on a clock) and labelling items that you have counted as a class with numbers.

There can be confusion around the terms 'numbers', 'numerals' and 'digits'. In Stage 1, use the term 'number' with learners when referring to a quantity or amount and when counting and making calculations. However, it is good practice to introduce learners to the words 'numbers', 'numerals' (a symbol that stands for a number) and 'digits' (the ten numerals 0, 1, 2, ... 9).

When counting forwards in ones, learners may believe that you must always start counting from 0 or 1 and end on 10. Therefore it is useful to ask learners to choose a number to start counting from or stop counting on once they have mastered counting from 1 to 10, for example: 'Start with 3 and end on 9.'

When counting objects, some learners may not understand that each number said is a label for each object and that an object must be counted only once, resulting in learners either under-counting or over-

counting. Organise objects so that they are easy to count, and count objects slowly and deliberately as a class, slightly moving each object once it is counted.

Supporting language awareness

Display the key words and discuss them with learners. Use and refer to these words throughout the lessons and encourage learners to use them too.

Knowing the number names is important for this unit. It is good practice to display numerals, the corresponding number words and a visual representation of each quantity around the classroom to expose learners to numbers in as many ways as possible.

Use every opportunity you can to count with learners, for example counting out resources.

Promoting Thinking and Working Mathematically

TWM.02 Generalising

Learners investigate counting forwards and discover that the number names come in the same order every time, no matter where they start counting from or which number they finishing counting on.

TWM.03 Conjecturing

Learners question why we need to be able to count, recognise and write numbers.

Success criteria

Learners can:
- read the numbers 1–10 represented as numerals
- write the numerals 1–10 as numerals
- recite the numbers 1–10 in order, starting or finishing on any number within that range
- count up to ten objects with one-to-one correspondence.

Note

For those teachers/schools following the CIPM Recommended Teaching and Learning Sequence, zero is not introduced until Term 1, Week 3 (Unit 16, LO1Np.01: Understand that zero represents none of something). Therefore, in this unit learners are working with numbers 1 to 10 and not zero. In subsequent units, however, learners are reciting, counting, reading and writing from zero.

Unit **3** Reading and writing numbers to 10

Lesson 1: **Counting to 10**

Learning objectives

Code	Learning objective
1Ni.01	Recite[, read and write] number names and whole numbers (from 1 [0] to 10 [20]).

Resources

ball or marble (per pair); 1–10 number cards from Resource sheet 1: 0–20 number cards (per class)

Revise

Use the activity *Counting rhymes* from Unit 3: *Reading and writing numbers to 10* in the Revise activities to make an initial assessment of each learner's knowledge of numbers to 10. Note any learners who may need additional practice and those who are already working at Stage 1.

Teach [SB] 🖵

- Discuss the picture in the Student's Book. Ask: **What are the children doing?** (counting to 10)
- [T&T] Ask: **Why do we need to count?** Take suggestions and elicit that we count to find 'how many' there are of something. Say: **When we think about possible answers to a mathematical question, we are conjecturing.**
- Explain that when we count, we use numbers.
- [T&T] Ask: **Can you find some numbers in our classroom?** Ask pairs to point out the numbers that they have found, addressing any misconceptions (e.g. confusing letters for numbers).
- Explain that when we count, we use numbers in a certain order.
- Display the 1–10 number track on **Slide 1**. Explain that it is a number track and that we can use it to help us to count. Demonstrate counting to 10 while pointing to each number on the number track, then say: **Count to 10 with me.**
- 👭 Ask: **What number would come next if I started counting from 3?** Explain that you can start counting from – or stop counting on – any number. Ask learners to choose numbers to start counting from and stop counting at and practise this as a class.
- Discuss the Guided practice example in the Student's Book.

Practise [WB]

- Workbook

Title: Counting to 10

Page: 14

- Refer to Activity 1 from the Additional practice activities.

Apply 👥 🖵 [TWM.02]

- Display **Slide 2**.
- Give each pair of learners a ball or marble.

- Learners roll the ball back and forth to each other, saying a number name each time they roll it (starting from 1, finishing at 10).
- [TWM.02] When learners are confident with this, they can choose different numbers to start counting from, discussing as a pair what number comes next and how they know this.

Review [TWM.02]

- Invite ten learners to the front of the class.
- Give each learner a single number card from 1 to 10.
- Line up the learners out of order, then ask them to say their number names in turn.
- Ask: **Are they counting?**
- [T&T] Ask: **What are they doing wrong? How can we make it right?**
- [TWM.02] Elicit that to count, the numbers must be in a special order. Ask the learners to help you to reorder the numbers, discussing which number comes next each time. They then say their number names in turn to check the order. Are there any mistakes in the order still? Ask learners to tell you how they know.
- Repeat if time allows.

Assessment for learning

- What number comes before/after…?
- Count forwards/backwards from…
- Do you have to start counting from 0 or 1/stop on 10? Why/why not?

Same day intervention
Support

- For learners who are struggling to count to 10, focus on counting to 5 first.
- Try taking turns to say numbers in order to prompt learners. For example, you say 'one', the learner says 'two', you say 'three' and so on.

Enrichment

- For learners who can already confidently count to 10, focus on asking questions such as **Can you think of a time when your mum/dad/teacher needed to count?** to promote discussion about this topic.

Lesson 2: **Reading numbers to 10**

Learning objectives

Code	Learning objective
1Ni.01	Recite, read [and write] number names and whole numbers (from 1 [0] to 10 [20]).

Resources

1–10 number cards from Resource sheet 1: 0–20 number cards (one card per learner); large book or scarf (per class)

Revise

Use the activity *Counting rhymes* from Unit 3: *Reading and writing numbers to 10* in the Revise activities to consolidate counting to 10.

Teach [SB]

- Discuss the picture in the Student's Book. Tell learners that your friend lives at number 6.
- [T&T] Ask: **How could you work out which house that is?** Allow learners to give suggestions. Praise any suggestions that mention counting as a strategy, and elicit that being able to read the numbers on the door would be a useful skill to have.
- Ask learners to point out any numbers that they can read in the classroom. Make a list of these numbers on the board, saying each number as you write it, and encourage learners to say these numbers as you point to them. Fill in any numbers from 1 to 10 that have not already been written on the board.
- [T&T] Ask: **What is different about number 10 when compared to the numbers below 10?** Discuss how some numbers have one digit and some have more than one digit. If appropriate, expand on the term 'digit' explaining how the numbers 0 (which stands for nothing) to 9 can also be called 'digits', and how these digits are used to make up other numbers, such as 10.
- ℗ Write the numbers from 1 to 10 on the board, out of order. Ask: **Can you tell me which numbers these are?** This should check whether learners are reading the numbers or just reciting them in order.
- Discuss the Guided practice example in the Student's Book. Explain to learners that they must count how many stars/planets there are and draw a line to the numeral that matches that amount.

Practise [WB]

- Workbook

Title: Reading numbers to 10

Page: 15

- Refer to Activity 2 from the Additional practice activities.

Apply 👥 🖥

- Display **Slide 1**.
- Ask learners to find a number card from 1 to 10 that you placed under their chairs before the lesson began.
- Call out a number and an action (for example, hop around the classroom).
- Everyone with that number card must do the action that number of times. Repeat for the other numbers.

Review

- Hold up large 1–10 number cards, one by one, slowly revealing the number from behind a book or scarf.
- Learners call out the number when they have worked out what it is.

Assessment for learning

- Tell me what number this is.
- Read this number.
- Point to the number 10 / 4 / 7.

Same day intervention
Support

- Ask learners to find numbers of personal significance (such as their age) on a number track and to read those numbers to you.
- Concentrate on one or two other numbers until learners recognise them too.

Number – Integers and powers/Counting and sequences

Unit **3** Reading and writing numbers to 10

Lesson 3: **Writing numbers to 10**

Learning objectives

Code	Learning objective
1Ni.01	Recite, read and write number names and whole numbers (from 1 [0] to 10 [20]).

Resources

mini whiteboard and pen (per learner); 1–10 number cards from Resource sheet 1: 0–20 number cards (per pair)

Revise

Choose an activity from Unit 3: *Reading and writing numbers to 10* in the Revise activities.

Teach [SB] 🖵 [TWM.03]

- [T&T] Direct learners to the picture in the Student's Book.
- [TWM.03] Ask: **What does the picture show? Why are the numbers on the list important? What might happen if you wrote the wrong number on the list?** Discuss this with learners, allowing them to draw the conclusion that being able to write numbers helps you to show information and give clear instructions.
- Give learners mini whiteboards and pens and tell them that they will be writing the numbers from 1 to 10.
- Display **Slide 1** and count from 1 to 10.
- Display **Slide 2** asking learners to watch carefully as you trace over the numeral 1, explaining what you are doing in the process. Then referring to the formation of the numeral on the slide, ask the learners to write the numeral on their whiteboard before holding up their whiteboards to show their numbers.
- Talk about the features of the numeral – is it curved or made of straight lines? Does it resemble anything familiar?
- Repeat above displaying **Slides 3–11** in turn.
- Now call out the numbers from 1 to 10 and ask learners to write them carefully on their whiteboards and then hold them up.
- Discuss the Guided practice example in the Student's Book.

Practise [WB]

- Workbook

Title: Writing numbers to 10

Page: 16

- Refer to Activity 2 from the Additional practice activities.

Apply 👥 🖵

- Display **Slide 12**.
- Give each pair of learners a set of 1–10 number cards and a mini whiteboard and pen.
- Learner 1 picks a number card and reads it to learner 2, making sure not to show learner 2 the number on the card.
- Learner 2 writes the number on the mini whiteboard.
- They check that the written number matches the number card.
- Learners swap roles and repeat.

Review

- Ask a volunteer to write any number from 1 to 10 on the board.
- Can the rest of the learners work out which number they have written?
- Repeat several times.

Assessment for learning

- Write the number…
- Where do you start when you write the number…?

Same day intervention

Support

- Focus on mastering the formation of numbers of personal significance or numbers that are easier to form (such as 1, 7).

Enrichment

- Learners who are already proficient at number formation could make a large 1–10 number track to display in the classroom.

Number – Integers and powers/Counting and sequences

Lesson 4: **How many?**

Learning objectives

Code	Learning objective
1Ni.01	Recite, read and write number names and whole numbers (from 1 [0] to 10 [20]).
1Nc.01	Count objects from 1 [0] to 10 [20], recognising [conservation of number and] one-to-one correspondence.

Resources

large number cards 1–7 from Resource sheet 1: 0–20 number cards or numbers 1–7 written on A4 sheets of card (per class); ten interlocking cubes (per pair); mini whiteboard and pen (per pair); 1–10 number cards from Resource sheet 1: 0–20 number cards (per group); container of objects for counting (per group)

Revise

Use the activity *Recognising numbers* from Unit 4: *Reading and writing numbers to 10* in the Revise activities to assess what learners already know about counting objects.

Teach [SB] [TWM.03]

- Direct learners to the picture in the Student's Book. **[TWM.03]** Ask: **How could the girl find out how many blocks there are?** Agree that she could count them. Ask: **Why might she want to know how many there are?** Take suggestions from the learners, encouraging them to refer to examples from their lives when they have wondered or needed to know how many there were.
- [T&T] Say: **Talk to your partner about how the girl could make counting them easy.**
- Explain that we count objects to find out 'how many', and we count by saying one number name for each object we count and that we must say the numbers in order.
- Demonstrate by asking five learners to come to the front of the class.
- Count the learners, pointing clearly to each one as you say each number. Now ask another two learners to join the learners at the front in a group rather than a line and count them again, deliberately over-counting or under-counting. Encourage learners correct your mistakes.
- Ask: **How can we make it easier to count accurately?** Elicit that you could line up the learners and point to each one as you count. Count them again with the rest of the class.
- Now give each learner in the line a 1–7 number card (in order) to check your counting. Say: **Number 7 is at the end of the line and the numbers start at 1 and are in order, so there must be 7 people.**
- Ask: **What would happen if we said a number name while we weren't pointing to an object we were counting?**
- Discuss the Guided practice example in the Student's Book.

Practise [WB]

- Workbook

Title: How many?

Page: 17

- Refer to Activity 2 from the Additional practice activities.

Apply 👥 🖥

- Display **Slide 1**.
- Extend by increasing the number of cubes each time (one handful, two handfuls, all).

Review

- Arrange learners into small groups and give each group some 1–10 number cards.
- Learners pick one card each and collect that amount of a given object (e.g. pencils) and lay them out on their table.
- They must all check that they have the correct amount of objects before putting their hands up to show that they are ready.

Assessment for learning

- How could you make counting easier?
- Why must we touch/point to each object as we say each number?

Same day intervention

Support

- Give learners objects that have been numbered 1–10 to count, to practise saying the number names in order as they point to each object.

Enrichment

- Ask a group of learners to pretend that they are making an advert about how to make counting objects easier, thinking about skills such as touching each object as you count them. They can use any countable objects to demonstrate this. They perform their advert for the rest of the class.

Number – Integers and powers/Counting and sequences

Additional practice activities

Activity 1 👥 ⚠2

Learning objectives
• Recite numbers in order to 10

Resources
chalk (per class); bean bag or similar for the Challenge 3 variation (per learner)

What to do
• Use chalk to draw ten circles ('stepping stones') outside. Each circle must be large enough for a learner to jump into.
• Make the stepping stones into a number track by writing a number from 1 to 10 in each one, starting on 1 and ending on 10. (You could ask learners to do this if they are learning how to write numbers.)
• Ask learners to jump along the numbers, counting in ones.

Variations
1 Draw only five circles and number them from 1 to 5.

3 [TWM.02] Ask learners to throw a bean bag and start jumping and counting from the number that their bag lands on, thinking carefully about which number comes next. Ask: **Does the same number always come after 5 even if you start counting from different numbers? Why do you think that is?**

Activity 2 👥 or 👥 ⚠2

Learning objectives
• Read and write numerals from 1 to 10
• Count from 1 to 10 objects, using one-to-one correspondence (variation)

Resources
small writing pad (per group or pair); plastic plates or bowls (for variation) (per group or pair); plastic fruit/buttons/small toys (for variation) (per group or pair)

What to do
• One learner in each group or pair is the waiter or shopkeeper and has an 'order' pad.
• The other members of the group must decide what they want and how many. The waiter writes down their orders (for example, 4 apples, 7 buttons). If learners are not confident enough to write words, they can draw very simple pictures of the objects, as long as they write down the correct number.

• Learners swap roles and repeat.
• [TWM.03] Ask: **Why do you need to be able to read and write numbers to take orders in a café?** Discuss as a group and come up with a list of other scenarios in which you would need to be able to read or write numbers.

Variations
1 Challenge 1 learners should take their turn as the waiter/shopkeeper with a Challenge 2 or 3 learner as support or in an adult-led group.

3 The waiter/shopkeeper brings the order, counting the objects onto the plates and giving them to the customers to count and check.

Number – Integers and powers/Counting and sequences

Unit 4: Reading and writing numbers to 20

Collins International Primary Maths Recommended Teaching and Learning Sequence: Term 2, Week 6

Learning objectives

Code	Learning objective
1Ni.01	Recite, read and write number names and whole numbers (from 0 to 20).
1Nc.01	Count objects from 0 to 20, recognising conservation of number and one-to-one correspondence.

Unit overview

In this unit, learners build on their knowledge of the numbers 0–10 by extending their ability to recognise, count and write numbers to 20. They recite the numbers to 20 forwards and backwards, eventually using this skill to count up to 20 objects. They read and write numbers both as numerals and as words, enabling them to label different amounts to 20.

Prerequisites for learning

Learners need to:
- recognise the numbers 0–10
- be able to count up to ten objects accurately
- be able to count on and back at any point from 0 to 10.

Vocabulary

number, numeral, count, how many, forwards, backwards

Common difficulties and remediation

While learners should now be very confident when reciting, reading and writing numbers from 0 to 10, some may find it difficult to grasp the numbers 11 to 20. Although some of the numbers (fourteen, sixteen, seventeen, eighteen and nineteen) follow a pattern that is easy to pick up, eleven, twelve, thirteen, fifteen and twenty don't fit this pattern. It can help to focus on these non-typical numbers one by one, pointing out that, for example, 11 is not 'oneteen', but eleven. Playing lots of games that involve counting will get learners into the habit of saying these numbers, making it easier to match their names with the numerals.

In this unit, writing the word for each number is taught, as well as the numeral. Learners may find this difficult if they struggle with reading and writing in general, especially as some of the number words are quite long. Encourage them to read and write the smaller number words (0–10) accurately, and the longer words will follow as it is often just a case of adding 'teen' on the end of a word that they can already write.

While learners will be proficient at counting up to ten objects, counting up to 20 objects can be confusing as there are far more objects to keep organised. Remind learners to line up the objects carefully and to leave a gap between adjacent objects so that they don't get confused when counting. In subsequent units they will learn how to count in groups of twos and fives, which will help them to count larger quantities more quickly.

Supporting language awareness

Display the key words for each lesson and discuss them with learners. Use and refer to these words throughout the lesson and encourage learners to use the words too, prompting them when necessary.

Displaying numbers written as numerals alongside their names (as well as a pictorial representation of the quantity/amount) is good practice and helps learners to read and write the number names.

Promoting Thinking and Working Mathematically

TWM.02 Generalising

Learners investigate counting forwards and backwards and discover that the number names come in the same order every time, no matter where they start counting from or which number they finishing counting on.

Success criteria

Learners can:
- recite, read and write the numbers 0–20 represented as numerals
- begin to read and write the numbers 0–20 represented as words
- count up to 20 objects accurately.

Number – Integers and powers/Counting and sequences

Lesson 1: **Counting to 20**

Learning objectives

Code	Learning objective
1Ni.01	Recite, read [and write] number names and whole numbers (from 0 to 20).

Resources

set of 0–10 number cards from Resource sheet 1: 0–20 number cards (per pair)

Revise

Use the activity *Counting circle* from Unit 4: *Reading and writing numbers to 20* in the Revise activities.

Teach [SB] [TWM.02]

- Direct learners to the picture in the Student's Book. **[T&T]** Ask: **What numbers do you recognise on the board game?** Now ask: **What number is the red counter on? What about the blue counter?**
- Tell learners that they are going to learn to count to 20. Ask: **Does anybody know which number comes after ten when we count forwards?** Elicit that it is 11. Do the same for the next two numbers, then practise counting forwards from different starting numbers to 13 together. Repeat this process for the next three numbers, then the final four numbers until you can count to 20 together as a class. During this process, point out the pattern that many of the numbers follow but also note any numbers that do not follow the pattern.
- Ask learners to choose starting numbers in the range 0–19 and count forwards as a class from these numbers.
- Now try counting backwards. 🖉 Ask: **If my starting number is 18 and I want to count backwards, what will the next number be?** Count backwards as a class starting from different numbers in the range 10–20.
- **[TWM.02]** Ask: **When I count forwards/backwards in ones, will the numbers ever be in a different order? Why/why not?** Discuss this, encouraging learners to recognise that numbers are always in a fixed position when we count forwards or backwards in ones and the order will never change. Remind learners that proving that the same thing will happen every time is called *generalising*.
- Discuss the Guided practice example in the Student's Book.

Practise [WB]

- Workbook

Title: Counting to 20

Page: 18

- Refer to Activity 1 from the Additional practice activities.

Apply 👥 🖥️

- Display **Slide 1**.
- Give pairs of learners a set of 0–10 number cards.
- They decide if they are going to count forwards or backwards. They pick a number card and together they count either forwards to 20 from that number or backwards from 20 until they reach that number.

Review

- Ask two volunteer learners to come to the front to stand either side of you.
- Say a number from 1 to 19. The learner to your right says the number that comes before it and the learner to your left says the number that comes after it.
- Repeat this with different learners and starting numbers.

Assessment for learning

- What number comes before/after 18?
- Count forwards from 0 to 20/3 to 17.
- Count backwards from 20 to 0/15 to 4.
- If I count forwards in ones, will the number after 13 always be 14? Why/why not?

Same day intervention
Support

- Take learning the numbers 11–20 slowly with learners who are struggling with this, allowing them first to be confident to count to 13 and back again, progressing a few numbers at a time until they can reach 20.

Number – Integers and powers/Counting and sequences

Lesson 2: **Reading numbers to 20**

Learning objectives

Code	Learning objective
1Ni.01	Recite, read [and write] number names and whole numbers (from 0 to 20).

Resources

Resource sheet 1: 0–20 number cards (shuffled) (per learner); Resource sheet 6: 0–20 number word cards (shuffled) (per learner)

Revise

Use the activity *Number race* from Unit 4: *Reading and writing numbers to 20* in the Revise activities.

Teach [SB]

- Direct learners to the picture in the Student's Book. Ask: **Does anybody know which numbers the children are wearing?** Note which learners can already identify these numbers. Now ask: **What do you think the words say?** Explain that, as well as being written in numerals, each number can be written as a word and that the words that match the numerals are underneath each child.

- Write the numerals 0–10 on the board and tell learners that, because they can already recognise these numerals, they will be able to recognise the numerals in any number because there are no more numerals to learn. Explain that the numbers 10–20 are all two-digit numbers. Write the numbers 10–20 on the board and ask: **What do you notice about these numbers?** Elicit that they all have 1 in the tens position and that the digit in the ones column follows the same pattern as 0–9, until we reach 20.

- Now write the matching word next to each number from 10 to 20, pointing out how some of the numbers from 10 to 20 follow the pattern of starting with the related 0–10 number, followed by teen (e.g. sixteen). Pay close attention to any numbers that do not follow this pattern and those that almost follow it but deviate slightly (e.g. thirteen and fifteen).

- Now write on the board a jumble of numerals from 10 to 20 and number words from 10 to 20 (about five of each) and ask learners to come to the front and point to the numeral or word that matches a number that you say.

- Discuss the Guided practice example in the Student's Book.

Practise [WB]

- Workbook

Title: Reading numbers to 20

Page: 19

- Refer to Activity 1 from the Additional practice activities.

Apply 👥 🖥

- Display **Slide 1**.

- Give each pair of learners a shuffled set of 0–20 number cards and a shuffled set of 0–20 number word cards.

- Learners work together to match the numerals to the words until all their cards are in pairs.

Review

- Give each learner a 0–20 numeral or number word card.

- Call out questions or instructions, for example: **Who's got 13?/Anyone with 15 or 16 jump up and down three times**.

- Learners should regularly swap their cards with other learners.

Assessment for learning

- What does this numeral say?
- What does this number word say?
- Point to 15 on the number line.
- Find the word that says twelve on the board.

Same day intervention

Support

- Allow learners to focus on reading the numerals for 11–20 before they try to read the associated words.

Enrichment

- Send small groups of learners to look for examples of numbers from 11–20 around the school. They could record these by taking photographs. When they are back, they share what they have found with the class.

Number – Integers and powers/Counting and sequences

Lesson 3: **Writing numbers to 20**

Number – Integers and powers/Counting and sequences

Learning objectives

Code	Learning objective
1Ni.01	Recite, read and write number names and whole numbers (from 0 to 20).

Resources

mini whiteboard and pen (per learner); Resource sheet 1: 0–20 number cards (per learner); Resource sheet 6: 0–20 number word cards (per learner)

Revise

Use the activity *Number race* from Unit 4: *Reading and writing numbers to 20* in the Revise activities.

Teach 🆂🅱

- Direct learners to the picture in the Student's Book. Ask: **How will the postman know which flat to deliver the letter to?** Elicit that they must find the flat number on the envelope.
- Discuss with learners the importance of being able to write numerals to 20. **[T&T]** Ask: **Can you think of a reason that you might have to write a number down?** Make a list of these situations on the board (e.g. on a shopping list, writing your age on a form).
- Go through the formation of the numbers 0–10 with the learners, asking volunteers to write these numbers on the board. Correct any problems with orientation of numbers or formation.
- Now ask: **Can anyone tell me how to write 11?** Repeat for the other numbers to 20. Learners draw on their knowledge of number recognition from Lesson 3 or they could use number cards for support.
- Now write the number names for 0–20 next to their matching numerals, discussing their spelling with the learners throughout. Remind learners of the 'teen' pattern when you reach these numbers.
- Discuss the Guided practice example in the Student's Book.

Practise 🆆🅱

- Workbook

Title: Writing numbers to 20

Page: 20

- Refer to Activity 2 from the Additional practice activities.

Apply 👤 🖥

- Display **Slide 1**.
- Give each learner a mini whiteboard and pen, a set of 0–20 number cards and a set of 0–20 number word cards.
- Learners pick a number card, find the matching word card, turn both face down and write them on their whiteboards. They turn the cards face up to check their work, then pick another number card and repeat.

Review

- Write 41 on the board. 🅿 Ask: **Have I written 14? Why/why not?** Elicit that teen numbers always begin with 1 and that it is important to get the two digits in the correct order.
- Write 31, 71, 81 and 51 on the board and invite learners to come up to the board to re-write them as teen numbers.

Assessment for learning

- How do you write the number 13 in numerals?
- How do you write the word for thirteen?
- Write the number and the word for your house number.

Same day intervention

Support

- Let learners focus on writing numbers and number names of personal significance (e.g. the number of their house).

Enrichment

- Learners who already have good number formation and are more familiar with the numbers 11–20 could construct a classroom number track, writing the numbers 1–20 on large cards to display in the classroom.

Lesson 4: **Counting and labelling objects to 20**

Learning objectives

Code	Learning objective
1Ni.01	Recite, read and write number names and whole numbers (from 0 to 20).
1Nc.01	Count objects from 0 to 20, recognising conservation of number and one-to-one correspondence.

Resources

dish of up to 20 countable objects, e.g. marbles/counters/cubes (per pair); five sticky labels (per pair); 20 building blocks (per class)

Revise

Use the activity *Counting out 20* from Unit 5: *Reading and writing numbers to 20* in the Revise activities.

Teach [SB] [📊]

- Direct learners to the picture in the Student's book. **[T&T]** Ask: **Why does it help us to know how many sweets are in the jar?** Take suggestions (e.g. if we want to share them, we know how many people we have enough sweets for). Ask: **How do you think somebody found out how many sweets there were in the jar?** Agree that someone must have counted them.

- 🗐 Ask: **What things do we need to remember to do when we're counting objects?** Make a list on the board (line up the objects carefully, touch each object as we count, say one number name for each object). Tell the learners that they will be counting up to 20 objects today so it is important to organise them into a line before counting them and to take care not to lose your place when counting. Suggest moving each object as it is counted rather than just touching it to help with this.

- Display the **Tree tool** and put 13 apples on the tree. Ask a volunteer learner to count the apples, prompting them to move them into a line first, then move each apple off the tree as they count it. Write '13' and 'thirteen' on the board. Now move the apples all over the screen and ask: **How many apples are there now?** Check by counting the apples again, reminding learners that there are still 13 apples and that, unless you add more or take some away, that amount won't change even when they are moved.

- Repeat the counting with different amounts of apples to 20. Use some amounts below 10 to help to address any problems that less able learners may still have with one-to-one correspondence.

- Discuss the Guided practice example in the Student's Book.

Practise [WB]

- Workbook

Title: Counting and labelling objects to 20

Page: 21

- Refer to Activity 2 from the Additional practice activities.

Apply [👥] [🖥]

- Display **Slide 1**.

- Give each pair a dish of countable objects (e.g. counters/cubes/marbles) and some stickers. Ensure that each pair's dish contains different amounts to 20.

- The pairs count their objects together and label the dish with a sticker saying the amount in numerals and words.

- They then swap their dish with another pair and repeat the process.

- Remind learners that the last learner to count the objects may have counted incorrectly so the label already on this dish may not be correct.

Review

- Explore different ways of organising 20 building blocks to make them easier to count. Try putting them into groups of twos, fives and tens, then counting each group in ones before moving onto the next group and continuing to count.

- Invite learners to come to the front of the class to help you organise the objects before you count them together.

Assessment for learning

- What things do you need to remember to do to help you to count?

- How many building blocks are there? Show me how you would count them.

- How would you organise these counters to make them easier to count?

Same day intervention

Support

- Let learners practise counting up to ten objects first to remind them about one-to-one correspondence.

Enrichment

- Give learners responsibility for counting out classroom objects during the lesson (e.g. checking that there are enough pencils).

Number – Integers and powers/Counting and sequences

Additional practice activities

Activity 1

> **Learning objectives**
> - Count forwards and backwards starting from any number from 0 to 20
> - Read the numerals 1–20

Resources

Resource sheet 7: Snakes and ladders game (per group); Resource sheet 9: 1–6 spinner (per group); pencil and paper clip for the spinner (per group); counter (per learner); large squared paper (per learner) (for variation)

What to do

- Look at the 'Snakes and ladders' board together and count the squares with the learners. Explain that when it is their turn, they will move their counter forwards the amount shown on the spinner. If they land on a ladder they can go up it. If they land on a snake they have to go down it. The winner is the first person to the finish (square 20).
- As learners play the game, count forwards with them when they go up a ladder, touching each square that

they bypass as you count. When they go down a snake, do the same but count backwards.
- Ask learners what number they have landed on every time it is their turn to encourage them to read the numerals.

Variations

2 Learners make their own board game using large squared paper and writing the numbers 1–20 in each square.

3 Ask Challenge 3 learners to make up different rules for the game. These can be as challenging as they like (e.g. you must move back two spaces when you land on a snake's head). They then teach the rest of their group how to play the game with their rules.

Activity 2 :

> **Learning objectives**
> - Read, write and count numbers from 0 to 20
> - Count up to 20 objects

Resources

A4 card (per learner); trays of paint in different colours (per group); items for printing, e.g. sponges (per group); coloured pens (per group); cut out shapes or stickers (per group)

What to do

- Give each learner a number between 1 and 20.
- Learners make a display card for that number, writing the numeral, the number word and printing that many prints on the card, counting them carefully to check that they print the correct amount.

Variations

1 Give Challenge 1 learners a card on which you have already printed an amount of shapes to 20 (you may wish to keep amounts to 10 and below to start with for these learners). They must count 'how many', then write the numeral and number word that match the amount.

2 Instead of printing, give learners cut-out shapes or stickers to count out for their display cards, giving them more opportunities to explore conservation of number if an adult moves the shapes around before they are stuck down.

3 Ask learners to make a full set of these cards (e.g. 1–10 or 10–20) to make a display garland.

Additional practice activities

Activity 3 👥 ⚠2

Learning objective
• To read, write and recognise the numbers 11–20

Resources
tray of sand/shaving foam/flour/other sensory material (per pair); 11–20 number cards (per pair); paintbrush, water and cardboard (optional – per pair)

What to do
• One learner in each pair thinks of a number from 11–20. They use their finger to write the number in the tray of sand.
• Their partner must find the matching number card and say the number that they think their partner has written. If the number match they swap roles and repeat the activity.

Variations
1 Learners should choose a number card and copy the number from the card into the sand rather than attempting this from memory if they need the support.

2 If you don't have any sensory materials available for this activity, learners could use their finger to draw the number on the palm of their partner's hand or use a paintbrush and water to paint the number onto cardboard.

Activity 4 👥👥 ⚠2

Learning objective
• Count forwards and backwards in the range of 11–20

Resources
A bag containing 20 red counters and 10 blue counters (per group); 1-20 number track or line with the numbers 11–20 highlighted (optional – per group)

What to do
• Each group sits in a circle. One pupil from each group starts to count slowly from 11–20. While they are doing this, the rest of the group pass the bag around the circle, taking one counter each as it is passed around. If a red counter is taken, the starting pupil continues to count forwards (until they reach 20 – if they get this far they re-start the count backwards to 11). If a blue counter is taken, they stop counting and the pupil with the blue counter continues the count.
• Continue the game until there are no counters left in the bag.

Variation
1 Learners who need support can use a number track of line with the numbers 11–20 highlighted to help them to keep track of the count.

Number – Integers and powers/Counting and sequences

Unit 5: Addition as combining two sets

Learning objectives

Collins International Primary Maths
Recommended Teaching and
Learning Sequence: Term 1, Week 4

Code	Learning objective
1Ni.02	Understand addition as: [- counting on] - combining two sets.

Unit overview

In this unit, learners begin to learn about addition by combining two sets of objects. They begin by counting 'how many' in each set, then counting all of the objects together, and progress to solving addition number sentences by representing the two numbers with objects. They learn the terms 'augend', 'addend', 'equals' and 'sum', and are introduced to simple addition number sentences.

Prerequisites for learning

Learners need to:
• be able to count to 10
• recognise the numbers 1–10
• be able to count up to ten objects accurately.

Vocabulary

add, augend, addend, sum, total, equals, altogether number sentence, count, part–whole diagram, plus

Common difficulties and remediation

Learners require plenty of practical activities, combining groups of objects to reinforce the concept of addition.

Ensure that learners can reliably count up to ten objects before attempting to teach them to combine groups of objects. If they have not yet mastered this skill, they will not reach the correct answer, which they may find disheartening.

Supporting language awareness

Display the key words for each lesson and discuss them with the learners. Use and refer to these words throughout the whole lesson and encourage learners to use the words too, prompting them when necessary.

It is important to teach the proper terms for addition (augend, addend, sum, equals) from a young age as this ensures that learners will have a good understanding of addition as they get older. Display these words, together with practical examples (e.g. a group of four objects with the word 'augend' above it + a group of five objects with the word 'addend' above it = a group of nine objects with the word 'sum' above it).

Use words relating to addition in everyday situations, for example: **There are six learners at this table and three at that table – that equals nine learners altogether**.

Promoting Thinking and Working Mathematically

TWM.04 Convincing
Learners apply what they have learned about counting two sets of objects to find out 'how many' in total by using this skill to solve simple addition number sentences.

Success criteria

Learners can:
• solve an addition by combining and counting two sets of objects (to 10).

Number – Integers and powers

Unit 5 Addition as combining two sets

Lesson 1: **Combining sets**

Learning objectives

Code	Learning objective
1Ni.02	Understand addition as: [- counting on] - combining two sets.

Resources

up to ten soft toys (per class); two dishes of five counters – a different colour in each dish (per learner)

Revise

Use the activity *Add the apples* from Unit 5: *Addition as combining two sets* from the Revise activities.

Teach [SB]

- Direct learners to the picture in the Student's Book. Ask: **How many flowers are there in each bunch? How many flowers are there in the vase?** Elicit that there are three flowers in the first bunch, four flowers in the second bunch and seven flowers in the vase.
- Explain that the number of flowers in the vase is the same as the number in the two bunches of flowers together. Demonstrate this by counting both sets of flowers together to prove that there are seven flowers in the two bunches.
- Tell learners that they are going to be finding out 'how many altogether' by counting two sets of objects together. Set out five soft toys on one side of a table and two soft toys on the other side. Ask: **How many toys are there in each set?** Take answers. **[T&T]** Ask: **How can we find out how many toys there are altogether?** Elicit that you could put the toys together in one big group and count them again. Demonstrate how to line up the toys to make them easier to count. It is helpful to give learners counters to match the soft toys so that they can copy the process, discussing what they are doing with a partner throughout.
- Repeat this with different starting amounts, keeping the total at 10 or less. Invite learners to come to the front to physically count the toys and then move them into one set before counting them altogether.
- Discuss the Guided practice example in the Student's Book.

Practise [WB]

- Workbook

Title: Combining sets

Page: 22

- Refer to Activity 1 from the Additional practice activities.

Apply 👤 🖥

- Display **Slide 1**.
- Give each learner two dishes of five counters – a different colour in each dish.
- Learners take any number of counters from each dish. They count how many are in each set, then line up all the counters and count them again.

Review

- Write two numbers (such as 7 and 2) on the board.
- Learners use their objects to count out sets to match the numbers, then combine the sets and count them to find how many altogether. Throughout the process, ask: **Why are you doing that? Why are you counting out two sets of objects? What will happen when you put the objects together? What will you find out?**
- Repeat with different starting numbers.

Assessment for learning

- How many are in each set?
- How many altogether?
- How could you work out how many there are altogether in these two sets?
- How can we make it easier to count all the objects from the two sets altogether?

Same day intervention

Support

- Begin with totals of up to 5 until learners are used to the process of counting the sets, combining the sets, then counting the objects altogether.

Enrichment

- Encourage learners to draw on their estimation skills from previous units to estimate how many there will be altogether before counting.

Lesson 2: **Part–whole diagrams**

Learning objectives

Code	Learning objective
1Ni.02	Understand addition as: [- counting on] - combining two sets.

Resources

chalk (per class) or six skipping ropes (per class); up to ten blue and ten yellow building blocks (per class); up to ten counters in two different colours (per learner); Resource sheet 10: Part–whole diagram (per learner)

Revise

Use the activity *How many fingers?* from Unit 5: *Addition as combining two sets* in the Revise activities.

Teach [SB]

- Direct learners to the picture in the Student's Book. Explain that this is another way of organising two sets to find out how many there are altogether and that it is called a part–whole model.

- Point out the two 'part' sections and say: **The amounts that go in the 'part' sections make up the amount that goes in the 'whole' section when you add them together. Two parts make one whole.** Count the cubes in each 'part' section with the learners, then count both sets together and point out that the total is in the 'whole' section of the diagram.

- Use chalk to draw a part–whole diagram on the floor (or use skipping ropes to set one out). Put five blue building blocks and three yellow building blocks on the floor next to the model. Put the blue blocks inside the bottom right 'part' section and the yellow blocks inside the bottom left 'part' section. Invite learners to come and count how many blocks are in each set. Then count how many altogether, moving each block into the 'whole' section as you count them. Repeat this with different starting numbers of blocks, encouraging learners to help you to set up the model and count the blocks each time. As you work through different examples, change from moving the blocks from the two 'part' sections into the 'whole' section, and instead write the total number (as a numeral) in the 'whole' section. Throughout this process, keep asking the learners: **What goes in this part of the diagram? What number matches this amount? How can we use this diagram to add the amounts/ numbers together?**

- Remember to show the learners strategies to help them, for example counting the larger set of objects first, then continuing to count the smaller set.

- Discuss the Guided practice example in the Student's Book.

Practise [WB]

- Workbook

Title: Part–whole diagrams

Page: 23

- Refer to Activity 2 from the Additional practice activities.

Apply [icons]

- Display **Slide** 1. Give each learner Resource sheet 10: Part–whole diagram and any number of counters in two different colours (from 3 to 10).

- Learners put their counters inside the 'part' sections of the diagram (for example, red in one, blue in the other). They count how many are in each set, then count how many there are altogether by moving each counter into the 'whole' section as they count it.

- Learners can swap their counters with another learner and repeat the activity, if time allows.

Review

- Write the numbers 4 and 3 on the board and set out the skipping ropes as an empty part–whole diagram (or draw one on the floor with chalk).

- [T&T] Ask: **How could we use a part–whole diagram to find out what 4 and 3 make if you add them together?** Elicit that you could put four objects in one 'part' section, three objects in the other 'part' section, then count how many there are altogether. Do this and write the total (7) on a piece of paper (or with chalk) in the 'whole' section.

- Repeat, using different starting numbers.

Assessment for learning

- Show me how you would arrange the counters in a part–whole diagram.
- What number goes in the 'whole' section?
- How many counters are there in each set?
- Which set would you count first?

Same day intervention

Support

- Give learners smaller amounts to work with (less than 5).

Unit 5 Addition as combining two sets

Lesson 3: **Writing addition number sentences**

Learning objectives

Code	Learning objective
1Ni.02	Understand addition as: [- counting on] - combining two sets.

Resources

up to ten interlocking cubes in two different colours (per learner); mini whiteboard and pen (per learner)

Revise

Use the activity *How many fingers?* (variation) from Unit 5: *Addition as combining two sets* in the Revise activities.

Teach [SB]

- Look at the picture in the Student's Book. Ask: **What does this picture show? Has anybody seen these symbols (+ and =) before? Do you know what they mean?** Explain that the picture shows an addition – two amounts being added together to find out how many altogether. Count the children in the two sets with the learners, then count both sets together to find out how many.
- Write + on the board. Explain that it is the add (or plus) sign and comes between two numbers or amounts to show that you must add them together.
- Write = on the board and explain that it is the equals sign and comes after the amounts or numbers that have been added together and before the total. It means that the amounts that have been added are equal to, or 'the same as', the total. Draw an example of an addition, using three stars and one star, on the board to demonstrate how the signs work.
- Now write the addition as 3 + 1 = 4, explaining to learners that you can use numbers in place of the amounts of objects or pictures. Show that this is the same as the stars addition.
- Point to 3 and say: **This is called the augend. It is the first number in the addition.** Point to 1 and say: **This is called the addend. We are adding it to the augend.** Point to the 4 and say: **This is called the sum of the augend and the addend. Another word for sum is total. It tells us how many there are altogether.**
- On the board, draw an addition using four triangles and three triangles, including + and =. Point to the four triangles and ask: **What is the augend?** Elicit that it is 4, and write 4 under the triangles. Now point to the three triangles and ask: **What is the addend?** Elicit that it is 3 and write 3 under the triangles. Read the number sentence: **4 + 3 =.** Count up all the triangles to find the answer and draw triangles to match, then write the answer (7) in place. Repeat with different starting numbers.
- Discuss the Guided practice example in the Student's Book.

Practise [WB]

- Workbook

Title: Writing addition number sentences

Page: 24

- Refer to Activity 1 (variation) from the Additional practice activities.

Apply 👤 💻

- Display **Slide 1**.
- Give each learner up to five interlocking cubes in one colour, up to five interlocking cubes in a different colour and a mini whiteboard and pen.
- Learners put the cubes into two coloured sets. They count how many are in each set and write this as a number sentence using + and =. They count how many cubes altogether to find the total.
- Learners can swap their cubes with a partner and repeat the activity if time allows.

Review

- Ask eight learners to come to the front and divide them into a set of five and a set of three.
- Ask a volunteer to come to the front and write the number sentence.
- Remember to discuss which number is the augend, which is the addend and to refer to adding as finding the sum of the numbers.

Assessment for learning

- What does + mean?
- What does = mean?
- Which number is the augend/addend?
- Write a number sentence to show this.

Same day intervention

Support

- Provide learners with + and = cards to put between groups of objects for practice before asking them to write number sentences.

Enrichment

- Learners write addition number sentences using the numbers 1–5 then solve them using groups of objects.

Lesson 4: **Using sets of objects to solve additions**

Learning objectives

Code	Learning objective
1Ni.02	Understand addition as: [- counting on] - combining two sets.

Resources

set of 0–5 number cards from Resource sheet 1: 0–20 number cards (per learner); Resource sheet 10: Part–whole diagram (per learner); ten counters (per learner); mini whiteboard and pen (per learner); set of large 0–10 number cards (per class); large + and = cards (per class)

Revise

Use the activity *Add the apples* (variation) from Unit 5: *Addition as combining two sets* in the Revise activities.

Teach [SB] [TWM.04]

* Look at the picture in the Student's Book. Point to the number sentence and ☞ ask: **Who can remember what this is called?** Elicit that it is a number sentence. Discuss the features of number sentences, encouraging learners to name the add and equals signs and point out the augend, addend and the sum. If learners are still struggling with any aspect of this, write several addition number sentences on the board and repeat this section of **Teach** from Lesson 3.
* Point out the part–whole diagram in the Student's Book and explain that it has been used to solve the addition. Draw a blank part–whole diagram on the board along with the following number sentence: 7 + 2 =.
* [TWM.04] Ask: **How would you use the part–whole diagram to solve this number sentence? What do you think you would put in each part? What would the 'whole' section tell you?** Explain to learners that by doing this, they are providing evidence to support their answers. Say: **We call this *convincing*.** Demonstrate how to draw dots in the 'part' sections to match the numbers in the number sentence. Then count how many dots there are altogether with the learners (9). Write 9 in the 'whole' section, and then write 9 as the total for the number sentence. Repeat this with different starting numbers.
* Discuss the Guided practice example in the Student's Book.

Practise [WB]

* Workbook

Title: Using sets of objects to solve additions

Page: 25

* Refer to Activity 1 (variation) or 2 (variation) from the Additional practice activities.

Apply 👤 🖥

* Display **Slide 1**.
* Give each learner a set of 0–5 number cards, a part–whole diagram resource sheet, ten counters and a mini whiteboard and pen.
* Learners choose two number cards and write them on their mini whiteboards as an addition number sentence without an answer.
* They use their counters and part–whole diagram sheets to solve the addition and write the total at the end of number sentence.
* Repeat with a different pair of number cards.

Review [TWM.04]

* Make a human number sentence with large number and + and = cards (e.g. 6 + 2 =). Volunteer learners hold up each card.
* ☞ Ask: **Who is holding the add/equals sign? Who has the augend/addend?**
* Ask a volunteer learner to draw a part–whole diagram on the board to solve the addition.
* Invite the class to discuss the process throughout, relating it to earlier lessons in this unit. Ask questions such as: **What sign shows us that we must put these two amounts together? Why are we writing numbers in different parts of this diagram?** and **How else could we have solved this addition?**

Assessment for learning

* How would you use a part–whole diagram to solve this addition?
* Count out counters to match the numbers in this addition.
* Which of these part–whole diagrams matches the number sentence?

Same day intervention
Support

* Keep totals to 5 or less for learners who are struggling with this concept or method.

Number – Integers and powers

Additional practice activities

Activity 1

Learning objectives
- Combine two sets of objects to find how many altogether
- Combine two sets of objects to solve a number sentence

Resources
paper plate or circle of paper (per learner); paper in different colours for cutting out shapes (per learner); scissors (per learner); glue stick (per learner); set of 1–5 number cards from Resource sheet 1: 0–20 number cards (per learner)

What to do
- Ask learners to think of two food items that go well together as a meal.
- They pick two number cards (such as 2 and 5) and cut out that number of food items (such as two burgers and five chips). Alternatively, you could already have these cut out for them to use.

They stick the food onto the paper plates in two clear sets.
- Learners count how many of each food they have, to check, and write the numbers above each set. They then count how many pieces of food are on the plate altogether and write the total amount on the bottom of the plate.

Variations
1 Keep totals to 6 or less for Challenge 1 learners (give them two sets of 1–3 number cards).

2 Give each learner an addition number sentence (e.g. 6 + 3 =) to solve, using this activity and method.

Activity 2

Learning objective
- Use a part–whole model to find 'how many altogether'

Resources
approximately six straws or pieces of string (per learner); glue stick and glue spreaders (per group); white sticky labels or paper cut-out squares – any amount from 3 to 10 (per learner); red and blue coloured pencils (per learner)

What to do
- Give each learner a piece of paper, some glue and some straws or string to divide it into a rectangular part–whole diagram, and sticky labels or paper cut-out shapes (any amount from 3 to 10).
- Learners use the straws or string and glue to set out an empty part–whole model on their paper.
- They colour some of their labels red and the rest of their labels blue.

- They arrange the labels in the part–whole model by colour, count each set, then count how many altogether and write the total in the last circle.

Variations
2 Adapt this activity as the unit progresses – learners could write numbers in the diagram sections instead of using labels or use this activity to solve addition number sentences.

3 Challenge 3 learners may wish to try putting number cards in each part of the part–whole model rather than objects. They could add two numbers to the 'parts' and count out counters for the 'whole'.

Number – Integers and powers

Unit 6: Addition as counting on

Collins International Primary Maths Recommended Teaching and Learning Sequence: Term 1, Week 5

Learning objectives

Code	Learning objective
1Ni.02	Understand addition as: - counting on [- combining two sets].
1Nc.04	Count on in ones, [twos or tens, and count back in ones and tens,] starting from any number (from 0 to 10 [20]).

Unit overview

In this unit, learners continue to learn about addition, this time learning how to count on to add two amounts. They begin by adding objects one at a time to an established set and progress to using number tracks and number lines to support counting on. They continue to construct and solve simple addition number sentences and think about real-life situations when they would need to count on to add two amounts.

Prerequisites for learning

Learners need to:
- recognise the numbers 1–10
- be able to count up to ten objects accurately
- be able to count on in ones to 10 from any starting number from 0 to 10
- have experience of writing and solving an addition number sentence
- know the components of a number sentence and what the operations + and = mean.

Vocabulary

add, augend, addend, sum, total, count on, number sentence

Common difficulties and remediation

Some learners still may not understand that you can count on from a given number and will want to start at 0. Giving the class lots of practice at counting on in ones from any number from 0 to 9 will help to prepare them for using this skill as they tackle addition. It also helps to use concrete objects such as interlocking cubes physically to demonstrate that they already have a given amount and must count on by a new amount of extra objects.

Some learners may wish to revert to the addition technique used in Unit 5 (combining groups). However, it is important that they experience both aggregation and augmentation in order to gain a fuller understanding of addition, so do check that all learners are using the counting-on technique throughout the lessons. Explain to learners that we need to learn different ways of solving mathematics problems so that we can use the most effective and efficient way to work out the solution to a problem.

Supporting language awareness

Display the key words for each lesson and discuss them with learners. Use and refer to these words throughout the lesson and encourage learners to use the words too, prompting them when necessary.

It is important to teach the proper terms for addition (augend, addend, sum, equals) from a young age as this ensures that learners will have a good understanding of addition as they get older. Display these words, together with practical examples (such as a group of four objects with the word 'augend' above it + a group of five objects with the word 'addend' above it = a group of nine objects with the word 'sum' above it).

Use words relating to addition and counting on in everyday situations, for example: **There were four learners sitting at the table. Two more come to sit with them. How many learners are sitting at the table now?**

Promoting Thinking and Working Mathematically

TWM.03 Conjecturing
Learners ask: **How many objects are there if I add X more?** They discover that by counting on X more, they find the sum of those objects.

TWM.04 Convincing
Learners apply what they have learned about counting on to add two amounts by using this skill to solve simple addition number sentences.

TWM.07 Critiquing
Learners draw comparisons between approaching addition by counting on and by combining sets. They learn how one technique may work better than the other in different situations.

TWM.08 Improving
Learners refine their counting on technique by using number tracks and number lines for support, moving on from counting on with concrete objects.

Success criteria

Learners can:
- solve an addition by counting on (to 10).

Unit 6 Addition as counting on

Lesson 1: **Adding more**

Number – Integers and powers/Counting and sequences

Learning objectives

Code	Learning objective
1Ni.02	Understand addition as: - counting on [- combining two sets].
1Nc.04	Count on in ones, [twos or tens, and count back in ones and tens,] starting from any number (from 0 to 10 [20]).

Resources

ten building blocks (per class); two sets of 1–9 interlocking cubes – each set a different colour (per learner)

Revise

Use the activity *Adding more birds* from Unit 6: *Addition as counting on* in the Revise activities.

Teach [SB] [TWM.07]

- Direct learners to the picture in the Student's Book. Explain to learners that the girl has bought some more pet fish. Ask: **How many fish does she already have in her bowl? How many more fish has she bought?** Elicit that she already has four fish and has bought two more. **[T&T]** Ask: **How could we find out how many fish she has now?** Learners will probably suggest combining the groups, as in Unit 5. Praise this suggestion but explain that they will be learning another way to add amounts today.
- Ask a learner to count out four building blocks to represent the fish in the bowl and two building blocks to represent the fish in the bag. Make a tower from the four bricks.
- Say: **We know that there were four fish to start with, so we can start on four and count on two more.** Touch the tower of four and say: **four…,** then add the remaining two bricks one at a time, saying: **five, six** as you do so. Explain that the last number you say when you have finished adding the blocks and counting is the total.
- **[TWM.07]** Ask: **What was different about the way we worked out how many altogether in this lesson? What was the same?** Elicit that instead of putting two groups together, we found a starting number then added more. **[T&T]** Ask: **Do you think this method is easier or more difficult than combining groups? When do you think it would be useful to use adding more instead?** Remind learners that when they compare the usefulness of methods, they are *critiquing*. Explain that this can be quicker than combining groups so is useful if you need to find out the answer fast.
- Repeat the block activity with different starting numbers. Show learners how it is easiest to start with the larger number as then there are fewer numbers to count on.
- Discuss the Guided practice example in the Student's Book.

Practise [WB]

- Workbook

Title: Adding more

Page: 26

- Refer to Activity 1 from the Additional practice activities.

Apply 👤 🖥️

- Display **Slide 1**. Give each learner two towers of interlocking cubes (with any amount from 1 to 5 in each set). Each set must be a different colour.
- Learners identify the taller tower and count how many cubes there are in it. They then add the cubes from the second tower one by one, counting on in ones from the original number in the first tower.
- Learners then swap their towers of cubes with another learner and repeat.

Review [TWM.03]

- Write two numbers (such as 6 and 3) on the board.
- **[T&T]** Ask: **What will you find out if you add the two amounts? How would you do that?** Encourage learners to talk through the process of adding more. Learners use interlocking cubes to make two towers, then count on from the larger number.
- Ask: **What is the number sentence?**
- Spend time reminding learners of how to write a number sentence to match the amounts of cubes, pointing out each operation and reminding them of what it does, and each feature of the number sentence (the augend, the addend and the sum).

Assessment for learning

- Which number/amount is larger?
- How would you use cubes to count on?
- Which number should you start from when you count on? Why?

Same day intervention

Support

Give learners practice at counting on in ones from any given number to 10 before they use counting on to add amounts.

Lesson 2: **Adding more to solve additions**

Learning objectives

Code	Learning objective
1Ni.02	Understand addition as: - counting on [- combining two sets].
1Nc.04	Count on in ones, [twos or tens, and count back in ones and tens,] starting from any number (from 0 to 10 [20]).

Resources

Resource sheet 8: 1–5 spinner (per learner); pencil and paper clip for the spinner (per learner); ten counting objects such as interlocking cubes (per learner); mini whiteboard and pen (per learner)

Revise

Use the activity *Adding more birds* from Unit 6: *Addition as counting on* in the Revise activities.

Teach SB [TWM.04]

- Direct learners to the picture in the Student's Book. Tell learners that there is a group of three children and that two more children want to be in the group. Point out the matching number sentence, saying: **Three children add two more children equals ...**

- Remind learners that they have been counting on to add more. Encourage them to do this with the children in the Student's Book picture. Ask: **How many children will there be in the group when the other two have joined?** Elicit that there will be five children.

- ✍ **[TWM.04]** Write 5 + 3 = on the board and ask five learners to sit at a table. Ask three more learners to stand to one side. Say: **There are five learners at the table. Three more want to sit down. How can we use counting on to find out how many will be at the table when the other learners are sitting down?** Take suggestions and guide learners through starting on 5 and counting on three more, moving each of the three learners to the table as you count on. Say: **Well done – you used what you know about counting on to *convince* me that the answer was correct.**

- Repeat this several times with different starting numbers, sometimes putting the smaller number first and reminding learners to start counting on from the larger number.

- Discuss the Guided practice example in the Student's Book.

Practise WB

- Workbook

Title: Adding more to solve additions

Page: 27

- Refer to Activity 1 from the Additional practice activities.

Apply 👤 🖥

- Display **Slide 1**.

- Give each learner a five-sided spinner from Resource sheet 8, a pencil and paper clip for the spinner, ten counting objects and a mini whiteboard and pen.

- Learners generate an addition by spinning the spinner twice and writing the two numbers it lands on in an addition number sentence.

- They use the counting objects to make both amounts, then count on from the larger amount by adding objects one at a time.

Review

- Write **7 + 2 =** on the board.

- Demonstrate to learners that as you already know that there are seven in the first group, you don't need to make the group from counters. You can just start on 7 and count on two more, only using counters for counting on. Do this by saying: **Put seven in your head.** Use your fingers as the two counters and hold up one finger for each number you count on as you say: **eight, nine.** Remind learners that the last number you say as you count is the total and write this at the end of the number sentence.

- Ask: **What are you doing when you count on to solve a number sentence?** Elicit that counting on is the same as adding more. Discuss the different ways that learners have counted on previously and apply them to the above number sentence.

Assessment for learning

- How would you use cubes to solve this addition by counting on?
- How would you use cubes to count on?
- Which number must you start from when you count on?

Same day intervention

Support

- Ask learners to count on one or two rather than counting on larger amounts.

Enrichment

- Ensure that Challenge 3 learners are counting on larger amounts (e.g. three, four or five).

Number – Integers and powers/Counting and sequences

Lesson 3: **Adding more with a number track**

Learning objectives

Code	Learning objective
1Ni.02	Understand addition as: - counting on [- combining two sets].
1Nc.04	Count on in ones, [twos or tens, and count back in ones and tens,] starting from any number (from 0 to 10 [20]).

Resources

set of large 1–10 number cards (per class); Resource sheet 8: 1–5 spinner (per learner); pencil and paper clip for the spinner (per learner); Resource sheet 4: 0–10 number track (per learner); counter (per learner)

Revise

Use the activity *Giant number line* (variation) from Unit 6: *Addition as counting on* in the Revise activities.

Teach 🆂🅱 📊 [TWM.08]

- Direct learners to the picture in the Student's Book. Ask: **If the boy is on 5 and wants to do two more jumps, which number will he land on?** Agree that he will land on 7.
- Use large 1–10 number cards to set up a number track on the floor and demonstrate jumping along a number track to count on from a given number. Invite learners to do the same, choosing different starting numbers and amounts of jumps each time.
- [TWM.08] Tell learners that using a number track can help to count on to add amounts together. Display the **Number track tool** and write **2 + 8 =** on the board. [T&T] Ask: **How do you think you would use the number track to help you to find the total?** Take suggestions. 🗣 Now ask: **Which number should we start from when counting on to solve an addition? Why?** Agree that we must start with the larger number so, in this case, we will start with 8. Circle 8 and model how to count on two more on the number track. Explain that the last number your finger touches when you finish counting on is the total.
- Repeat this several times with different additions, keeping the totals less than 10. Discuss different ways of monitoring how many you have counted on (for example, drawing a dot on every number you touch or using a counter). Invite learners to come to the front to try this.
- Discuss the Guided practice example in the Student's Book.

Practise 🆆🅱

- Workbook

Title: Adding more with a number track

Page: 28

- Refer to Activity 2 (variation) from the Additional practice activities.

Apply 👤 💻 [TWM.08]

- Display **Slide 1**.
- Give each pair a five-sided spinner, a pencil and paper clip for the spinner, Resource sheet 4: 0–10 number track and a counter.
- Learners use the spinner to generate two numbers to make an addition number sentence, which they write on the number track resource sheet.
- They place the counter on the larger number, then count on along the number track to find the total. They write the answer to complete the number sentence.
- They repeat with different starting numbers.
- Ask the learners to report back – did they find it easier or more difficult to add more using the number track than the other methods used to add more so far? Is there anything they were unsure of? Remind them that by using the number track, they are *improving* their counting on strategies.

Review

- Practise the mental counting on technique, using fingers, from **Review** in **Lesson 2**.
- Ask learners to use a number track to check their answers.

Assessment for learning

- Tell me what six add four equals, by counting on.
- How would you use a number track to find out how much six add four equals?
- Count on two more from five.
- Why is it easier to count on from the higher number?

Same day intervention
Support

- Give learners concrete objects (such as counters) to put on each number on their number tracks as they count on.

Lesson 4: **Adding more with a number line**

Learning objectives

Code	Learning objective
1Ni.02	Understand addition as: - counting on [- combining two sets].
1Nc.04	Count on in ones, [twos or tens, and count back in ones and tens,] starting from any number (from 0 to 10 [20]).

Resources

mini whiteboard and pen (per learner); ruler (per learner)

Revise

Use the activity *Giant number line* from Unit 6: *Addition as counting on* in the Revise activities.

Teach [SB] [📊] [TWM.08]

- Look at the picture in the Student's Book. Ask: **What number sentence would match?** Agree that it would be 5 + 3 = 8 because the boy is starting on 5 and will take three more jumps, to land on 8.
- Point out the dotted line 'jumps' to show where the boy is jumping to. Remind learners that you can draw 'jumps' on a number line to keep track of how many numbers you have jumped forwards or backwards. Remind learners that a number line is similar to a number track, but the numbers are below the line, whereas a number track is like a caterpillar, with numbers in its segments.
- [TWM.08] Remind learners that they have added more by counting on along a number track before. Ask Learners to talk you through the process, comparing it to a method that uses concrete objects. Explain that you can also count on to add more with a number line. Display the **Number line tool** showing 0–10. Write **3 + 6 =** on the board. 📝 Ask: **Which number would we start from if we wanted to count on to find the total?** Elicit that you must start from the larger number even when it is not the augend. Circle 6 on the number line and draw three jumps from 6 to 9.
- Number the jumps 1, 2 and 3 to demonstrate how to check that you have done enough jumps.
- Repeat this with different additions, keeping the totals less than 10.
- Discuss the Guided practice example in the Student's Book.

Practise [WB]

- Workbook

Title: Adding more with a number line

Page: 29

- Refer to Activity 2 from the Additional practice activities.

Apply [👤] [🖥]

- Display **Slide 1**.
- Give each learner a mini whiteboard, a pen and a ruler.
- Each learner uses their ruler to draw a number line on their whiteboard, labelling it from 0 to 10.
- They then use their number line to solve the additions on the slide.

Review [📊]

- Say: **I went into four shops in town. Then I went to a different town and went into two more shops. How many shops did I go into altogether?**
- Learners write a number sentence to match this on their mini whiteboards and hold them up. Provide counters, number tracks and number lines.
- Display the **Number line tool** showing 0–10 to check the answer.
- Ask: **Which method did you use to work out the answer? Would another method work better? Why?** Remind learners that when they compare methods to decide which worked best, they are *critiquing*.

Assessment for learning

- How would you use a number line to find out how much six add three equals?
- Count on four more from five on a number line.
- Why does it help to number the 'jumps' on a number line?
- Why is it easier to count on from the larger number?

Same day intervention

Support

- Some learners may need more time to consolidate counting on with a number track before they use number lines. Ensure that they have adult support with number lines.

Enrichment

- Ask learners to compare adding more on a number line to adding more on a number track. What is the same? What is different? Which method do they find easiest and why?

Number – Integers and powers/Counting and sequences

Additional practice activities

Activity 1

Learning objectives
- Count on more objects to find how many
- Count on more objects to solve a number sentence (variation)

Resources
shop counter – a table or shop role-play area (per group); plastic or paper cut-out fruit or food or toys (per group); set of 1–5 number cards from Resource sheet 1: 0–20 number cards (per group); a basket or cardboard box (per group); mini whiteboard and pen (per group) (variation only)

What to do
- Learners work in groups of three. One learner is the shopkeeper and the other two learners are customers.
- The two customers each pick one number card. The learner with the larger number goes first.
- The first learner asks for the amount shown on their number card of what the shop is selling (such as apples/toy cars/teddies). The shopkeeper counts them out in a line on the counter.
- The second learner asks for the amount shown on their number card of the same item. The shopkeeper counts out that many into a box.

- All three learners then count on from the starting number (the number in a line on the counter) as the shopkeeper adds the second amount of objects one by one from the box to the line on the counter.
- **[TWM.04]** Encourage learners to talk through the process, discussing what they are doing and why, which amount they must start with, and so on.
- Learners swap roles and repeat.

Variations
1 Give Challenge 1 learners two sets of 1–3 number cards initially to keep totals to 6 and under.

2 Learners write an addition number sentence on their mini whiteboards to match each shopping trip.

3 Encourage Challenge 3 learners to estimate how many there will be altogether before counting.

Activity 2

Learning objectives
- Make and use a number track
- Solve additions by counting on along a number track

Resources
ten lolly sticks (per group); ten balls of modelling clay (per group); tissue paper or standard paper in different colours (per group); scissors (per pair); 11 small paper cut-out circles (per group); glue stick (per group)

What to do
- Learners make two or three flowers each using the lolly sticks as stems and cutting out tissue or paper for the petals. For each flower, they write a number from 1 to 10 on a circle of paper and glue it to the centre of the flower.
- Arrange the flowers in a number line from 1 to 10 by putting the stems into the balls of modelling clay.

- One learner makes a bee by colouring the remaining small circle with black and yellow stripes.
- Give learners various additions (keeping the totals less than 10). They take turns to put the bee on the larger number, then move it from flower to flower to count on to find the total.

Variations
1 Challenge 1 learners should practise counting forwards and backwards along the number line before using it to solve additions.

2 Learners stick the flowers to the wall instead of standing them in balls of modelling clay and use chalk to draw the 'jumps' between each flower to replicate a number line.

Unit 7: Subtraction as take away

Collins International Primary Maths
Recommended Teaching and
Learning Sequence: Term 2, Week 1

Learning objectives

Code	Learning objective
1Ni.03	Understand subtraction as: [- counting back] - take away [- difference].

Unit overview

In this unit, learners are introduced to subtraction. First, they explore what happens when they physically remove objects from a group, then they cross out representations of objects from a group. This leads on to using this technique to solve simple subtractions. Learners then progress to visualising subtractions as part–whole diagrams, introducing the relationship between addition and subtraction. Learners explore subtraction in real-life situations and as number sentences.

Prerequisites for learning

Learners need to:
• recognise the numbers 0–10
• be able to count up to ten objects accurately
• be familiar with number sentences
• know what = means.

Vocabulary

take away, subtract, number sentence, minuend, subtrahend, equals, group, part-whole diagram

Common difficulties and remediation

As learners have previously learned about addition but have not experienced subtraction until this unit, some may struggle to adjust to taking objects away from a set rather than adding them or counting on. This is common when adjusting to learning about different operations and these learners will soon get used to subtraction strategies. Use words relating to subtraction as much as possible when teaching this unit and in classroom life (for example: **I had six pencils on my desk but two learners took one each – that makes four pencils left on my desk**) to help with the adjustment process.

When learners have removed objects from a group, they may forget which group they are supposed to count to find out how many objects are left. Remedy this by asking learners to remove the objects and put them somewhere away from the main group (for example, in a plastic bag or a small box or on their chair).

Supporting language awareness

Display the key words for each lesson and discuss them with learners. Use and refer to these words throughout the lesson and encourage learners to use the words too, prompting them when necessary.

It is important to teach the proper terms for subtraction (**minuend**: the starting number in a subtraction from which an amount is taken away; **subtrahend**: the amount or number that is taken away from the minuend) from a young age as this ensures that learners will have a good understanding of subtraction as they get older. Display these words together with practical examples where possible.

Promoting Thinking and Working Mathematically

TWM.04 Convincing
Learners apply what they have learned about taking away by using this skill to solve simple subtraction number sentences.

TWM.08 Improving
Learners refine their taking away technique by moving on from using concrete objects to pictures and diagrams.

Success criteria

Learners can:
• solve a subtraction by taking away (minuend 10 or less).

Number – Integers and powers

Unit **7** Subtraction as take away

Lesson 1: **Taking away objects**

Learning objectives

Code	Learning objective
1Ni.03	Understand subtraction as: [- counting back] - take away [- difference].

Resources

sheet of A4 paper (per pair); ten toy cars (or counters as an alternative) (per pair); set of 0–10 number cards from Resource sheet 1: 0–20 number cards (per pair); ten counters (per learner) (for the Workbook)

Revise

Use the activity *Taking away beads* from Unit 7: *Subtraction as take away* in the Revise activities.

Teach [SB] [.il]

- Direct learners to the picture in the Student's Book. Ask: **How many cakes were on the cake stand? How many are being taken away? How many are left?** Count the starting amount of cakes (5), the amount being taken away (2) and the remaining amount (3) with learners.
- Explain to learners that taking away is also called 'subtraction' and when you take an amount away from another amount, you are 'subtracting' it.
- [T&T] Ask: **How many cakes were there to start with? Were there more or less cakes left at the end? Why?** Discuss how taking away makes an amount less.
- Ask four volunteer learners to stand at the front of the class. Count them together. Now take one volunteer to the other side of the classroom. Say: **"There were four learners. I have taken one learner away. I am** *conjecturing* **– that means I am asking a mathematical question: I want to know how many learners are left now?**. Count the remaining learners and agree that there are three left.
- Repeat this with different starting amounts of learners, taking away a different amount of learners each time.
- Repeat this activity using the **Tree tool**, displaying one tree. Allow learners to come to the front to take away given amounts of birds from the tree.
- Discuss the Guided practice example in the Student's Book.

Practise [WB]

- Workbook

Title: Taking away objects

Page: 30

- Refer to Activity 1 from the Additional practice activities.

Apply [👥] [🖥]

- Display **Slide 1**.

- Give pairs of learners an A4 piece of paper (to use as a 'car park'), ten toy cars (you could use counters as an alternative) and a set of 0–10 number cards.
- Learners choose two number cards and identify the larger number. They drive that many cars into the 'car park'.
- They then look at the smaller number and drive that many cars off the paper. They count how many cars are left in the 'car park'.

Review

- Set up the **Apply** activity and use the numbers 3 and 0.
- [📢] Ask: **Which number is larger?** Elicit that it is 3 so learners must drive three cars into the car park.
- Ask: **How many cars do I need to drive out of the car park?** Elicit that you need to drive zero cars out again. Remind learners that zero means 'none', which means that you don't drive any cars out of the car park. Count how many cars are left and establish that there is still the same amount.
- Repeat several times to reinforce that, if you take zero away from an amount, the amount stays the same.

Assessment for learning

- Take three toy cars away from this group of seven. How many cars are left now?
- If you take some counters away from this group will there be more or less counters left?
- If I have four sweets and I give one away, how many sweets are left? Use counters to show me.

Same day intervention
Support

- Focus on taking away one object and counting how many are left before taking away more than one object.

Enrichment

- Give learners 'real-life' scenarios in which amounts are taken away and ask them to use counters to find out how many are left.

Lesson 2: Taking away to solve subtractions

Learning objectives

Code	Learning objective
1Ni.03	Understand subtraction as: [- counting back] - take away [- difference].

Resources

mini whiteboard and pen (per pair); ten interlocking cubes (per pair)

Revise

Use the activity *Taking away beads* from Unit 7: *Subtraction as take away* in the Revise activities.

Teach [SB]

• Direct learners to the picture in the Student's Book. Ask: **How many frogs were there to start with? How many frogs have been taken away? How many frogs are left?** Point out that because the frogs are just a drawing, you can't take one away in the same way that you can with real objects, so the ones that have been taken away have been crossed out. Discuss ways you could 'take away' drawn objects (for example, by covering some with your hands or a piece of paper or crossing them out).

• Point to the number sentence above the frogs. Ask: **What is different about this number sentence from the addition number sentences that you have been solving?** Elicit that there is a different operation sign. Write the calculation 6 – 4 = 2 on the board, point to the subtraction symbol and say: **This is the subtraction sign; this number sentence says 'six subtract four equals two'. Subtract means to take away.**

• Point to 6 and say: **The first number in a subtraction number sentence is called the minuend. This is our starting number.**

• Point to 4 and say: **The second number in a subtraction number sentence is called the subtrahend. We subtract that number from the minuend.**

• Point to 2 and say: **This is the amount after we have subtracted the subtrahend from the minuend.** If appropriate, tell the learners that the amount left is referred to as the 'difference'.

• Write various subtraction number sentences on the board, keeping the minuend at 10 or less. Solve them with the learners by first taking away with concrete objects, and then doing the same by drawing counters on the board and crossing out to take them away.

• Discuss the Guided practice example in the Student's Book.

Practise [WB]

• Workbook

Title: Taking away to solve subtractions

Page: 31

• Refer to Activity 1 from the Additional practice activities.

Apply [pairs icon] [screen icon]

• Display **Slide 1**.

• Learners, in pairs, solve the subtractions on the slide. Learner A uses interlocking cubes and learner B draws counters to match on their mini whiteboard and 'takes away' by crossing some of them out.

• Learners check to see if they you both get the same answer.

• Learners swap roles and repeat until all the subtractions are solved.

Review [TWM.04]

• On the board, write: 9 – 5 =.

• [TWM.04] [icon] Ask: **How many ways can you think of to solve this subtraction?** Take suggestions, working through each method with the learners. Ensure that you include different methods, using concrete objects, drawing objects and crossing some out, and drawing or using objects and covering some with your hands or paper. Remind the learners that demonstrating how they solved the problem is important because it shows evidence to *convince* everyone that the answer is correct.

Assessment for learning

• If you haven't got any objects, how can you solve 4 – 3?

• Show me how to solve 4 – 3 on a whiteboard/ piece of paper.

• Which number is the minuend/subtrahend?

• Which number is our starting number?

Same day intervention

Support

• Let learners continue to have more practice with concrete objects rather than drawing and crossing out if they are not yet confident with pictorial representations only.

Number – Integers and powers

Unit 7 Subtraction as take away

Number – Integers and powers

Lesson 3: **Taking away with part–whole diagrams**

Learning objectives

Code	Learning objective
1Ni.03	Understand subtraction as: [- counting back] - take away [- difference].

Resources

Resource sheet 10: Part–whole diagram (per pair); set of 0–10 number cards from Resource sheet 1: 0–20 number cards (per pair); ten counters (per pair)

Revise

Use the activity *Part–whole subtraction race* from Unit 7: *Subtraction as take away* in the Revise activities.

Teach 🔲 [TWM.08]

- Direct learners to the picture in the Student's Book. Discuss how there are four cherries to start with (in the top section of the diagram) and that one cherry has been taken away and that it has been put in the bottom left section of the diagram. Ask: **How many cherries are left?** Agree that three are left. Explain how the bottom right section of the diagram shows the remaining three cherries. Say: **Four cherries, take away one cherry, equals three cherries** while asking learners to point to the appropriate parts of the diagram in the Student's Book.
- Remember to remind the learners that taking away is subtraction, so when we take away an amount, we are subtracting it.
- [TWM.08] Draw a rectangular part–whole diagram on the board and ask a learner to give you a starting number between 2 and 10 (such as 7). Stick that amount of counters to the 'whole' section of the diagram with clear tape or modelling clay. Ask another learner to give you an amount to take away (such as 2). Move that many counters to the bottom left 'part' section, counting them with learners as you move each one. Ask: **Why are we moving these counters? What are we doing with them? What do we need to do to the counters that are left in the 'whole' section? What will that tell us?** Then move the remaining counters to the bottom right 'part' section. Say: **We have taken away two counters.** Ask: **How many counters were left?** Agree that there are five left and say: **Seven subtract five equals two.** Repeat several times with different amounts.
- Discuss the Guided practice example in the Student's Book.

Practise 🔲

- Workbook

Title: Taking away with part–whole diagrams

Page: 32

- Refer to Activity 2 from the Additional practice activities.

Apply 👥 🖥

- Display **Slide 1**.
- Give each pair of learners a rectangular part–whole diagram resource sheet, ten counters, a set of 0–5 number cards and a set of 6–10 number cards.
- Learners choose one number card from each pile. The larger number is their starting number and they count that many counters into the 'whole' section of their part–whole diagram.
- They count how many counters are left, while moving them into the remaining 'part' section of the diagram. They move the smaller number of counters into the bottom left 'part' section of the diagram.
- Repeat with different starting numbers.

Review

- Draw a part–whole diagram as in **Teach**.
- Ask learners to give you a starting amount (such as 9). Ask a learner to place that many counters in the 'whole' section of the diagram.
- Now ask: **If I need to take away zero counters, what should I do to the diagram?** Agree that you don't take away any counters.
- Move all of the counters to the bottom right section of the diagram. Point to the bottom left section and say: **I started with nine counters. I took away zero counters.** Point to the bottom right section and say: **There are nine counters left. Nine counters subtract zero counters equals nine counters.**
- Repeat this with different starting amounts to reinforce what happens when you take zero away.

Assessment for learning

- Show me how you would use a part–whole diagram to subtract 2 from 5.
- I am taking away 2 from these five counters. Where should they go on the part–whole diagram?
- Where does the starting amount go on the part–whole diagram?

Same day intervention

Support

- If learners are unable to be supported by an adult throughout this lesson, allow them to work with a peer of higher ability to boost understanding.

Lesson 4: **Solving subtractions with part–whole diagrams**

Learning objectives

Code	Learning objective
1Ni.03	Understand subtraction as: [- counting back] - take away [- difference].

Resources

Resource sheet 10: Part–whole diagram (per pair); ten counters (per pair); approximately six skipping ropes (per group); ten counters (per learner) (for the Workbook)

Revise

Use the activity *Part–whole subtraction race* (variation) from Unit 7: *Subtraction as take away* in the Revise activities.

Teach [SB] [TWM.04]

• Direct learners to the picture in the Student's Book and read out both the word problem and the matching number sentence. Look at the part–whole diagram and discuss how it has been used to solve the subtraction.

• On the board, write: 10 – 4 =. Remind learners of what the number sentence is asking them to do (subtract), of the names relating to each number in the subtraction (minuend and subtrahend) and of the function of the operations (subtract and equals).

• [TWM.04] Give each pair of learners a part–whole diagram resource sheet and ten counters and ask them to use their part–whole diagram to solve the subtraction on the board. Discuss the answer. Ask: **Did anyone get a different answer? Did anyone forget how to use the diagram or run into any other problems?** Repeat, using different minuends and subtrahends (keeping the minuend to 10 or less).

• Discuss the Guided practice example in the Student's Book.

Practise [WB]

• Workbook

Title: Solving subtractions with part–whole diagrams

Page: 33

• Refer to Activity 2 from the Additional practice activities.

Apply 👥 🖥 [TWM.04]

• Display **Slide 1**.

• Give groups of learners enough skipping ropes (about six) to mark out a rectangular part–whole diagram.

• As a group, they choose one of the number sentences on the slide to solve.

• The same number of learners as in the minuend stand in the 'whole' part of the diagram.

• The same number of learners as in the subtrahend take themselves away from the 'whole' section and move into the bottom left 'part' section.

• The remaining learners move into the bottom right 'part' section.

• The group tell you how many were left, to solve their chosen subtraction.

Review [TWM.08]

• Draw a part–whole diagram for 7 – 3 = 4 on the board. Underneath it, write the following number sentences: 7 – 6 = 1, 8 – 4 = 4, 7 – 3 = 4, 10 – 5 = 5, 7 – 1 = 6, 3 – 2 = 1.

• 🗣 Ask: **How can we work out which of these number sentences matches this part–whole diagram?** Take suggestions and discuss how to rule out number sentences, first by checking whether the starting amount matches the minuend, then by checking whether the second amount matches the subtrahend, and finally checking whether the final amount matches the total. Encourage learners to talk through the process, describing what they are doing and why each time and what each part of the diagram represents.

Assessment for learning

• Show me how you would use a part–whole diagram to solve 8 – 5 =.

• Which number sentence matches this part–whole diagram?

• I am taking away 5 from these eight counters. Where should they go on the part–whole diagram?

Same day intervention

Support

• If learners cannot be supported by an adult throughout this lesson, allow them to work with a peer of higher ability, to boost understanding.

Enrichment

• Ask learners to write the minuend and the subtrahend in the 'part' sections of their diagrams, then use counters to show the answer in the 'whole' section.

Number – Integers and powers

Additional practice activities

Activity 1

Learning objectives
- Subtract by taking objects away
- Identify when an amount of objects has been taken away from a group

Resources
ten countable objects, e.g. interlocking cubes/counters/small toys (per pair); set of 1–10 number cards from Resource sheet 1: 0–20 number cards (per pair)

What to do
- Learner A picks a number card (numbers 6–10 only) and counts out that many countable objects in front of them. They then turn around and close their eyes.
- Learner B picks a number card (numbers 1–5 only). They can choose to take that amount away from their partner's objects or not to take any objects away at all.

- Learner A opens their eyes and counts their remaining objects. They say how many are left and whether their partner took some objects away or not.
- Learners swap roles and repeat.

Variations
2 Learners write a subtraction number sentence, using their two chosen numbers, before continuing the activity as above.

3 Learner B first estimates whether they think items have been taken away or not by looking at their objects and deciding if it looks as if there are fewer than there were to start with.

Activity 2

Learning objectives
- Use a part–whole diagram to take away
- Use a part–whole diagram to solve subtractions (variation)

Resources
A3 sheet of paper (per group); brown crayon or pen (per group); ten cardboard cut-outs in the shape of sheep (per group); cotton wool or white yarn (per group); stick-on eyes or eyes made from small round sticky labels (per group)

What to do
- Draw a rectangular part–whole model on each A3 sheet of paper. Stick one in the middle of each group's table. Draw a gate with brown crayon between the 'whole' section and the part sections below it.
- Each learner makes one or two sheep using the cardboard cut-outs, cotton wool and stick on eyes.
- Give each group a starting amount. Learners line up that many sheep in the 'whole' section of the diagram.

- Now give them an amount to 'take away'. Learners work together to jump that many sheep over the fence and into the bottom left 'part' section.
- They jump the remaining sheep into the right 'part' section and count them to find out how many sheep were left.
- Repeat with different amounts of sheep each time.

Variations
1 Challenge 1 learners will benefit from acting out this scenario. Ask one of the learners in the group to be the farmer and the rest to be the sheep. The farmer directs how many sheep should jump over the gate according to the chosen number cards.

2 Give learners a subtraction number sentence to act out with the sheep and the part–whole diagram.

Number – Integers and powers

Unit 8: Subtraction as counting back

Collins International Primary Maths Recommended Teaching and Learning Sequence: Term 2, Week 2

Learning objectives

Code	Learning objective
1Ni.03	Understand subtraction as: - counting back [- take away - difference].
1Nc.04	Count [on in ones, twos or tens, and count] back in ones [and tens], starting from any number (from 0 to 10 [20]).

Unit overview

In this unit, learners continue to learn about subtraction, this time using the method of counting back. They start by revisiting counting back in ones and using this skill alongside concrete objects as a subtraction strategy. They then progress to solving subtractions by 'jumping' backwards along a line of objects, a number track and a number line. They continue to identify the components of a subtraction number sentence and work out the answer to subtraction problems as both number sentences and everyday situations.

Prerequisites for learning

Learners need to:
- recognise the numbers 0–10
- be able to count up to ten objects accurately
- be able to count backwards in ones from 10 to 0
- be familiar with the concept of subtraction
- be familiar with the components of subtraction number sentences.

Vocabulary

count back, subtract, number sentence, minuend, subtrahend, equals, number track

Common difficulties and remediation

Some learners may be resistant to learning a new subtraction method if they have become comfortable with the strategy of taking away. Remind them that the more strategies we have to solve mathematical problems, the easier it is and the better we understand it. Praise any efforts to use counting back as a strategy even if they are not immediately successful.

Counting back in ones as a method can be confusing as it is difficult to keep track of how many numbers you have counted back. Encourage learners to hold up fingers as they count back in ones to keep track of how many numbers they have said as they count.

'Jumping' back along a number line or track can confuse some learners as they may not understand

that the last number they land on is the answer, instead thinking that the last number they said as they counted the jumps is the answer. For this reason, it is important to teach this method alongside concrete objects first (see Lesson 2) to give learners the ability to deconstruct this method by manipulating the objects where necessary.

Supporting language awareness

Display the key words for each lesson and discuss them with learners. Use and refer to these words throughout the lesson and encourage learners to use the words too, prompting them when necessary.

It is important to teach the proper terms for subtraction (minuend, subtrahend) from a young age as this ensures that learners will have a good understanding of subtraction as they get older. Display these words together with practical examples where possible.

Promoting Thinking and Working Mathematically

TWM.04 Convincing
Learners apply what they have learned about counting back by using this skill to solve simple subtraction number sentences.

TWM.07 Critiquing
Learners try out different methods to count back to solve subtraction problems and compare them to find a method that works well for them and the purpose.

TWM.08 Improving
Learners refine their counting back technique by moving on from using concrete objects to number tracks and number lines.

Success criteria

Learners can:
- solve a subtraction by counting back (minuend 10 or less).

Number – Integers and powers/Counting and sequences

Unit **8** Subtraction as counting back

Lesson 1: **Counting back in ones to subtract**

Learning objectives

Code	Learning objective
1Ni.03	Understand subtraction as: - counting back [- take away - difference].
1Nc.04	Count [on in ones, twos or tens, and count] back in ones [and tens], starting from any number (from 0 to 10 [20]).

Resources

ten building blocks (per class); ten interlocking cubes (per pair); set of 0–10 number cards from Resource sheet 1: 0–20 number cards (per pair); mini whiteboards and pens (optional) (per learner)

Revise

Use the activity *Rocket launch* from Unit 8: *Subtraction as counting back* in the Revise activities.

Teach [SB]

- Direct learners to the picture in the Student's Book. Ask: **Is the rocket's countdown forwards or backwards?** Practise counting back from 10 to 0, then counting back in ones from any given number to 10 and ending on any given number between 0 and 9.
- Tell learners that they will be learning a new way to subtract this week. Explain that if you count back, you are subtracting.
- Draw five circles on the board and number them 1–5. Tell learners that you want to subtract 2. Model colouring in the two circles on the end, asking learners to check that you have coloured the correct amount. Then touch your finger on circle 5 and count back in ones, moving your finger as you count and stopping when you reach the first un-coloured circle. Say: **Five subtract two equals three**. Do this several times using different amounts.
- Now show learners how to use the same technique with objects by counting out a given number of building blocks, separating the amount that you want to subtract from the rest of the blocks slightly, then counting backwards from the starting number, touching each block as you go and stopping when you touch/count the first block in the original group of blocks.
- Discuss the Guided practice example in the Student's Book.

Practise [WB]

- Workbook

Title: Counting back in ones to subtract

Page: 34

- Refer to Activity 1 (variation) from the Additional practice activities.

Apply 👥 🖥

- Display **Slide 1**.

- Give pairs of learners a set of 0–10 number cards and ten interlocking cubes.
- Learners choose two number cards and identify the larger number. They count out that many cubes, then look at the smaller number and move that amount of cubes slightly apart from the rest of the line of cubes. They put their fingers on the last cube in the line and count back the smaller amount, stopping when they touch the first cube in the original group of cubes.
- Learners repeat with different starting numbers

Review [TWM.07]

- Say: **Yesterday I saw seven birds in my garden. Today I saw four fewer birds.**
- 🗣 Ask: **How can you work out how many birds I saw today by counting back?** Discuss this with learners, reminding them of the technique, discussing how they could draw the birds or use cubes or counters to represent the birds.
- [TWM.07] Learners work in pairs and choose to use mini whiteboards and pens or cubes to work out the answer. Did everybody get the same answer? Which method did they use and why? What did they do during each step of the process? Ask learners to *critique* the two methods, stating which worked best for them.

Assessment for learning

- What number would you land on if you counted back two numbers from 10?
- If you count back in ones from 6, what number comes next?

Same day intervention

Support

- Focus on counting back one or two numbers until learners are confident enough to count back further.

Enrichment

- Ensure that Challenge 3 learners have the opportunity to count back larger amounts (e.g. 7, 8 and 9).

Lesson 2: **Subtracting on a number track**

Learning objectives

Code	Learning objective
1Ni.03	Understand subtraction as: - counting back [- take away - difference].
1Nc.04	Count [on in ones, twos or tens, and count] back in ones [and tens], starting from any number (from 0 to 10 [20]).

Resources

ten interlocking cubes numbered 1–10 with stickers (per learner); Resource sheet 4: 0–10 number track (per learner)

Revise

Use the activity *Teddy number track subtraction* from Unit 8: *Subtraction as counting back* in the Revise activities.

Teach [SB] [TWM.04/08]

- Direct learners to the picture in the Student's Book. Point out that the kangaroo can jump back down the number track to find the answer to the subtraction. Ask: **How many jumps does the kangaroo need to do? Where must she start from? Where must she stop jumping? What will the answer be?**

- With reference to the picture in the Student's Book, explain how to count back along a number line to solve a subtraction. Remind learners that the starting number is 6 so she must start on that number. She needs to do two jumps because we are subtracting 2, and we cannot start counting until she has completed her first jump. Count two jumps down the number line with learners and remind them that the answer is the last number that you touch when you have completed all the jumps.

- Point out that you don't have to count backwards if you are using a number track because, as long as you count the jumps as you move your finger backwards, you will land on the correct number.

- Draw a 0–10 number track on the board. Write 9 – 3 = on the board. **[TWM.04/08]** Ask a learner volunteer to come forward and put their finger on the starting number. Ask the other learners to tell them how many jumps they must make to solve the subtraction. Repeat this with different subtractions and learners.

- Discuss the Guided practice example in the Student's Book.

Practise [WB]

- Workbook

Title: Subtracting on a number track

Page: 35

- Refer to Activity 1 from the Additional practice activities.

Apply [icons]

- Display **Slide 1**.

- Give learners ten interlocking cubes numbered 1–10 with small stickers. (Alternatively, some learners could use Resource sheet 4: 0–10 number track.)

- Learners use their cube number track to solve the subtractions displayed on the slide.

- Encourage less able learners to use this method, and then the counting back technique in Lesson 1 so they see the link between the two methods.

Review [TWM.04/08]

- On the board, write: 6 – 6 =.

- **[TWM.04/08]** Ask: **How would you solve this subtraction by using a number track?** Discuss answers and clear up any misconceptions.

- Give learners a 0–10 number track each and encourage them to count back with you. Ask: **What number did we land on?** Agree that we landed on 0. Ask: **What do you notice about the minuend and the subtrahend in the number sentence?** Elicit that both numbers are the same. **[T&T]** Ask: **If both numbers are the same in a subtraction, what is the outcome and why?**

- Repeat this with another subtraction in which the minuend and subtrahend are the same to see what happens. Carry out the process two more times to reinforce the rule that when the minuend and subtrahend are the same, the answer will always be 0.

Assessment for learning

- How would you use a number track to find the answer to 5 – 3?
- Which number would you start from on the number track if you were solving 5 – 3?
- Where on the number track must we stop?
- How many jumps must you make to solve 5 – 3?

Same day intervention

Support

- Link this method of counting back in ones by allowing learners to use the strategy from Lesson 1 after the strategy taught in this lesson so that they understand the relationship between jumping backwards and counting back in ones.

Number – Integers and powers/Counting and sequences

Lesson 3: **Subtracting on a number line**

Learning objectives

Code	Learning objective
1Ni.03	Understand subtraction as: - counting back [- take away - difference].
1Nc.04	Count [on in ones, twos or tens, and count] back in ones [and tens], starting from any number (from 0 to 10 [20]).

Resources

ruler – at least 0–10 cm (per pair); mini whiteboard and pen (per pair); string or rope, tape, chalk, 0–10 number cards in various sizes (per class)

Revise

Use the activity *Counting back race* from Unit 8: *Subtraction as counting back* in the Revise activities.

Teach SB [TWM.08]

- Direct learners to the picture in the Student's Book. Ask: **What is being used to solve the subtraction?** Agree that it is a number line.
- Point out that a ruler is being used as a number line in the picture. Hand out rulers to pairs of learners and draw a ruler on the board with divisions from 0 to 10. Show learners how to use a ruler as a number line by counting forwards and backwards along it, then by using it to solve a subtraction. Write 5 – 4 = on the board. Ask learners to find 5 on their rulers and put their finger on it then count back four numbers. Agree that the answer is 1. Repeat this with a different number sentence.
- Explain to learners that number lines are more useful than number tracks because to work out an addition or subtraction with a number track, you must be given a number track first, whereas you can easily draw or make your own number line or use a ruler.
- [TWM.08] Tell learners that they are going to *improve* their counting back technique by using a number line. On the board, show learners how to draw a number line. Explain that you could either write the numbers 0–10 along it or start on 0 on the left and finish on the minuend on the right. Write 7 – 3 = on the board then draw a 0–7 number line to match this number sentence. Ask a learner to come to the front to draw three jumps to count back from 7. Ask: **How will we know what the answer is?** Elicit that the answer is the number that you land on after you have finished counting back.
- Discuss the Guided practice example in the Student's Book.

Practise WB

- Workbook

Title: Subtracting on a number line

Page: 36
- Refer to Activity 1 from the Additional practice activities.

Apply 👥 🖥

- Display **Slide 1**.
- Give each pair of learners a ruler and a mini whiteboard and pen.
- Learners solve the subtractions on the slide. Learner A uses the ruler as a number line. Learner B draws a number line on the mini whiteboard and uses this.
- They check that they both got the same answer for the first subtraction then swap roles and repeat with the second number sentence.

Review [TWM.04/08]

- Make sure that you have the following located around the classroom where learners can see them: rope or string, tape, number cards, chalk, anything else that you feel could be used to construct a number line.
- Remind learners that they have been drawing their own number lines and using rulers as number lines.
- [TWM.04/08] [T&T] Ask: **Can you see anything else around the classroom that you could use to make a number line?** Discuss suggestions and add some of your own. Explore making number lines from skipping ropes, string, tape, number cards or drawing or writing them with chalk on the floor. Use your constructed number lines to solve subtraction number sentences, always keeping the minuend at 10 or less.

Assessment for learning

- How would you use a number line to find the answer to 9 – 5?
- Which number would you start from?
- How could you use a ruler as a number line?

Same day intervention

Support

- If learners struggle with the concept of using a ruler as a number line or drawing their own, give them a 0–10 number line to use to solve the subtractions.

Lesson 4: **Counting back to solve subtractions**

Learning objectives

Code	Learning objective
1Ni.03	Understand subtraction as: - counting back [- take away - difference].
1Nc.04	Count [on in ones, twos or tens, and count] back in ones [and tens], starting from any number (from 0 to 10 [20]).

Resources

set of 0–10 number cards from Resource sheet 1: 0–20 number cards (per pair); mini whiteboard and pen (per pair); Resource sheet 11: 0–10 number line (optional) (per pair)

Revise

Choose an activity from Unit 8: *Subtraction as counting back* in the Revise activities.

Teach SB

- Direct learners to the picture in the Student's Book. Ask: **What do you think the boy is doing?** Elicit that he is solving the subtraction.
- On the board, write: 10 – 7 =. Demonstrate to learners how to solve the subtraction by using the following mental strategy. Say: **We need to put ten in our heads and count back seven. We can use our fingers to keep track of how many we have counted back.** Touch your head and say: **Ten.** Now count back, saying: **Nine, eight, seven, six, five, four, three.** Hold one finger up for each number you say. Show the learners that you are holding up seven fingers so must stop counting. Explain that the last number you said when counting back is the answer. Repeat this with a different subtraction, this time asking learners to keep track of how many fingers you are holding up so they can tell you when to stop.
- Now repeat the process several times, inviting learners to copy you and use the strategy together. Remind them that if they want to check their answer, they can use a number line to do so.
- Discuss the Guided practice example in the Student's Book.

Practise WB

- Workbook

Title: Counting back to solve subtractions

Page: 37

- Refer to Activity 2 from the Additional practice activities.

Apply 👥 🖥

- Display **Slide 1**.
- Give pairs of a set of 0–10 number cards, a mini whiteboard and pen, and a 0–10 number line (optional).
- Learners pick two number cards and write them as a subtraction on their mini whiteboards, making sure to use the larger number as the minuend.
- Learner A uses the mental strategy to count back to find the answer. Learner B assists by keeping track of how many fingers learner A is holding up.
- Optional: They check the answer by using a number line.
- Learners swap roles and repeat.

Review

- Say: **I had seven cats but three of them ran away.**
- 🖐 Ask: **What number sentence would match that subtraction problem?** Write the number sentence on the board.
- Ask learners to work with a partner to find out how many cats are left. Encourage them to use the mental strategy from this lesson and to check their answer by using a number line.

Assessment for learning

- Show me how you would count back in your head to solve 6 – 4 =.
- How will you know when to stop counting back?
- How could you check your answer?

Same day intervention

Support

- Ask learners to only subtract up to 3 from a given number so that they don't get confused about how many fingers they are holding up.

Number – Integers and powers/Counting and sequences

Additional practice activities

Activity 1

> **Learning objective**
> • Solve a subtraction by using a number track or number line

Resources

rope or string (per group); ten pegs of the same colour (per group); one peg of a different colour (per group); ten socks labelled 1–10 with stickers or paper cut-out clothing labelled 1–10 (per group); set of 1–10 number cards from Resource sheet 1: 0–20 number cards (per group); mini whiteboard and pen (per group)

What to do

• Hang the string between two tables as though it is a washing line. Ask learners to order the socks from 1 to 10 and hang them on the washing line.

• Learners pick two 1–10 number cards and write them as a subtraction number sentence on their mini whiteboard making sure to put the larger number first (for example, 4 – 2 =).

• They work together to solve the subtraction by touching the different - coloured peg over the number

that matches the minuend and then 'jumping' back by the number that matches the subtrahend. They clip the different - coloured peg onto the last sock that they jump to as they count back.

• Learners repeat with different starting numbers.

• **[TWM.04]** Encourage group discussion throughout the activity: **Why are they 'jumping' back? What does the number they land on tell us? How would they solve the same subtraction on a drawn number line?**

Variations

2 Learners solve the subtraction by counting backwards from the minuend rather than by counting jumps.

3 Give Challenge 3 learners larger subtrahends as they will be more confident at keeping track of their place on a number line when counting back.

Activity 2 ![icon] ![2]

> **Learning objective**
> • Use a mental strategy to subtract by counting back

Resources

large number track on the floor made by taping large 0–10 number cards in place (per group); set of 0–10 number cards from Resource sheet 1: 0–20 number cards (per group)

What to do

• One learner stands on 10 and waits for instructions.

• Another learner chooses a 0–10 number card as the amount to subtract from 10.

• The group counts back from 10 by the number shown on the card, while the learner jumps back one space for each number that is said as the group counts. The group use their fingers to keep track of how many numbers they have counted.

• When the group have finished counting, they check that the learner on the number track is standing on the last number they said when they finished counting.

Variation

1 Use this activity to count backwards with learners who forget which order the numbers come in as they count back from 10. Ask them to stop counting on a different number each time. When they are confident with counting back, introduce instructions to subtract amounts (1, 2 or 3 only).

Unit 9: Subtraction as difference

Collins International Primary Maths
Recommended Teaching and
Learning Sequence: Term 2, Week 3

Learning objectives

Code	Learning objective
1Ni.03	Understand subtraction as: [- counting back - take away] - difference.

Unit overview

In this unit, learners explore subtraction as finding the difference between two numbers or amounts. They begin by comparing towers and lines of similar-sized objects to see how many more are in the taller tower/ longer line. They move on to using this technique to solve subtraction number sentences and word problems. Learners then draw representations of objects to solve subtractions in this way. Finally, they learn how to find the difference between two numbers on a number line by counting how many 'jumps' it takes to get from a given number to another.

This unit consolidates what learners know about subtraction number sentences and gives them another strategy to solve them.

Note

Learners need to be familiar with the concept of 'zero' in order to work through this unit and Unit 10. This is introduced in Unit 16, which comes before Unit 10 in the recommended sequence of work. If you are not following this sequence, we recommend that you teach the lesson 'Zero' from Unit 16 (p.185) prior to this unit even if you wish to leave the rest of Unit 16 until a later date.

Prerequisites for learning

Learners need to:
- recognise the numbers 0–10
- be able to count up to ten objects accurately
- be familiar with the concept of subtraction
- be familiar with the components of subtraction number sentences
- be able to identify which of two towers or lines is taller or longer than the other.

Vocabulary

difference, subtract, number sentence, minuend, subtrahend, equals, number line

Common difficulties and remediation

Some learners find it difficult to comprehend that 'finding the difference' is the same as subtracting, as they have become accustomed to seeing subtraction as taking something away. Teaching how to find the difference between two amounts first (Lesson 1) and then moving on to doing the same to solve subtraction number sentences (Lesson 2) reinforces the link, as does explaining that it is another strategy to help us to solve subtractions, just as taking away and counting back are.

Using a number line to find the difference may confuse some learners as they will be using it differently from the way they did in Unit 9: Subtraction as counting back, but for the same purpose. Give learners plenty of practice at this so that they understand that a question is asking them to subtract if it asks them to find the difference, count back or take away.

Supporting language awareness

Display the key words for each lesson and discuss them with learners. Use and refer to these words throughout the lesson and encourage learners to use the words too, prompting them when necessary.

It is important to teach the proper terms for subtraction (minuend, subtrahend, difference) from a young age as this ensures that learners will have a good understanding of subtraction as they get older. Display these words together with practical examples where possible.

Promoting Thinking and Working Mathematically

TWM.04 Convincing
Learners apply what they have learned about comparing sets of objects to find the difference by using this skill to solve simple subtraction number sentences.

TWM.08 Improving
Learners refine their strategy by moving on from using concrete objects to finding the difference on a number line.

Success criteria

Learners can:
- solve a subtraction by finding the difference between two given numbers (minuend 10 or less).

Unit (9) Subtraction as difference

Lesson 1: **Finding the difference**

Learning objectives

Code	Learning objective
1Ni.03	Understand subtraction as: [- counting back - take away] - difference.

Resources

ten interlocking cubes (per class); sheet of paper (per learner); ten stickers or paper cut-out shapes and a glue stick (per learner); set of 1–5 number cards from Resource sheet 1: 0–20 number cards

Revise

Use the activity *Hands up* from Unit 9: *Subtraction as difference* in the Revise activities.

Teach [SB]

- Direct the learners to the picture in the Student's Book. Ask: **How many books are in each pile? How could you work out how many more books are in the larger pile?** Take suggestions.
- Tell the learners that they will be learning how to find the difference between two amounts. Explain that 'finding the difference' means finding how many more there are in one amount than the other. Refer them back to the picture in the Student's Book, saying: **There are five more books in the larger pile than the smaller pile. The difference between the numbers of books is five.**
- Ask eleven learners to come to the front. Line seven of them up (standing) and line the remaining four up (kneeling) in front of the first four learners in the standing group. **[T&T]** Say: **I want to know the difference between the two groups of learners. How would you find that out?** Ask learners to feed back to the class, eliciting that you could count the standing learners who do not have a kneeling learner in front of them and that therefore the difference between seven and four is three.
- Now make towers of interlocking cubes representing the same amounts and show learners that you can work out the difference using cubes as well as people. Explain that when we are finding the difference with objects, we must line up the objects carefully so that we can see clearly how many more there are in the longer line. Repeat this several times with different starting amounts of learners/cubes.
- Discuss the Guided practice example in the Student's Book.

Practise [WB]

- Workbook

Title: Finding the difference

Page: 38

- Refer to Activity 1 from the Additional practice activities.

Apply 👤 🖥

- Display **Slide 1**.
- Give each learner a sheet of paper, ten stickers (or paper cut-out shapes and a glue stick) and a set of 1–5 number cards.
- Learners choose two number cards. They make a line from their stickers to match each amount. Ensure that learners position the two lines of stickers one above the other so that differences between the amounts of stickers are obvious.
- They count the extra stickers on the longer line to find the difference between the two amounts and write the answer at the bottom of the paper.

Review

- **[T&T]** Ask: **If my starting number was zero and my finishing number was four, what is the difference?**
- Demonstrate to learners that the problem with using objects to find the difference between 0 and another number is that you cannot make a line or tower of zero objects because zero means none.
- Make a tower of four cubes and show learners, then hold out an imaginary tower and say: **This is a tower of zero cubes.** Demonstrate counting the difference between them by counting all four cubes in the tower of four and say: **The difference between zero and four is four.**
- Repeat with 0 and other numbers less than 10 to reinforce that the difference between 0 and any number/amount will always be the same as the other number.

Assessment for learning

- What is the difference between 8 and 10?
- How could you use cubes to find the difference?
- Why must we line up our objects carefully when we compare them to find the difference?

Same day intervention
Support

- Let learners find small differences of 1 or 2 so that they don't become confused when counting larger amounts.

Lesson 2: **Subtracting by finding the difference**

Learning objectives

Code	Learning objective
1Ni.03	Understand subtraction as: [- counting back - take away] - difference.

Resources

20 building blocks (per class); paper (per learner); dishes of red and blue paint (per group)

Revise

Use the activity *Hands up* from Unit 9: *Subtraction as difference* in the Revise activities.

Teach [SB] [TWM.04]

- Direct learners to the picture in the Student's Book. Ask: **Can you find the difference between the amounts of windows that are lit up on the towers?** Work through the problem with learners, counting the extra windows that are lit up on one of the towers to find the answer. Now direct their attention to the number sentence and show them that the numbers in the number sentence are the same as the numbers of windows and the answer to the difference between the amounts. Explain that when you are asked to find the difference between two amounts, you are being asked to subtract the smaller amount from the larger amount.

- [TWM.04] Write 8 − 3 = on the board, saying: **Eight subtract three. That's the same as finding the difference between three and eight.** Ask: **How can I use blocks to find the difference and solve the subtraction?** Ask a volunteer to make the two towers of blocks, then count the extra blocks in the tallest tower while other learners talk them through the process and discuss the outcome.

- Now show learners how to find the difference without objects by drawing counters in lines on the board. Cross out the extra counters in the longer line and count them together.

- Discuss the Guided practice example in the Student's Book.

Practise [WB]

- Workbook

Title: Subtracting by finding the difference

Page: 39

- Refer to Activity 1 from the Additional practice activities.

Apply 🧑 🖥

- Display **Slide 1**.
- Give learners paper and two dishes of paint, one blue and one red.

- Learners solve the subtractions on the slide by making fingerprint lines to match the minuend and the subtrahend using one colour of paint. They go over the top of the extra prints in the longer line with the other colour of paint to find the difference and write the answer next to the lines.

Review

- Build a tower of nine blocks and a tower of six blocks.
- 🗣 Ask: **What number sentence would match these towers if we wanted to find the difference between the amounts in each tower?** Encourage learners to use the terms 'minuend', 'subtrahend', 'subtract', 'difference' and 'equals'. Elicit that the larger number must be the minuend, the smaller number must be the subtrahend and the difference between them is the answer. Find the difference with the learners and write 9 − 6 = 3 on the board.

Assessment for learning

- How would you find the answer to 8 − 3 = by finding the difference?
- How would you use blocks to find the difference to answer 8 − 3 =?
- Which amount of blocks must be the minuend/ subtrahend?

Same day intervention

Support

- Give learners more practice with using concrete objects to find the difference if drawing or printing the amounts causes them to lose focus.

Enrichment

- Give learners a set of 1–10 number cards to make their own subtraction number sentences to solve in this way.

Number – Integers and powers

145

Lesson 3: **Finding the difference on a number line**

Number – Integers and powers

Learning objectives

Code	Learning objective
1Ni.03	Understand subtraction as: [- counting back - take away] - difference.

Resources

Resource sheet 11: 0–10 number line (per learner); set of 1–10 number cards from Resource sheet 1: 0–20 number cards (per pair); Resource sheet 11: 0–10 number line (per pair); two counters (per pair)

Revise

Use the activity *Find the difference towers* from Unit 9: *Subtraction as difference* in the Revise activities.

Teach [SB] [📊] [TWM.08]

• Direct learners to the picture in the Student's Book. Ask: **Which flower did the bee start his journey from? Which flower has he ended up on? How many flowers has he landed on?** Tell learners that the amount of flowers the bee has landed on is the difference between his starting number and his finishing number.

• [TWM. 08] Tell learners that they will be using number lines and number tracks to find the difference between numbers, *improving* their strategies for this skill. Hand out 0–10 number lines. Display the **Number line tool** (0–10) and circle the numbers 3 and 8. Show learners how to put their fingers on the starting number (3) and count along the numbers from 3 to 8, starting from the number that is next to 3 and including 8. Ask: **How many numbers were there?** Elicit that there were five numbers, so 5 is the difference between 3 and 8. Repeat this with different pairs of numbers, asking learners to count along with you on their number lines.

• [TWM. 08] Give learners different strategies to use with their number lines, such as crossing off each number that they count, drawing jumps or dots, or placing a sticker on each number counted to make checking their answer easier. Encourage learners to describe and discuss the strategies – which do they find the easiest? Are they confused by any strategy? How is crossing out numbers the same as drawing a dot above each number? Is the outcome the same for every strategy?

• Discuss the Guided practice example in the Student's Book.

Practise [WB]

• Workbook

Title: Finding the difference on a number line

Page: 40

• Refer to Activity 2 from the Additional practice activities.

Apply 👥 🖥

• Display **Slide 1**.

• Give each pair a set of 1–10 number cards, a number line and two counters.

• Learners pick two numbers, place a counter on each of the numbers and find the difference by counting the numbers between them and the last number.

Review [📊] [TWM.08]

• Discuss what would happen if you needed to find the difference between 0 and any number. Ask: **Where would you start counting the difference from?** Elicit that it is the same as finding the difference between any two numbers, so you would start by counting the numbers between.

• [TWM.08] Ask learners to demonstrate finding the difference between 0 and various numbers to 10 on the **Number line tool**.

Assessment for learning

• What is the difference between 6 and 9?

• How would you use a number line to find the difference between 6 and 9?

• Point to the number I would count first if I wanted to find the difference between 6 and 9.

• Which number would I stop counting on if I was finding the difference between 6 and 9 on a number line?

Same day intervention

Support

• Give learners small stickers to put on each number between their two numbers on their number line and the larger number. They can then count how many stickers to find the difference.

Lesson 4: **Subtraction as difference on a number line**

Learning objectives

Code	Learning objective
1Ni.03	Understand subtraction as: [- counting back - take away] - difference.

Resources

set of 1–10 large number cards (per class); 20 large stickers (per class); 0–10 number cards from Resource sheet 1: 0–20 number cards (per learner); Resource sheet 11: 0–10 number line (per learner); small stickers or a pencil (per learner)

Revise

Use the activity *Find the difference towers* from Unit 9: *Subtraction as difference* in the Revise activities.

Teach ⬚ [TWM.08]

- Direct learners to the picture in the Student's Book. Ask: **How is the number line being used to solve the subtraction?** Elicit that it is being used to find the difference. Ask: **What is the difference between 3 and 8?** Count the jumps and agree that the difference is 5. Say: **8 – 3 = 5.**
- Remind learners that 'find the difference' is another way to say 'subtract', so when you find the difference between two numbers, you are subtracting the smaller number from the larger number.
- [TWM.08] On the board, write: 6 – 2 =. Invite 6 learners to come to the front of the class and give them each a large number card from the range 1–6 to hold. They stand in order like a human number line. Ask a learner to come to the front to show the class where to start counting from and where to stop to find the difference. Count the numbers to find the difference with the learners. ⬚ Ask: **What could we do to these numbers on our number lines to make it easier to find the difference?** Elicit that you could cross out or place a sticker on each number as you count to find the difference. Do this now, putting a sticker on each learner whose number has been counted. Count the learners with stickers to check the answer.
- Repeat several times with different subtractions within 10.
- Discuss the Guided practice example in the Student's Book.

Practise ⬚

- Workbook

Title: Subtraction as difference on a number line

Page: 41

- Refer to Activity 2 from the Additional practice activities.

Apply 👤 🖥

- Display **Slide 1**.
- Learners use a 0–10 number line to find the difference to solve the subtractions on the slide.
- Provide small stickers to stick on the numbers, or pencils to cross out numbers as support.

Review

- Say: **Leo goes to two clubs every week after school. Munisha goes to four clubs. What is the difference between the number of clubs they go to?**
- ⬚ Ask: **What number sentence would match that subtraction problem?** Write the number sentence on the board.
- Ask learners to work with a partner and use their number lines to find the difference.

Assessment for learning

- Show me how you would find the difference on a number line to solve 8 – 5 =.
- Which numbers do you count if you are finding the difference between 7 and 10?
- How could you check your answer?

Same day intervention

Support

- Give learners interlocking cubes to make towers with to check their answers after using a number line.

Enrichment

- Give learners a subtraction word problem to solve by finding the difference on a number line.

Number – Integers and powers

147

Number – Integers and powers

Additional practice activities

Activity 1

Learning objectives
- Use objects to find the difference between two amounts
- Use objects to solve a subtraction by finding the difference (variation)

Resources
box of building blocks or interlocking cubes (per pair); set of 1–10 number cards from Resource sheet 1: 0–20 number cards (per pair); small pieces of paper and pencil (per pair)

What to do
- Pairs of learners pick two 1–10 number cards.
- Using building blocks or interlocking cubes, they each make a tower to match the amount on their card. The towers must be placed next to each other.
- Together, they count the additional blocks in the taller tower to find the difference, then write the difference on a piece of paper that they put in front of their towers.
- They repeat with two other starting numbers.
- Each group can build their own 'city' of towers, between the pairs of which they have found the

difference. Challenge 1 learners can do this as an adult-led activity in a group rather than in pairs.
- [TWM.04] Ask pairs of learners to report back to the class, showing them their starting numbers and how they used the blocks to find the difference, explaining what they are doing and why throughout.

Variations
1 When learners have made their two towers, ask them to take the extra blocks off of the tallest tower, count them, then replace them with the same number of blocks in a different colour so that they have a clearer visual representation of the difference.

2 As above, but learners write a subtraction number sentence on their paper using their two numbers, filling in the answer when they have found the difference.

Activity 2

Learning objectives
- Find the difference between two numbers on a number line
- Solve a subtraction by finding the difference between two numbers on a number line

Resources
two small toy animals (per pair); sheet of blue paper (per pair); ten small balls of modelling clay (per pair); set of 1–10 number cards from Resource sheet 1: 0–20 number cards (per pair)

What to do
- Pairs of learners set up ten 'stepping stones' (the balls of modelling clay) going across the blue paper that represents a river. They use the 1–10 number cards to number the stepping stones in order.
- Learners choose any two numbers from 1 to 10 and place one animal on each of the stepping stones representing those numbers.

- They find the difference by counting the amount of stepping stones between the two animals and the second stepping stone with an animal on it. Learners must remember to start counting from the stepping stone after the first animal, not from the first animal.
- Learners repeat with different pairs of numbers.

Variations
2 Learners write a subtraction number sentence to match the two numbers that they have chosen and find the answer as above.

3 Ask Challenge 3 learners to look at the difference (the stepping stones) between the two animals and estimate how much the difference is before they count.

Unit 10: Addition and subtraction to 10 (A)

Learning objectives

Collins International Primary Maths
Recommended Teaching and
Learning Sequence: Term 2, Week 4

Code	Learning objective
1Ni.04	Recognise complements of 10.
1Ni.05	[Estimate,] add and subtract whole numbers (where the answer is from 0 to 10 [20]).
1Nc.04	Count on in ones, [two or tens,] and count back in ones [and tens], starting from any number (from 0 to 10 [20]).

Unit overview

In this unit, learners are introduced to complements of and to 10. They use concrete objects to find pairs of amounts that equal 10 and begin to memorise these as number facts. They then learn the addition and subtraction facts for these pairs of numbers, using the addition and subtraction strategies that they have learned in previous units, to check their results. Learners then move on to finding complements for other numbers to 10 (e.g. 7) and apply addition and subtraction to these in the same way.

Note

Learners need to be familiar with the concept of 'zero' in order to work through this unit and Unit 9. This is introduced in Unit 16, which comes before Units 9 and 10 in the recommended sequence of work. If you are not following this sequence, we recommend that you teach the lesson 'Zero' from Unit 16 (p.185) prior to this unit even if you wish to leave the rest of Unit 16 until a later date.

Prerequisites for learning

Learners need to:
- recognise the numbers 0–10
- be able to count up to ten objects accurately
- be familiar with the concepts of addition and subtraction
- have a variety of addition and subtraction strategies that they are able to use to work out and check an answer.

Vocabulary

complements, add, addition, subtract, subtraction, equals, number sentence, total

Common difficulties and remediation

Learners require plenty of practical activities in which they can split ten (or fewer) objects into two groups, changing the amount of objects in each group every time. This allows learners to experience and therefore understand the different complements of and to 10.

Some learners may grasp the concept of complements of and to 10 quickly but will struggle to relate this to addition and subtraction. It can help to give them addition and subtraction operations cards to put between two groups of objects (for example, 6 cubes + 4 cubes = 10 cubes) so that they can see examples of this with concrete objects before moving on to abstract representations.

Likewise, trying to see the link between addition and subtraction can confuse some learners, so using objects and operations cards as above, and moving the objects and cards into different positions to create a subtraction number sentence from an addition number sentence (for example, 6 + 4 = 10 could convert to 10 – 6 = 4 easily in this way), can help to reinforce this idea.

Supporting language awareness

Display the key words for each lesson and discuss them with learners. Use and refer to these words throughout the lesson and encourage learners to use the words too, prompting them when necessary.

It is important to teach the proper terms for addition (augend, addend, sum) and subtraction (minuend, subtrahend, difference) from a young age, as this ensures that learners will have a good understanding of addition and subtraction as they get older. Display these words together with practical examples where possible.

Promoting Thinking and Working Mathematically

TWM.01 Specialising
Learners work out whether two amounts or numbers are complements of a number to 10.

TWM.04 Convincing
Learners write addition and subtraction number sentences to match complements of and to 10 to demonstrate their understanding of how the pairs of numbers can be manipulated and how addition and subtraction are related.

Success criteria

Learners can:
- find and identify complements of 10
- find and identify complements to 10
- write addition and subtraction number sentences for complements of and to 10
- check the answers to addition and subtractions using a variety of previously learned strategies.

Number – Integers and powers/Counting and sequences

Unit **10** Addition and subtraction to 10 (A)

Lesson 1: **Making 10**

Learning objectives

Code	Learning objective
1Ni.04	Recognise complements of 10.
1Nc.04	Count on in ones, [two or tens,] and count back in ones [and tens], starting from any number (from 0 to 10 [20]).

Resources

ten interlocking cubes (per learner); mini whiteboard and pen (per learner)

Revise

Choose an activity from Unit 10: *Addition and subtraction to 10 (A)* in the Revise activities.

Teach 🔲 🖥

- Direct learners to the picture in the Student's Book. Ask: **Why do you think the children in each pair are holding hands? Look at the numbers on their T-shirts. Do you notice anything?** Take suggestions, then explain that the numbers on the children's T-shirts all add up to 10 for each pair. Give pairs of learners ten interlocking cubes and ask each pair to check one of the pairs of numbers, using the cubes to ensure that they add up to 10.
- Explain that pairs of numbers that add up to 10 are called 'complements of 10' and that knowing these pairs is helpful to us when we add or subtract. Display **Slide 1** and talk through the different complements of 10, asking learners to match each set of cubes on the slide with their own cubes.
- Tell learners that an easy way to check complements of 10 is on our fingers. Hold up ten fingers and ask learners to do the same. Put down one finger and count along the fingers that are still up. Say: **One, two, three, four, five, six, seven, eight, nine fingers that are up and one finger that is down makes ten fingers altogether.**
- Ask volunteer learners to come to the front to hold up an amount of fingers up to 10, while the rest of the class work out how many fingers must be down to make ten fingers altogether.
- [T&T] Ask: **If I hold up ten fingers, what number must I add to it to make 10? What about if I hold up zero fingers?** Remind learners that 10 and 0 and 0 and 10 are also complements of 10.
- Discuss the Guided practice example in the Student's Book.

Practise 🔳

- Workbook

Title: Making 10

Page: 42

- [TWM.01] Refer to Activity 1 from the Additional practice activities.

Apply 👤 🖥 [TWM.01]

- Display **Slide 2**.
- Give each learner ten interlocking cubes and a mini whiteboard and pen.
- Challenge learners to find as many different pairs of numbers that make 10 by breaking the line of cubes into two parts and counting how many cubes are in each part. Tell the learners that by doing this, they are specialising - finding examples that fit the criteria of totalling 10.
- They record their pairs of numbers on mini whiteboards.

Review

- Give each learner a set of ten interlocking cubes.
- Call out numbers from 0 to 10. Learners hold up the amount of cubes that make 10 when added to your starting number.

Assessment for learning

- What number makes 10 when added to 8?
- Tell me two numbers that make 10.
- Are 5 and 7 totals of 10? How do you know?
- I'm holding up six fingers. How many more fingers make 10?

Same day intervention

Support

- Let learners work with an adult to make a poster showing the different complements of 10, using their findings with their cubes. They could use ten paper cut-out shapes to show each pair of amounts/numbers on the poster to help them to memorise them.

Enrichment

- Write several pairs of numbers on the board, some of which total 10, some of which total other numbers. Learners must identify the pairs that do not total 10 and work out their totals to check this.

Lesson 2: **Addition and subtraction facts for 10**

Learning objectives

Code	Learning objective
1Ni.04	Recognise complements of 10.
1Ni.05	[Estimate,] add and subtract whole numbers (where the answer is from 0 to 10 [20]).
1Nc.04	Count on in ones, [twos or tens,] and count back in ones [and tens], starting from any number (from 0 to 10 [20]).

Resources

Resource sheet 10: Part–whole diagram (per learner); ten counters (per learner); mini whiteboard and pen (per pair); operations cards –, +, = from Resource sheet 13: Operations cards (per pair)

Revise

Choose an activity from Unit 10: *Addition and subtraction to 10 (A)* in the Revise activities.

Teach [SB]

- Direct learners to the picture in the Student's Book. Ask them to count the brown teddies, then the pink teddies. Remind learners that they have been finding complements of 10 and that 8 and 2 are a complement of 10 because the two numbers make 10 when added together.

- Give each learner a part–whole diagram resource sheet and ten counters. Draw a part–whole diagram on the board. Fill in your diagram with 10 in the whole section and 8 and 2 in the part sections. Explain to learners that they can use part–whole diagrams to help them to write additions and subtractions for complements of 10. Point out that 10 is the whole number and that 8 and 2 are the parts that make up the whole.

- Ask learners to put eight counters and two counters in the part sections of their diagrams. Say: **8 + 2 = 10** and write this on the board. Encourage learners to check this by counting their eight counters into the whole section of the diagram, then counting on the extra two counters into the whole section. Now ask them to move the counters back to their starting positions and demonstrate that you could do this the other way around, starting with 2 and adding eight more counters. Write 2 + 8 = 10 on the board.

- Now show learners how to make subtractions. Ask them to put ten counters into the whole section of the diagram, then move eight of those counters to one of the part sections to subtract them. Show learners that there are two counters left, so 10 – 8 = 2. Write this on the board and demonstrate that you could also use the same numbers to make 10 – 2 = 8.

- Discuss the Guided practice example in the Student's Book. Ask: **What other addition/ subtraction could be the answer here?**

Practise [WB]

- Workbook

Title: Addition and subtraction facts for 10

Page: 43

- Refer to Activity 1 (variation) from the Additional practice activities

Apply 👥 🖥 [TWM.04]

- Display **Slide 1**.

- Give each pair a mini whiteboard and pen, a set of operations cards (+, –, =) and ten counters.

- Learner A writes a pair of numbers that are complements of 10 on the whiteboard.

- **[TWM.04]** Learner B makes an addition or subtraction number sentence to match the pair of numbers, using counters and the operations cards. Learners must check that the components of the number sentence total 10 to prove that learner A's pair of numbers are complements of 10.

- Learners swap roles and repeat.

Review

- Divide learners into groups of five.

- The first learner in each group writes two numbers that are complements of 10 on a mini whiteboard. They pass it to the next learner who must write a matching addition or subtraction number sentence.

- The whiteboard is passed down the line until four different number sentences that match the number pair have been written.

- The number sentences must be checked in order to prove that the pair of numbers are complements of 10. Encourage learners to use a part–whole diagram and counters to check their answers if they need to.

Assessment for learning

- Write four different number sentences to match this part–whole diagram.
- What is the missing number in this number sentence?

Same day intervention

Support

- If learners are struggling with using part–whole diagrams, allow them to use ten interlocking cubes for support instead.

Number – Integers and powers/Counting and sequences

Number – Integers and powers/Counting and sequences

Lesson 3: **Making numbers to 10**

Learning objectives

Code	Learning objective
1Ni.05	[Estimate,] add and subtract whole numbers (where the answer is from 0 to 10 [20]).
1Nc.04	Count on in ones, [two or tens,] and count back in ones [and tens], starting from any number (from 0 to 10 [20]).

Resources

five interlocking cubes (per pair); modelling clay (per pair) (optional); mini whiteboard and pen (per pair); ten counters (per pair); set of 3–9 number cards from Resource sheet 1: 0–20 number cards (per pair); 0–9 number cards from Resource sheet 1: 0–20 number cards (one card at random per learner); ten interlocking cubes (per learner) (for the Workbook)

Revise

Choose an activity from Unit 10: *Addition and subtraction to 10 (A)* in the Revise activities.

Teach 🆂🅱 [TWM.01/02]

- Direct learners to the picture in the Student's Book. Ask: **What is each set of pictures showing? What number is each set making?** Elicit that they show different ways of making one number – they are the complements of 6.
- Explain that as well as the complements of 10, there are complements for each number to 10. Write 5 on the board and give pairs of learners five interlocking cubes. **[TWM.01/02] [T&T]** Ask: **Can you think of a pair of numbers that make 5 when added together?** Take suggestions and write the complements of 5 on the board.
- ⓟ Ask: **Can you think of a way to organise these 8 learners into two groups?** Invite further learners to the front to decompose the group of 8 into two groups in different ways.
- Remind learners that each pair of numbers will work either way around when added together (for example, 7 and 1, and 1 and 7). Write the complements of 8 on the board as they are found.
- Discuss the Guided practice example in the Student's Book.

Practise 🆆🅱

- Workbook

Title: Making numbers to 10

Page: 44

- Refer to Activity 2 from the Additional practice activities.

Apply 👥 🖥

- Display **Slide 1**.
- Give learners some modelling clay (optional), a set of 3–9 number cards and ten counters.
- Learners make two 'cakes' from the modelling clay (alternatively they could draw two 'cakes'

on a mini whiteboard). They pick a number card and count out that many counters. They share the counters between the cakes to decorate them. They count how many counters are on each cake and record it as a complement of their number on their mini whiteboard.
- Learners repeat until no more complements can be found for that number, then start again with a new number.

Review [TWM.01]

- Give each learner a 0–9 number card.
- Learners walk around the room looking for someone who has the number that makes 9 when added to their number.
- When they have found a partner, they sit together with their cards displayed in front of them. Use these number pairs to make a list of the complements of 9 on the board. Are there any missing? If so, write number pairs on the board and ask: **Which of these number pairs are complements of [X]?** Learners work this out using countable objects if necessary.

Assessment for learning

- How many ways can you make 6 with interlocking cubes?
- What number makes 6 when added to 4?
- How many different pairs of numbers that make 6 can you write down?
- Do 4 and 3 make 7? Why/Why not?

Same day intervention

Support

- Start learners off with the numbers 3, 4 and 5 as there will be fewer complements to find than for numbers 6 to 9.

Enrichment

- Challenge learners to find out how many pairs of numbers they can find that total 8 or 9.

Lesson 4: **Addition and subtraction facts to 10**

Learning objectives

Code	Learning objective
1Ni.05	[Estimate,] add and subtract whole numbers (where the answer is from 0 to 10 [20]).
1Nc.04	Count on in ones, [two or tens,] and count back in ones [and tens], starting from any number (from 0 to 10 [20]).

Resources

Resource sheet 10: Part–whole diagram (per pair, then per learner); ten counters (per pair, then per learner); ten interlocking cubes (per pair); 0–9 number cards from Resource sheet 1: 0–20 number cards (one at random per learner)

Revise

Choose an activity from Unit 10: *Addition and subtraction to 10 (A)* in the Revise activities.

Teach [SB] [TWM.04]

- Direct learners to the picture in the Student's Book and discuss what it shows. Explain to learners that it shows a complement of 7 both in cubes and in a part–whole diagram, and it shows the related addition and subtraction facts.
- Give pairs of learners a part–whole diagram resource sheet and ten counters and show them how to use the counters on the diagram to add and subtract to get the associated addition and subtraction facts for 7.
- **[TWM.04]** Ask: **Who can think of another complement of 7?** Take a suggestion and ask learners to make this on their part–whole diagrams, then use their counters on the diagram to work out the matching addition and subtraction facts. Write these on the board as learners volunteer them. Encourage discussion about the number sentences and why and how the same numbers are used. Ask: **What happens if we add 4 to 3? What happens if we subtract 3 from 7? How is that similar to the addition number sentence?**
- Repeat this with an even starting number and explore what happens when the two numbers in the complement of that number are the same (e.g. when 6 is made of 3 and 3). Demonstrate to learners that there is only one matching addition and one matching subtraction because both the parts of the number are the same.
- Discuss the Guided practice example in the Student's Book.

Practise [WB]

- Workbook

Title: Addition and subtraction facts to 10

Page: 45

- Refer to Activity 2 from the Additional practice activities.

Apply 👥 👨‍👩‍👧 🖥 [TWM.04]

- Display **Slide 1**.
- Give each group a target number of 10 or less (such as 9).
- Each pair within the group picks a complement of that number and represents it using interlocking cubes, then using counters on a part–whole diagram.
- **[TWM.04]** They write the two matching additions and two matching subtractions on the part–whole diagram sheet. Encourage learners to discuss the outcome of each number sentence and to show how the number sentences relate to one another using the part–whole diagram.
- At the end of the activity, learners can walk around the room looking at all of the groups' work to see the different complements of each number and the related addition and subtraction facts.

Review

- Give each learner a 0–9 number card.
- Write an addition on the board (such as 5 + 2 = 7).
- Ask all learners with the numbers that are the complements in that number sentence to hold up their cards. Remind learners that complements are the parts, not the whole, so if they have the largest number in the addition number sentence, it is not a complement. Ask learners for a related addition fact and the related subtraction facts.
- Repeat this with different number sentences.

Assessment for learning

- Tell me a pair of numbers that make 7.
- How many number sentences can you write for this complement of 7?
- Write a subtraction number sentence to match this addition (or vice versa).

Same day intervention
Support

- Ask learners to write one addition and one subtraction only.

Additional practice activities

Activity 1

Learning objectives
- Identify and use complements of 10
- Relate addition and subtraction to complements of 10 (variation)

Resources
ten counters (per pair); set of 0–10 number cards from Resource sheet 1: 0–20 number cards (per pair)

What to do
- Give pairs of learners ten counters and a shuffled set of 0–10 number cards.
- [TWM.01] Learners choose a starting number from 0 to 10. They then pick a number card and use their counters to work out if their starting number and the number they have picked are complements of 10. They do this by lining up their 10 counters, counting along them until they reach their starting number and moving these counters to one side, then checking if the remaining amount matches the

number they have picked. If it does not match, the numbers are not complements of 10.
- They continue to do this with the same starting number and different number cards until they find a pair of numbers that are complements of 10.
- Learners repeat with a new starting number.

Variations

1 Give Challenge 1 learners two different number cards–they must work out which of these numbers matches is a complement of 10 with their starting number.

 Learners write an addition and/or subtraction sentence to match the pairs of numbers that are complements of 10.

Activity 2

Learning objectives
- Know the complements to (or of) 10 and the related subtraction facts
- Know the complements to (or of) 10 and the related addition facts (variation)

Resources
blindfold (per pair); ten counters or other countable items to be used as 'treasure' (per pair); set of 0–10 number cards from Resource sheet 1: 0–20 number cards (per pair); mini whiteboard and pen (per pair)

What to do
- Learner A lines up a given number of counters to 10 (for example, 8) and puts on a blindfold.
- Learner B chooses a number card (less than or equal to learner A's number) and takes away that amount of counters.

- Learner A removes the blindfold, counts the remaining counters and uses their knowledge of complements to 10 to work out how many counters have been taken.
- Learners write a number sentence to match on their mini whiteboards.

Variation

 Learners write a matching addition fact for the subtraction fact.

Unit 11: Addition and subtraction to 10 (B)

Learning objectives

Collins International Primary Maths
Recommended Teaching and
Learning Sequence: Term 2, Week 5

Code	Learning objective
1Ni.05	Estimate, add and subtract whole numbers (where the answer is from 0 to 10 [20]).
1Nc.04	Count on in ones, [twos or tens,] and count back in ones [and tens], starting from any number (from 0 to 10 [20]).

Unit overview

In this unit, learners consolidate what they have learned about adding and subtracting numbers to 10. They hone their estimation skills by drawing on their knowledge of complements to 10 then check the accuracy of their estimates by counting on or back, both with objects and as a mental strategy. They apply addition and subtraction to real-life situations and pick the strategy from other addition and subtraction units that they feel the most comfortable with to solve these problems. This unit should leave learners proficient in their knowledge of addition and subtraction to 10 and ready to move on to working with numbers to 20.

Prerequisites for learning

Learners need to:
- recognise the numbers 0–10
- be able to count up to ten objects accurately
- be able to count on and back at any point from 0 to 10
- be familiar with the concepts of addition and subtraction
- be familiar with the concept of estimation
- have a variety of addition and subtraction strategies that they are confident to use.

Vocabulary

add, addition, augend, addend, sum, subtract, subtraction, minuend, subtrahend, difference, take away, equals, estimate, number sentence, count on, combine sets

Common difficulties and remediation

It is still helpful to learners to provide concrete objects for them to use to check their estimated answers as this helps to consolidate their understanding of numbers as amounts and vice versa. It also assists them in developing a deeper understanding of the strategies that they are using.

Some learners may still not have grasped that estimation is about using what you know to make your best guess, rather than guessing at random. Reminding learners of what they know already can inform their guesses (e.g. familiar patterns of objects to 10 and complements to 10).

Some learners may not be confident when choosing a strategy to use to find the answer to an addition or subtraction. Remind them of the strategies they have learned and ask them which one they like using the best and why.

Supporting language awareness

Display the key words for each lesson and discuss them with learners. Use and refer to these words throughout the lesson and encourage learners to use the words too, prompting them when necessary.

It is important to teach the proper terms for addition (addend, augend, sum) and subtraction (minuend, subtrahend, difference), from a young age as this ensures that learners will have a good understanding of addition and subtraction as they get older. Display these words, together with practical examples where possible.

Promoting Thinking and Working Mathematically

TWM.04 Convincing
Learners estimate an answer using known facts for support, then check the answer.

TWM.07 Critiquing
Learners work out answers to additions and subtractions using previously learned strategies, comparing which ones work best for them and choosing the best strategy to answer a given problem.

Success criteria

Learners can:
- provide a good estimate of the answer to an addition or subtraction to 10
- solve an addition or subtraction using different methods
- relate addition and subtraction to real-life situations.

Lesson 1: **Estimating an answer**

Learning objectives

Code	Learning objective
1Ni.05	Estimate, add and subtract whole numbers (where the answer is from 0 to 10 [20]).
1Nc.04	Count on in ones, [twos or tens,] and count back in ones [and tens], starting from any number (from 0 to 10 [20]).

Resources

mini whiteboard and pen (per learner); ten counters (per pair); Resource sheet 11: 0–10 number line (per pair)

Revise

Use the activity *Higher or lower* from Unit 11: *Addition and subtraction to 10 (B)* in the Revise activities.

Teach [SB]

- Direct learners to the picture in the Student's Book. Show learners the addition and the cubes that relate to it. Explain that the children in the picture are estimating the answer to the addition 5 + 3 and, today, this is what learners will be doing.
- Remind learners that estimating means using what you already know to make a good guess. Point to the cubes in the Student's Book and remind learners that they can use their knowledge of objects in patterns to help them to estimate how many there are altogether. Now point to the number sentence and explain that you can also estimate without seeing objects to match each of the numbers.
- Ask learners to point to 5 in the number sentence and tell learners that, as they can see they are adding something to 5, they know the answer must be more than 5, so 0, 1, 2, 3, 4 or 5 would not be a good estimate.
- Now point to 3 and tell learners that as the addend is less than 5 and we already know that 5 + 5 = 10, the answer must be less than 10, so 10 would not be a good estimate either. Ask: **Who can think of a good estimate for the answer?** Take suggestions, reminding learners that the answer must be between 6 and 9 if a number below 6 or over 9 is given.
- Solve the addition together by counting on from the larger number (i.e. 5). Use objects to count on, then ask: **How else can you count on?** Elicit that you could do it in your head or use a number line.
- Repeat this process for two more additions, then write a subtraction on the board. Explain how to make a good estimate by using what is already known (for example, if the minuend is 6, the answer can't be more than 6. If the subtrahend is nearly as large as the minuend, the answer/difference must be very small. Learners also make use of their knowledge of complements to 10). Count back to find the answer and repeat twice more with other subtractions to 10.
- Discuss the Guided practice example in the Student's Book.

Practise [WB]

- Workbook

Title: Estimating an answer

Page: 46

- Refer to Activity 1 from the Additional practice activities.

Apply 👥 🖥 [TWM.04]

- Display **Slide 1**.
- Give each learner a mini whiteboard and pen, and each pair ten counters and a number line.
- Learners write down an estimate to the addition they see on the slide. They compare this with their partner's estimate. The learners explain why they think they have made a good estimate and which facts they used to make it.
- They then work out the answer together by counting on, using the counters or number line for support if they wish. The winner is the learner whose estimate was closer to the answer. If both estimates were as close to the answer as each other, it is a draw.
- Learners repeat this for the subtraction on the slide.

Review

- Ensure each learner has a mini whiteboard and pen.
- Write various additions and subtractions to 10 on the board. Learners estimate each answer and write a number on their whiteboard to indicate their estimate.

Assessment for learning

- Estimate the answer to 6 − 2 = / 4 + 2 =.
- Why have you chosen 6 as an estimate?
- How can you check whether your estimate was correct?
- How could you count on/back to find the answer?

Same day intervention
Support

Give learners cubes to assist them with both estimating answers and counting on or back to find the answers.

Lesson 2: **Choosing how to solve an addition**

Learning objectives

Code	Learning objective
1Ni.05	Estimate, add [and subtract] whole numbers (where the answer is from 0 to 10 [20]).

Resources

Resource sheet 11: 0–10 number line (per learner); ten counters (per learner); Resource sheet 10: Part–whole diagram (per learner); ten interlocking cubes (per learner); mini whiteboard and pen (per learner)

Revise

Use the activity *Addition and subtraction machines* from Unit 11: *Addition and subtraction to 10 (B)* in the Revise activities.

Teach 🆂🅱 [TWM.07]

- Direct learners to the picture in the Student's Book. Indicate all the different ways there are to solve an addition like the one shown in the book. Remind learners that estimating the answer first is useful and explain that when we solve the addition, our estimate should be close to our answer. This helps us to know whether we have solved it correctly.

- [TWM.07] Write an addition on the board and ask: **How many ways can you think of to solve this addition?** Ask learners to think about strategies (such as counting on or combining groups) and what they could use to help them (for example, mental strategies, or objects, number lines/tracks, part–whole diagrams). Ask: **Why would you choose that method?** Make a list on the board. Remember to include recalling known facts (e.g. complements to 10) and remind learners that it is important to estimate the answer first using what you already know from the number sentence as clues.

- Go through the list, asking volunteer learners to come to the front of the class to demonstrate how to use each method shown. Address any misconceptions and practise any methods with which learners are less secure, until they are more confident. Praise learners who refer to each part of the addition by the correct terms (augend, addend and sum).

- [TWM.07] Discuss when you would use each method and when it would be more difficult to use certain methods or equipment – for example, if you didn't have any objects available to count on with, you could use a ruler as a number line and use that to count on instead. Remind learners that they are *critiquing* methods to find the one that works best.

- Discuss the Guided practice example in the Student's Book.

Practise 🆆🅱

- Workbook

Title: Choosing how to solve an addition

Page: 47

- Refer to Activity 2 from the Additional practice activities.

Apply 👤 🖥 [TWM.08]

- Display **Slide 1**.

- [TWM.08] Learners solve the additions using the method of their choice. Ensure that equipment to support them with this (such as number lines, counters) is readily available.

- When learners reach the red addition (4 + 3 =), they must use a method they are less confident in.

Review [TWM.07]

- 🗣 Ask: **Which strategies did you find the most difficult to use during this lesson?**

- Go through any strategies mentioned by learners, addressing any problem areas and asking learners whether a different method would be easier or more useful and why.

Assessment for learning

- Tell me two different ways that you could solve this addition.
- What is the answer to 5 + 3 = ?
- I solved this addition by combining two sets of objects. How else could I solve it?
- Estimate the answer to 5 + 3 = then choose how to solve it to check your estimate.

Same day intervention
Support

- Rather than allowing learners to use one strategy only, give them adult support to use different strategies, if possible.

Enrichment

- Challenge learners to always use the strategy that they find the most difficult so that they improve their skills.

Number – Integers and powers/Counting and sequences

Lesson 3: Choosing how to solve a subtraction

Learning objectives

Code	Learning objective
1Ni.05	Estimate[, add] and subtract whole numbers (where the answer is from 0 to 10 [20]).

Resources

Resource sheet 11: 0–10 number line (per learner); ten counters (per learner); Resource sheet 10: Part–whole diagram (per learner); ten interlocking cubes (per learner); mini whiteboard and pen (per learner)

Revise

Use the activity *Addition and subtraction machines* from Unit 11: *Addition and subtraction to 10 (B)* in the Revise activities.

Teach [SB] [TWM.07]

- Direct learners to the picture in the Student's Book. Point out that, like Lesson 2, it shows lots of different ways to solve a number sentence, but that this time the number sentence is a subtraction.
- Write a subtraction on the board and ask: **How many ways can you think of to solve this subtraction?** Ask learners to think about strategies (such as counting back, finding the difference or taking away) and what they could use to help them (for example, mental strategies, or objects, number lines/tracks, part–whole diagrams). Make a list on the board. Remember to include recalling known facts (complements to 10) and remind learners that it is important to estimate the answer first using what you already know from the number sentence as clues.
- Go through the list, asking volunteer learners to come to the front of the class to demonstrate how to use each method shown. Address any misconceptions and practise any methods with which learners are less secure, until they are more confident. Praise learners who refer to each part of the subtraction by the correct terms (minuend, subtrahend and difference).
- [TWM.07] Ask learners when they would choose to use each method and when it would be more difficult to use certain methods or equipment – for example, if they didn't have a part–whole diagram available they could count back in their heads, using their fingers to keep track, or use a ruler as a number line to find the difference.
- Discuss the Guided practice example in the Student's Book.

Practise [WB]

- Workbook

Title: Choosing how to solve a subtraction

Page: 48

- Refer to Activity 2 from the Additional practice activities.

Apply [TWM.07]

- Display **Slide 1**.
- [TWM.07] Learners solve the subtractions, using a different method each time. Ensure that equipment to support them with this (such as number lines, counters) is readily available.
- When they have finished solving each subtraction, they tell a partner how they did it, talking them through the method they used and why they chose it.

Review [TWM.07]

- Ask: **Which strategies did you find the most difficult to use during this lesson?**
- Go through any strategies mentioned by learners, addressing any problem areas and asking learners whether a different method would be easier or more useful and why.

Assessment for learning

- Tell me two different ways that you could solve this subtraction.
- What is the answer to 9 – 2 = ?
- I solved this subtraction by counting back. How else could I solve it?
- Estimate the answer to 9 – 2 = then choose how to solve it to check your estimate.

Same day intervention

Support

- Rather than allowing learners to use one strategy only, give them adult support to use different strategies, if possible.

Enrichment

- Challenge learners to always use the strategy that they find the most difficult so that they improve their skills.

Lesson 4: **Addition and subtraction in real life**

Learning objectives

Code	Learning objective
1Ni.05	Estimate, add and subtract whole numbers (where the answer is from 0 to 10 [20]).

Resources

photocopied real-life addition and subtraction sheet (see Apply) (per learner); Resource sheet 11: 0–10 number line (per learner); ten counters (per learner); Resource sheet 10: Part–whole diagram (per learner); ten interlocking cubes (per learner); ten soft toys (per class)

Revise

Choose an activity from Unit 12: *Addition and subtraction to 10 (B)* in the Revise activities.

Teach [SB]

• Direct learners to the picture in the Student's Book. Say: **Three children have a drink of fruit juice. Two children have a drink of water. How many drinks is that altogether?** Agree that there are five drinks in total, checking this by counting on from three and touching each of the bottles of water as you count on.

• Explain that we use addition and subtraction in real life all the time, just like in the picture. For example, we need to know how many of something we need in total, or how many people will be left when some leave, or how many snacks we need to buy altogether. Tell learners that they have all the skills to solve real-life additions and subtractions because they can use any of the methods that they have learned, but first they have to work out whether they need to add or subtract in each situation.

• Say: **I invited eight friends to my party, but three of them can't come. How many friends can come to my party?** Now ask: **Do I need to add or subtract to work out the answer?** Elicit that you need to subtract because you are taking something away from the starting number of 8. Write it as a subtraction on the board: 8 – 3 = and ask: **How can I solve the subtraction?** Allow volunteer learners to demonstrate different strategies for this.

• Say: **I bought four ice creams for my family, but then my three cousins joined us. How many ice creams do I need to buy altogether?** ♫ Now ask: **Do I need to add or subtract to work out the answer?** Elicit that you must add as you are adding more to a starting number. Ask: **How can I solve this addition?** Allow volunteer learners to demonstrate their strategies for this.

• Discuss the Guided practice example in the Student's Book.

Practise [WB]

• Workbook

Title: Addition and subtraction in real life

Page: 49

• Refer to Activity 1 from the Additional practice activities.

Apply 👤 🖥 [TWM.07]

• Display **Slide 1**.

• Give each learner a photocopied sheet on which is written (or two similar, more appropriate/relevant problems, making sure to edit the slide accordingly):

1. **Ben has ☐ _____ . Shan gives him ☐ more. How many does he have now?**
2. **Shan had ☐ _____ . Ben took ☐ away. How many does he have now?**

• Learners fill in the sheets with the numbers and objects of their choice then solve the problems using the methods of their choice.

• Be sure to have part–whole diagrams, counters, number lines and interlocking cubes available for learners to use.

• [TWM.07] Invite learners to share their problem and how they solved it with the class. They should say why they chose the method, what answer it gave them, and whether they would choose a different method next time.

Review

• Ask learners to work in pairs or small groups to come up with a real-life addition or subtraction based on ten soft toys, at the front of the class.

• Invite pairs/groups to come to the front and use the soft toys to act out their additions and subtractions. Invite the other learners to try to solve them in their pairs/groups as they are acted out.

Assessment for learning

• Do you need to add or subtract to solve that problem?

• How would you choose to solve the problem? Why?

Same day intervention
Enrichment

• Ask learners to write their own real-life problems. Ask them to think about situations in which they or someone they know had to add or subtract and use this scenario as the basis of their problem.

Number – Integers and powers/Counting and sequences

Additional practice activities

Activity 1

Learning objectives
- Estimate the answer to an addition or subtraction
- Relate addition and subtraction to real - life scenarios

Resources

a few sets of ten items – e.g. ten toy cars, ten counters, ten books (per group)

What to do

- Learners work in groups of three.
- Learner A is the shopkeeper and has a variety of items (such as ten toy cars, ten cubes, ten books) in their shop, in groups of 10.
- Learners B and C are the customers. They choose what sort of items they would like (such as cars) and each ask for up to 5 of that item (for example, learner B asks for four cars, learner C asks for one car).
- **[TWM.04]** Learner A estimates how many they would like, explaining their reasoning for this estimate, then counts that many out.

- The three learners work out the answer by counting on and check how close the estimate was.
- Learners swap roles and repeat.

Variation

 Adapt for subtraction as follows: Learners B and C are returning up to ten items (such as seven cars) to the shop. They decide how many they want to give back to the shopkeeper (an amount less than 10 that is also less than the amount of items they are starting with, for example, five cars). Learner A estimates how many items they will have left when they have returned some of them (for example, three cars). Learners count back together to check the answer.

Activity 2

Learning objectives
- Use different strategies to solve an addition
- Use different strategies to solve a subtraction (variation)

Resources

ten interlocking cubes (per group); Resource sheet 10: Part–whole diagram (per group); ten counters (per group); Resource sheet 11: 0–10 number line (per group); two sets of 0–5 number cards from Resource sheet 1: 0–20 number cards (per group); mini whiteboard and pen (per group)

What to do

- Learners work in groups of three, or as pairs in groups of six.
- Each learner (or pair) chooses two number cards to make an addition (such as 4 + 2 =). They estimate the answer then use one of the pieces of equipment provided to work out the answer, using counting on or combining two sets as their strategy.
- They record the number sentence, estimate and answer on a mini whiteboard, then pick a different

two number cards and use a different piece of equipment or strategy to find the answer.

Variations

1 Challenge 1 learners should do this activity in an adult-led group, with the adult supporting them while they use each strategy.

2 Adapt for subtraction by asking learners to choose between counting back, finding the difference or taking away, using the equipment already listed. Use one set of 0–10 number cards and another set of 0–5 number cards instead of two sets of 0–5 number cards for this.

3 Challenge 3 learners should choose the strategy which they find the most difficult so that they practice this until they are confident.

Number – Integers and powers/Counting and sequences

Unit 12: Addition and subtraction to 20 (A)

Collins International Primary Maths
Recommended Teaching and
Learning Sequence: Term 3, Week 2

Learning objectives

Code	Learning objective
1Ni.04	Recognise complements of 10.
1Ni.05	Estimate, add [and subtract] whole numbers (where the answer is from 0 to 20).
1Ni.06	Know doubles up to double 10.
1Nc.04	Count on in ones, [twos or tens, and count back in ones and tens,] starting from any number (from 0 to 20).

Unit overview

This unit builds on what learners have already learned and mastered about addition, this time extending the totals to up to 20. They add a one-digit number to a two-digit number, learning to put the two-digit number first then count on. They learn the strategies of making 10 then adding more and using near doubles when solving additions. They also revisit estimation, estimating an answer to an addition using what is already known (for example, that 12 + 5 must be more than 10, less than 20 and more than 15) before solving the addition using a strategy that they feel confident with.

Note

Learners need to be familiar with the concept of doubling in order to work through this unit and Unit 13. This is introduced in Unit 14, which comes before Units 12 and 13 in the recommended sequence of work. If you are not following this sequence, we recommend that you teach Unit 14 before Units 12 and 13 so that learners have a full understanding of the strategies used in these units.

Prerequisites for learning

Learners need to:
- recognise the numbers 0–20
- be able to count up to 20 objects accurately
- be able to solve additions to 10 confidently
- have a variety of addition and subtraction strategies that they are able to use to calculate and check an answer
- recognise doubles to 10.

Vocabulary

addition, add, more, double, estimate, one-digit, two-digit, complements, ten, more

Common difficulties and remediation

It can be difficult to comprehend how to use what you already know to estimate an answer. Remind learners to look carefully at the numbers and to think about

the addition facts to 10 that they know. If you are adding a two-digit number and a one-digit number, it is not possible for the answer to be 10 or less. If you are adding 2 and 16, you already know that the answer is more than 16. Encourage learners to spend time carefully looking at the numbers in an addition number sentence when making an estimate of the answer. Also remind them that they should always use what they know to help them work out the answer to a problem.

Making 10 and adding more can feel quite laborious at this stage, but it is worth persevering with as it is a very useful strategy for adding larger numbers in later stages. Practising with countable objects will help.

Supporting language awareness

Display the key words for each lesson and discuss them with learners. Use and refer to these words throughout the lesson and encourage learners to use the words too, prompting them when necessary.

It is important to teach the proper terms for addition (addend, augend, sum) from a young age as this ensures that learners will have a good understanding of addition as they get older. Display these words, together with practical examples where possible.

Promoting Thinking and Working Mathematically

TWM.07 Critiquing
Learners think about which strategies would or would not be appropriate to use to solve an addition.

TWM.08 Improving
With near doubles and making 10, learners add more strategies to their addition methods to refine their addition techniques.

Success criteria

Learners can:
- use a variety of methods, including counting on, to solve additions with answers to 20
- provide a reasonable estimate of an answer to an addition.

Number – Integers and powers/Counting and sequences

Lesson 1: **Addition facts to 20**

Learning objectives

Code	Learning objective
1Ni.05	Estimate, add [and subtract] whole numbers (where the answer is from 0 to 20).
1Nc.04	Count on in ones, [twos or tens, and count back in ones and tens,] starting from any number (from 0 to 20).

Resources

Resource sheet 12: 0–20 number line (per learner); 20 counters (per learner); mini whiteboard and pen (per group)

Revise

Use the activity *Counting on race* from Unit 12: *Addition and subtraction to 20 (A)* in the Revise activities.

Teach [SB] [TWM.07]

- Direct learners to the picture in the Student's Book. Ask: **Can you estimate how many children there are in the large group? ... small group? ... altogether?** Take suggestions. Now ask learners to count how many children there are in each group. Write the answers (14 and 4) on the board. Ask: **What problems might you have if you wanted to add the groups together?** Take suggestions.
- Ask: **Which would take longer, counting to 4 or counting to 14?** Explain that when adding a one- and a two-digit number together, it is quicker and easier if you start with the larger number, the two-digit number and count on the smaller number, the one-digit amount. If you do it the other way around it may take a long time and you could easily get confused and miscount.
- **[TWM.07]** On the board, write: 12 + 4 =. Ask: **How could we count on to get the answer?** Elicit that you count, using counters or a number line, or count on in your head, using your fingers for support. Demonstrate all three techniques with the learners, asking them to describe the process of each method and to discuss whether they feel each method would be a good choice and why. Remind them that they are *critiquing* the methods.
- Repeat with different numbers, asking learners to check that you are starting with the larger number, the two-digit number.
- Discuss the Guided practice example in the Student's Book.

Practise [WB]

- Workbook

Title: Addition facts to 20

Page: 50

- Refer to Activity 1 from the Additional practice activities

Apply 👤 🖥 [TWM.07]

- Display **Slide 1**.
- Learners use counting on from the two-digit number to solve the additions on the slide.
- **[TWM.07]** They choose to count on using counters, a number line or mentally, depending on their ability level and which method they feel confident in using. Allow learners time to discuss their choice of method, how they got their answer and whether they think the method was a good choice.

Review

- Write on the board a different one- and two-digit addition for each group to solve.
- 🗣 Ask: **Which addition do you think will have the largest answer?**
- The groups write their answers on mini whiteboards and share them with the rest of the class.

Assessment for learning

- Which number would you start with in this addition? Why?
- What different strategies could you use to solve this one- and two-digit number addition?
- Show me two ways of counting on to solve 14 + 5 =.

Same day intervention
Support

- Learners are likely to feel most comfortable using concrete objects (counters) to solve these additions. To help them to progress when using counters, start on the larger number but don't count out counters to match it – only use counters to count on the smaller amount, from the starting number.

Lesson 2: **Making 10 and adding more**

Learning objectives

Code	Learning objective
1Ni.04	Recognise complements of 10.
1Ni.05	Estimate, add [and subtract] whole numbers (where the answer is from 0 to 20).
1Nc.04	Count on in ones, [twos or tens, and count back in ones and tens,] starting from any number (from 0 to 20).

Resources

mini whiteboard and pen (per learner); two ten frames from Resource sheet 2: Ten frames (per learner); 5–9 number cards from Resource sheet 1: 0–20 number cards (per learner); 20 counters (per learner)

Revise

Use the activity *Ten and some left over* from Unit 12: *Addition and subtraction to 20 (A)* in the Revise activities.

Teach [SB] [TWM.08]

- **[TWM.08]** Direct learners to the picture in the Student's Book. Ask: **How many butterflies are there? How could you organise them to make counting easier?** Try out some of the learners' ideas, counting the butterflies each time, then explain that making ten in one of the ten frames would make it easier to count the two amounts together. Tell the learners that by doing this, you are *improving* on your existing methods for solving addition problems.
- Remind learners that they know the complements for 10 and make a list of these on the board with the learners.
- Explain that knowing the complements for 10 can help you to add two amounts when the answer is more than 10, because you can make 10 first then add more.
- Display the **Ten frame tool** or draw two ten frames on the board. Draw eight dots in one and three dots in the other. Remind learners that adding 2 to 8 makes 10.
- Cross out two of the dots from the second ten frame and draw them in the first ten frame. Say: **I have made 10. Now add on what is left in the second ten frame to find the answer**. Ask a learner to do this for you, then write 8 + 3 = 11 and 10 + 1 = 11.
- Repeat with different numbers.
- Discuss the Guided practice example in the Student's Book. Ask: **What other addition could be the answer here?** (10 + 1)

Practise [WB]

- Workbook

Title: Making 10 and adding more

Page: 51

- Refer to Activity 2 from the Additional practice activities.

Apply 👤💻

- Display **Slide 1**.
- Give each learner a mini whiteboard and pen, two ten frames, a set of 5–9 number cards and 20 counters.
- Learners pick two number cards and write an addition on their whiteboards (such as 9 + 5 =).
- They put counters on their ten frames to match the numbers, then move counters from the second ten frame to make ten in the first ten frame. They then write a new number sentence and the answer (i.e. 10 + 4 = 14).
- Learners repeat with different numbers.

Review

- Write 14 on the board.
- Learners put ten counters in one ten frame and four counters in the other. They try to find different combinations of numbers to make 14 by moving the counters between the ten frames.

Assessment for learning

- How would you use ten frames to add these two numbers?
- What must you add to the first number to make 10?
- What number is left over?
- How much is 10 add the amount that is left over?

Same day intervention

Support

- Ensure that learners only have to add 1 or 2 when they have made 10.

Number – Integers and powers/Counting and sequences

163

Unit 12 Addition and subtraction to 20 (A)

Lesson 3: **Near doubles**

Number – Integers and powers/Counting and sequences

Learning objectives

Code	Learning objective
1Ni.05	Estimate, add [and subtract] whole numbers (where the answer is from 0 to 20).
1Ni.06	Know doubles up to double 10.
1Nc.04	Count on in ones, [twos or tens, and count back in ones and tens,] starting from any number (from 0 to 20).

Resources

mini whiteboard and pen (per pair); 20 interlocking cubes (per pair)

Revise

Use the activity *Double or near double?* from Unit 12: *Addition and subtraction to 20 (A)* in the Revise activities.

Teach SB

- Direct learners to the picture in the Student's Book. Ask: **How many kittens are in each basket? How many kittens are there altogether? How could you use doubling to help you to work out how many kittens there are altogether?**
- Remind learners that they know doubles to 10 and briefly revise these with the class.
- Tell learners that a pair of numbers that are very close together are called 'near doubles' and that today they will be adding near doubles together. Explain that 6 is only one more than 5, and they know that double 5 is 10 so they can estimate that the answer must be 'about 10'. As 6 is only one more than 5, they can add on one more.
- Use the **Tree tool** to demonstrate, putting five apples on one tree and six apples on the other. Move the sixth apple off the second tree, then ask what double 5 is. Establish that it is 10, then add the one apple on its own to make 11.
- Show learners a different way to solve the problem. Show the first tree with three apples and the second tree with four apples (3 + 4). Say: **If we remove one apple from the second tree, we now have 3 + 3. You know that 3 + 3 = 6. So now you need to count on one to get the answer.** (7) Point out that either method will give the same answer so learners should use whichever one they find most suitable.
- Repeat with more examples before discussing the Guided practice example in the Student's Book.

Practise WB

- Workbook

Title: Near doubles

Page: 52

- Refer to Activity 2 from the Additional practice activities.

Apply 👥 💻 [TWM.08]

- Display **Slide 1**.
- Pairs of learners work together to solve the additions on the slide.
- They use interlocking cubes to make towers for both numbers, then remove a cube to make a double. They count on from the double to add the 1 for the cube that they removed and record their answers on a mini whiteboard.
- Ask pairs of learners to show the class how they solved one of the additions on the slide. Ask: **Why are you removing one of the cubes from one of the towers? How does that help you to find the answer? Would that have worked if the numbers had been further apart? Why/why not?**

Review

- Write 7 + 5 = on the board and ask learners to make towers to match.
- 🔖 Ask: **Is this a near double?** Elicit that there is more than 1 between the two numbers, but they are still quite close together.
- Demonstrate how to remove the extra two cubes from the tower of 7 then ask: **What is double 5?** Agree that it is 10, then add on the extra 2 (12).
- Repeat this for 3 + 5 =.

Assessment for learning

- What double would you use to work out the answer to 5 + 6?
- Show me how to use doubles to solve 2 + 3 =. What is the answer?
- Explain to me how your would work out the answer to 4 + 5.

Same day intervention

Enrichment

- Ask learners to spot real - life examples of near doubles in objects in the school environment (e.g. a table with 6 children sitting at it and a table with 5 children sitting at it). They should add these pairs of amounts together using the near doubles technique from this lesson.

Lesson 4: **Addition and estimation to 20**

Learning objectives

Code	Learning objective
1Ni.05	Estimate, add [and subtract] whole numbers (where the answer is from 0 to 20).
1Nc.04	Count on in ones, [twos or tens, and count back in ones and tens,] starting from any number (from 0 to 20).

Resources

Resource sheet 12: 0–20 number line (per pair); 20 interlocking cubes (per pair); 20 counters (per pair); mini whiteboard and pen (per pair)

Revise

Choose an activity from Unit 12: *Addition and subtraction to 20 (A)* in the Revise activities.

Teach 🔲 [TWM.07]

- Direct learners to the picture in the Student's Book. **[TWM.07]** Ask: **Which way do you think the girl should choose to solve the addition?** Discuss the suggestions, pointing out that she could put the larger number first and count on or make 10 then add the rest. Explain that the numbers aren't close enough together to use the near doubles strategy.

- Tell learners that before they solve the Student's Book addition, they need to estimate the answer using what they already know about the numbers. 🗣 Ask: **What do you already know about these numbers?** Elicit that they know that 8 and 2 make 10, so the answer must be more than 10, and that both numbers are less than 10, so the answer must be less than 20. Take suggestions of estimates and write them on the board, then solve the addition using one of the agreed strategies.

- Repeat this with different additions with answers to 20. Estimate the answer first, pointing out what you already know about the numbers, then decide which strategy would be most appropriate. Be sure to use a range of different strategies during this part of the lesson.

- Discuss the Guided practice example in the Student's Book.

Practise 🔲

- Workbook

Title: Addition and estimation to 20

Page: 53

- Refer to Activity 1 from the Additional practice activities.

Apply 👥 🖥 [TWM.07]

- Display **Slide 1**.
- Provide 0–20 number lines, interlocking cubes and counters for learners to use for support and a mini whiteboard and pen.

- Pairs of learners solve the additions on the slide. They estimate the answer first, then must each pick a different method to solve each addition (e.g. one puts the larger number first and counts on, while the other uses the making 10 and adding more or near double method).

- The learners must decide between them whose method was the most appropriate for the addition or the easiest way off finding the answer and why.

Review

- Ask: **Were any of your estimates in this lesson very different from the answer?** Allow learners to give you examples and look at the number sentences that these relate to by writing them on the board. Go through each one, reminding learners to estimate based on what they know. Find a list of 'it must be' and 'it can't be' statements for each number sentence (e.g. 'It must be more than 10 because one number has two digits' and 'It can't be more than 14 because double 7 is 14 and both of these numbers are less than 7').

Assessment for learning

- What do you know about the numbers in 9 + 5 = that can help you to estimate the answer?
- Which strategy would you use to solve 9 + 5 =? Why?
- Show me two ways to solve 9 + 5 =.

Same day intervention
Support

- Suggest to learners which strategy they could use to solve each addition.

Number – Integers and powers/Counting and sequences

Additional practice activities

Activity 1

Learning objectives
- Add single- and two-digit numbers
- Estimate first to solve a simple addition (variation only)

Resources
two or three sets of large 0–5 number cards (per class); two or three sets of large 10–15 number cards (per class)

What to do
- Give pairs of learners a set of 0–5 number cards, a set of 10–15 number cards and a mini whiteboard and pen each. Learner A shuffles the single-digit number cards and learner B shuffles the two-digit number cards.
- Both learners pick a number from the set of cards that they have shuffled and hold it up. They must add the numbers together as quickly as they can by putting the two-digit number first then counting on. When they know the answer, they write it on their whiteboard. They check the answer by counting on together. Whoever wrote the correct answer first wins a point.
- Repeat several times with different number cards.

Variations

1 Provide counters for Challenge 1 learners for support for this activity.

2 Both learners estimate the answer. They write their estimates on their whiteboards then solve the addition by putting the two-digit number first and counting on. The learner with the closer estimate gets a point.

3 Challenge learners to work out the answer by counting on in their heads. They can then check their answer using a 0–20 number line.

Activity 2

Learning objectives
- Solve an addition by making 10 then adding more
- Solve an addition by using doubles (variation only)

Resources
strips of red and blue paper of suitable length to make paper chains (per learner); glue stick (per learner)

What to do
- Ask learners to solve 7 + 6 = by the method below.
- Learners count out red strips of paper to match the larger number (7). They count out blue strips of paper to match the smaller number (6).
- They make a paper chain from the red paper strips to match the larger number (7).
- They put the 6 strips of blue paper next to it, then add blue paper links to the existing chain to make it a chain of 10 links.
- The learners then make the remaining blue strips into a separate paper chain.
- They add the remainder to 10 to find the answer.

Variation

 Learners use doubling to solve the addition by making a red paper chain of 6 and a blue paper chain of 6 then adding the extra red link to one of the chains.

Unit 13: Addition and subtraction to 20 (B)

Learning objectives

Collins International Primary Maths Recommended Teaching and Learning Sequence: Term 3, Week 3

Code	Learning objective
1Ni.05	Estimate, add and subtract whole numbers (where the answer is from 0 to 20).
1Nc.04	Count on in ones, twos or tens, and count back in ones and tens, starting from any number (from 0 to 20).

Unit overview

In this unit, learners extend and consolidate their work on addition and subtraction. They count back in ones to subtract a one-digit number from a two-digit number that is 20 or less. Note that at this stage, learners are not expected to solve subtraction calculations to 20 that involve regrouping of ones, such as 14 – 8 or 12 – 5.

They learn and recall addition and subtraction number facts to 20, using part–whole diagrams and number families (i.e. 4 + 3 = 7, 3 + 4 = 7, 7 – 4 = 3, 7 – 3 = 4) to help them. Finally, they learn about 'equal statements', recognising that the expressions on either side of the equals sign must have the same (equal/equivalent) value (for example, 6 + 4 = 15 – 5).

Note

Learners need to be familiar with the concept of doubling in order to work through this unit and Unit 12. This is introduced in Unit 14, which comes before Units 12 and 13 in the recommended sequence of work. If you are not following this sequence, we recommend that you teach Unit 14 before Units 12 and 13 so that learners have a full understanding of the strategies used in these units.

Prerequisites for learning

Learners need to:
- recognise the numbers 0–20
- be able to count up to 20 objects accurately
- be able to solve additions to 10 confidently
- have a variety of addition and subtraction strategies that they can use to calculate and check an answer
- be able to use a part–whole diagram to solve additions and subtractions.

Vocabulary

addition, add, subtract, subtraction, equals, number families, tens, ones, count back, part–whole diagram

Common difficulties and remediation

It can be a difficult for some learners to comprehend the concept of finding equal statements, as they are used to seeing just one number as the answer after the equals sign. Remind learners that they know lots of addition and subtraction facts to 20 and that these can help them to make equal statements about numbers. Try writing a number on the board and asking learners to think of as many ways of making that number, using addition and subtraction, as possible. You can then pick any of these statements to put either side of the equals sign.

Supporting language awareness

Display the key words for each lesson and discuss them with learners. Use and refer to these words throughout the lesson and encourage learners to use the words too, prompting them when necessary.

It is important to teach the proper terms for addition (augend, addend, sum) and subtraction (minuend, subtrahend, difference) from a young age as this ensures that learners will have a good understanding of subtraction as they get older. Display these words together with practical examples where possible.

Promoting Thinking and Working Mathematically

TWM.02 Generalising
Learners find many different ways of making a number to 20 and use two of these number sentences to create an equal statement.

TWM.08 Improving
Learners continue to refine their addition and subtraction skills by using different strategies.

Success criteria

Learners can:
- find and recall number facts to 20 and apply these to addition and subtraction
- write an equal statement for a number to 20
- subtract a one-digit number from a two-digit number that is 20 or less (no regrouping).

Number – Integers and powers/Counting and sequences

Lesson 1: **Subtraction facts to 20**

Learning objectives

Code	Learning objective
1Ni.05	Estimate, add and subtract whole numbers (where the answer is from 0 to 20).
1Nc.04	Count on in ones, twos or tens, and count back in ones and tens, starting from any number (from 0 to 20).

Resources

mini whiteboard and pen (per learner); Base 10 equipment – 1 ten and 9 ones (per learner)

Revise

Use the activity *Subtracting the ones* from Unit 13: *Addition and subtraction to 20 (B)* in the Revise activities.

Teach 📖 📊

- Direct learners to the picture in the Student's Book. Ask: **Why do you think the child is subtracting 3 from 7 when the subtraction she needs to answer is 17 – 3?** Take suggestions then explain that you can subtract one-digit numbers from two-digit numbers by subtracting the subtrahend from the ones digit.
- Write 15 – 2 = ? on the board. Ask a learner to use the **Base 10 tool** to make 15. Say: **Because 2 is less than the ones digit, we can take it away from the ones digit and leave the tens digit as it is because it will stay the same.** Ask a learner to take two blocks away from the ones. Ask: **How many ones are left?** Agree that there are three. Say: **10 add three ones is 13.** Count the blocks to check by starting on 10 and counting on the ones. Repeat this with different starting amounts to 20.
- Now demonstrate how to use a number line to subtract from the ones digit. Write 19 – 5 = on the board. Use the **Number line tool** to show learners how to start at 9 and jump back five places. Remind them that the tens digit will stay the same so as 9 – 5 = 4, 19 – 5 = 14. Repeat this with different starting numbers to 20 making sure not to use examples that require regrouping of ones.
- Ask: **How else could you subtract from the ones digit?** Elicit that you could count back in your head, using your fingers for support if you needed to. Practise this skill with learners.
- Discuss the Guided practice example in the Student's Book.

Practise 📓

- Workbook

Title: Subtraction facts to 20

Page: 54

- Refer to Activity 1 from the Additional practice activities.

Apply 👤💻

- Display **Slide 1**.
- Learners use Base 10 equipment to solve the subtractions on the slide, subtracting from the ones digit only. They record their answers on mini whiteboards.
- Afterwards, ask: **[T&T] Did you find it easy to solve the subtractions using this method? Why/why not? Did you prefer to use Base 10 or a number line to do it?** Remind learners that by adding this method to their existing subtraction strategies, they are *improving*.

Review 📊

- Write 16 – 6 = on the board.
- 🗣 Ask: **If we are taking away from the ones digit, what will the answer be?** Elicit that 6 – 6 = 0, therefore 16 – 6 = 10. Practise this with different numbers in which the subtrahend is the same as the ones digit in the minuend. If necessary, use the **Base 10 tool** and/or **Number line** tool to support learners.

Assessment for learning

- What is the ones digit in 16?
- 16 – 4 = ? Which digit are we subtracting from?
- Show me how to take away from the ones digit to solve 16 – 4 =.

Same day intervention
Support

- Let learners use Base 10 equipment throughout, even to support their calculations in the Workbook.

Enrichment

- Ask Challenge 3 learners to use a number line to subtract from the ones digit rather than Base 10 equipment once they have understood the basic idea behind this.

Lesson 2: **Number facts to 20 – part–whole diagrams**

Learning objectives

Code	Learning objective
1Ni.05	Estimate, add and subtract whole numbers (where the answer is from 0 to 20).
1Nc.04	Count on in ones, twos or tens, and count back in ones and tens, starting from any number (from 0 to 20).

Resources

Resource sheet 10: Part–whole diagram (per pair); 20 counters (per pair); set of 13–20 number cards from Resource sheet 1: 0–20 number cards (per pair)

Revise

Use the activity *Number facts* from Unit 13: *Addition and subtraction to 20 (B)* in the Revise activities.

Teach 🅂🄱

- Direct learners to the picture in the Student's Book. Ask: **What do you think the part–whole diagrams show?** Take suggestions then tell the learners that they show different ways of making 16.
- Give pairs of learners a part–whole diagram resource sheet and 20 counters. Ask them to place 16 counters in the whole section, then move them (copying your lead on the board) to match the first diagram in the Student's Book. Say: **There are 12 counters in this part and four counters in this part – that makes 16 counters altogether.**
- Move all the counters back to the whole section, then move them to match the second diagram in the Student's Book. Ask: **How many counters are in each part?** Agree that there are 14 counters and two counters and that 14 and 2 together make 16.
- Move the 16 counters back into the whole section. Ask pairs of learners to find a different way to split the 16 counters between the parts, using their own part–whole diagram sheets. When everyone has done this, invite learners to the front to show the class how they split the 16 counters.
- Discuss anything that learners notice, for example some may notice that these number facts are similar to the number facts to 10, as often just the second digit is split. Refer them back to what they learned about this in Lesson 1. Some learners may offer additions or subtractions to match the numbers. This will be investigated in Lesson 3, but feel free to explore this in this lesson if appropriate.
- Discuss the Guided practice example in the Student's Book. Ask: **What other addition could be the answer here?**

Practise 🅆🄱

- Workbook

Title: Number facts to 20 – part–whole diagrams

Page: 55

- Refer to Activity 2 from the Additional practice activities.

Apply 👥💻

- Display **Slide 1**.
- Give pairs of learners a set of 13–20 number cards, 20 counters and a part–whole diagram resource sheet.
- Learners pick a number and use the counters to find as many different ways of making that number as possible. When they run out of ways to make it, they pick a new number and start again.

Review

- Give each learner a 0–20 number card.
- Write a number from 0 to 20 (e.g. 18) in the whole part of a part–whole diagram on the board.
- Write a number that makes up part of 18 in one of the parts (e.g. 12).
- Learners who have the number card that makes up the other part of the starting number hold their number card up.

Assessment for learning

- How many different ways can you think of to make 17?
- How would you use your part–whole diagram to find ways to make 17?
- What number would you add to 13 to make 17?

Same day intervention

Support

- This is a good opportunity for learners to practise their number facts to 10 if they are not already secure with these. When they have done so, they can use these to help them with number facts to 20, by using the strategies for subtraction taught in Lesson 1.

Enrichment

- Ask learners to predict what they think some of the number facts for 20 will be, based on what they know about the number facts for 10. They can check these using counters.

Number – Integers and powers/Counting and sequences

Unit **13** Addition and subtraction to 20 (B)

Number – Integers and powers/Counting and sequences

Lesson 3: **Number families**

Learning objectives

Code	Learning objective
1Ni.05	Estimate, add and subtract whole numbers (where the answer is from 0 to 20).

Resources

mini whiteboard and pen (per pair); Resource sheet 10: Part–whole diagram (per pair); 10 counters (per pair)

Revise

Use the activity *Number facts* from Unit 13: *Addition and subtraction to 20 (B)* in the Revise activities.

Teach [SB]

- Direct learners to the picture in the Student's Book. Ask: **What can you tell me about the three numbers in the part–whole diagram?** Elicit that if you add two of the numbers together, they make the third number, and if you subtract one of the numbers from the largest number, it equals the remaining number.

- Explain to the learners that three numbers like this are called a 'number family' and that each number family of three numbers includes four number sentences, each using all three numbers – two additions and two subtractions. Ask learners to look at the number sentences beside the part–whole diagram to check that all of them use all three of the numbers. Ask volunteers to check these by using a number line or counters.

- Write a number up to 10 (such as 7) on the board and ask learners to tell you a pair of numbers that equal 3 when added together. Write these three numbers in a part–whole diagram. Ask: **Who can tell me a number sentence to match these three numbers?** Learners who need support with this could use a part–whole diagram and counters, as for Lesson 2. List the four number sentences on the board.

- Repeat for different sets of three numbers extending to numbers to 20 if appropriate (i.e. 8 + 4 = 12, 4 + 8 = 12, 12 – 8 = 4, 12 – 4 = 8).

- Discuss the Guided practice example in the Student's Book.

Practise [WB]

- Workbook

Title: Number families

Page: 56

- Refer to Activity 2 from the Additional practice activities.

Apply 👥 🖥

- Display **Slide 1**.

- Pairs of learners choose a number to 10 and choose two smaller numbers that total that number when added together. They each write them in a part–whole diagram on a mini whiteboard.

- Learner A writes two addition number sentences to match the numbers. Learner B writes two subtraction number sentences to match the numbers.

- They check that they are correct using a part–whole diagram resource sheet and counters, then pick another three numbers, swap roles and repeat.

Review

- ✍ Write $6 + \square = 10$, $\square + 6 = 10$, $10 - 6 = \square$, $10 - \square = 6$ on the board and ask: **What is the missing number in this number family?**

- Show learners how you can work out the missing number by counting on from 6 until you get to 10 and seeing how many fingers you are holding up, or by subtracting 6 from 10 and seeing what number you end up on.

- Write another set of number family sentences with a number missing on the board and ask learners to find the missing number.

Assessment for learning

- What four number sentences match the number family 7, 5 and 2?
- If $7 - 5 = 2$, what number must you add to 5 to get 7?
- Write a number family and matching number sentences for 8.

Same day intervention
Enrichment

- Encourage learners to work with numbers to 20.

Lesson 4: **Equal statements**

Learning objectives

Code	Learning objective
1Ni.05	Estimate, add and subtract whole numbers (where the answer is from 0 to 20).

Resources

10 counters (per pair); Resource sheet 11: 0–10 number line (per pair); mini whiteboard and pen (per pair); = sign card from Resource sheet 13: Operations cards (per pair)

Revise

Choose an activity from Unit 13: *Addition and subtraction to 20 (B)* in the Revise activities.

Teach [SB] [TWM.01]

- Direct learners to the picture in the Student's Book. Ask: **What is the answer to the addition? What is the answer to the subtraction? Why do you think they have = between them?** Agree that both of the calculations equal 9. Explain that the = sign is between them because they both equal the same amount. Tell learners that when two calculations have the same answer, they are called 'equal statements'.

- Ask learners to tell you as many different ways of making 5 as they can, using addition and subtraction. Learners can use number lines or counters for support if they wish. Write a list of these on the board. Pick any two and write them next to each other with = between them, such as 9 − 4 = 3 + 2. Say: **The equals sign means that they are the same. They both equal the same number.** Demonstrate that you could do this with any of the calculations on the board because they all equal 5.

- Repeat this with different target numbers. Include one example in which the statements are not equal. 🗣 Ask: **Are these equal statements? Why/ why not?**

- Explain that to make an equal statement, you don't have to use one addition and one subtraction – you could use two additions or two subtractions, as long as they both equal the same number. Demonstrate this using other examples. **[TWM.01]** Challenge pairs of learners to find equal statements for a target number, using their knowledge of number pairs or number families to do so. Say: **By finding equal statements for a target number, you are *specialising*.**

- Discuss the Guided practice example in the Student's Book.

Practise [WB]

- Workbook

Title: Equal statements

Page: 57

- Refer to Activity 2 (variation) from the Additional practice activities.

Apply 👥 🖥 [TWM.02]

- Display **Slide 1**.

- Pairs of learners choose a target number from 0 to 10. They each secretly write a calculation that equals that number, on a mini whiteboard. They put their whiteboards next to each other and put an = sign card in between them. If they have both written the same calculation, one of them must erase it and write a different one.

- They check their calculations using counters or a 0–10 number line.

Review

- Write six different calculations, each on a separate mini whiteboard. There must be three sets of answers that are the same, but keep them mixed up so that they are not next to each other. Challenge learners to change the positions of the whiteboards and place = sign cards between them until they have created three equal statements.

Assessment for learning

- Write an equal statement to match 9.
- I have written 6 + 3 =. What could you write next to make it an equal statement?
- Is 4 + 3 = 7 − 2 an equal statement? Why/ why not?
- What is an equal statement?

Same day intervention
Enrichment

- Encourage learners to work with numbers to 10.

Number – Integers and powers/Counting and sequences

Additional practice activities

Activity 1 👤 🔺2

Learning objective
• Solve a subtraction up to 20 by subtracting from the ones digit

Resources
black paper rectangles to make 'flats' (two per learner); 20 yellow 'window' squares (per learner); black felt-tipped pen or crayon (per learner); glue stick (per pair)

What to do
• Learners choose a number from 15 to 19 (e.g. 16) and a number from 1 to 5 (e.g. 3).
• They use the black paper and yellow 'window squares' to make 'flats' to match their starting number. The tens flat has ten yellow windows (to represent the lights being on) in a vertical line. The ones flat has that many (e.g. 6) yellow windows in a vertical line. They stick the two flats next to each other on a piece of paper.
• Learners then subtract their one-digit number from the ones flat, by colouring that many (e.g. 3) windows black in the ones flat.
• They count how many yellow windows are left to find the answer.

Activity 2 👤 🔺2

Learning objective
• Find number families to 10 (20 for the variation)

Resources
card cut-out house shape (per learner); four small paper 'window' squares (per learner); paper 'door' rectangle (per learner); glue stick (per pair)

What to do
• Give each learner a house cut-out, a paper rectangle and four paper squares.
• Learners choose a number to 10 (e.g. 9) and find two numbers that make that many when added together (such as 6 and 3). They can use a part–whole diagram and counters for support if they need to.
• They write the numbers on the paper rectangle (the 'door') – the starting number top centre and the two numbers that make that number on either side of the bottom. They stick the 'door' onto the house.
• They then generate two additions and subtractions, using the numbers on the door (such as 6 + 3 = 9, 3 + 6 = 9, 9 – 6 = 3, 9 – 3 = 6) and write these on each square of paper ('windows').
• Learners stick the 'windows' to the house.
• You could use the houses (and the flats from Additional practice activity 1) for a display to create a number families village.

Variations
1 Learners choose a number to 5 and create a number family,

2 Give learners two paper square 'windows' instead of four. Learners write the target number in the roof of the house (for example, 17). Instead of finding number families, learners write equal statements for this number, one on each 'window' (such as 6 + 3 and 10 – 1). They write = on the door and stick it between the windows on the house.

3 Learners choose a number to 20 and create a number family.

Number – Integers and powers/Counting and sequences

Unit 14: Doubling

Collins International Primary Maths Recommended Teaching and Learning Sequence: Term 3, Week 1

Learning objectives

Code	Learning objective
1Ni.06	Know doubles up to double 10.

Unit overview

In this unit learners are introduced to doubling, first exploring what happens when you double amounts of objects up to and including 5 and then to 10, and then progressing to doubling numbers by using a number line. They use this skill to solve additions and apply it to real-life situations.

Prerequisites for learning

Learners need to:
- recognise the numbers 0–20
- be able to count up to 20 objects accurately
- be able to add together two numbers which total 20 or less.

Vocabulary

double, number line

Common difficulties and remediation

Some learners may not understand immediately that to double amounts, you must make another amount of exactly the same size (for example, to double three marbles you need another three marbles) and may initially try to add another one marble or another amount at random. Using a ten frame to organise the objects to ensure that both amounts match can help, as can using a mirror to double amounts.

Supporting language awareness

Display the key words for each lesson and discuss them with learners. Use and refer to these words throughout the lesson and encourage learners to use the words, prompting them when necessary.

Refer to doubling in everyday classroom situations, for example: **I have two helpers, but I need double that amount. How many helpers do I need altogether?**

Promoting Thinking and Working Mathematically

TWM.03 Conjecturing
Learners discover that doubling an amount means making the same amount again.

TWM.07 Critiquing
Learners learn how to double by using different methods and compare these to discover which method is the most appropriate.

Success criteria

Learners can:
- double amounts to 10
- double numbers to 10.

Number – Integers and powers

Unit **14** Doubling

Lesson 1: **Doubling amounts to 5**

Learning objectives

Code	Learning objective
1Ni.06	Know doubles up to double 10.

Resources

two sheets of A4 paper (per pair); ten toy cars (per pair); set of 0–5 number cards from Resource sheet 1: 0–20 number cards (per pair)

Revise

Use the activity *Double the monkeys* from Unit 14: *Doubling* in the Revise activities.

Teach [SB]

- Direct learners to the picture in the Student's Book. Ask: **What do you notice about the legs of all the animals?** Elicit that they all have the same number of legs on both sides of their bodies. Say: **The animals legs have been doubled. What do you think 'doubled' means?** Take suggestions. Tell learners that by beginning to understand that doubling means 'making the same amount again'.

- Explain that doubling an amount means putting two 'lots' of the same amount together. Ask learners to hold up one pencil. Ask: **If you double the number of pencils you are holding, how many will that be?** Ask learners to hold up another pencil and say: **Double 1 is 2.**

- Refer to the image in the Student's Book again, counting the number of legs on one side of the cat, then doubling the number by 'counting on' all the legs on the other side. Do this for all the animals.

- Now show learners how to use their fingers to make doubles, for example holding up three fingers on one hand and then holding up three fingers on the other hand. Count how many fingers that is altogether and say: **Double 3 is 6.**

- [T&T] Ask: **What do you think double zero would be?** Take suggestions and then explain that double zero is zero because if you double nothing you still have nothing.

- Discuss the Guided practice example in the Student's Book.

Practise [WB]

- Workbook

Title: Doubling amounts to 5

Page: 58

- Refer to Activity 1 from the Additional practice activities.

Apply 👥 🖥

- Display **Slide 1**.
- Give pairs of learners two sheets of paper, a set of 0–5 number cards and ten toy cars.
- Learners put two sheets of paper on the floor to represent car parks.
- They pick a number card and drive that many cars into the first car park.
- They drive the same number of cars into the second car park, then count all of the cars to find double their starting number.

Review

- Call out a number from 1 to 5.
- Learners hold up the number of fingers that is double that number.

Assessment for learning

- What is double…?
- How would you use objects to work out double…?
- How could you use your fingers to find double…?

Same day intervention

Enrichment

- Ask learners to find examples of doubles on their bodies (e.g. eyes, legs, fingers, toes) and to work out the starting number and double it (e.g. double five toes equals ten toes).

Number – Integers and powers

Lesson 2: **Doubling amounts to 10**

Learning objectives

Code	Learning objective
1Ni.06	Know doubles up to double 10.

Resources

shuffled set of 0–10 number cards from Resource sheet 1: 0–20 number cards (per learner); 20 counters (per learner); mini whiteboard and pen (per learner)

Revise

Use the activity *Double the monkeys* from Unit 14: *Doubling* in the Revise activities.

Teach [SB] [bar chart icon]

• Direct learners to the picture in the Student's Book. Ask: **How many players are there in the red team?** Count together to establish that there are nine players. Point out that there are also nine players in the yellow team. Count all the players together and say: **Double 9 is 18**.

• Remind learners that doubling means making twice as much. Revise doubles to double 5 by calling out numbers to 5 and asking learners to use their fingers to double the number.

• Now use the **Tree tool** to show learners doubles to double 10. Put ten birds on one side of the screen, then the same number on the other side. Then count how many there are altogether. Explain that we don't have enough fingers to check doubles of numbers over 5, so we can use objects instead, until we start to recall the doubles to double 10 without help.

• Call learners to the front of the class to use the **Tree tool** to make doubles of numbers up to double 10.

• Discuss the Guided practice example in the Student's Book.

Practise [WB]

• Workbook

Title: Doubling amounts to 10

Page: 59

• Refer to Activity 1 from the Additional practice activities.

Apply [person icon] [monitor icon]

• Display **Slide 1**.

• **[TWM.03]** Learners pick a 0–10 number card, count out that many counters, then count out the same number again. They combine the two sets of counters, then count them all to find double the number. They record their answers on mini whiteboards.

Review

• Ask a learner to come to the front to be your assistant.

• Ask the class to count how many times you clap. Clap six times.

• [speech icon] Ask: **How many claps would double that amount be altogether?** Some learners may begin to recall doubling facts and be able to tell you without counting.

• Ask your assistant learner to double the claps by clapping the same number of times as you clapped. Count on from 6 as they do it, as a class.

• Repeat with different numbers of claps.

• Say: **You can have doubled amounts of objects and you have doubled claps. Can you think of anything else we can double?** Take suggestions from the learners, each time asking them to explain what doubling means and how they would double their suggestions.

Assessment for learning

• What is double...?

• How would you use counters to work out double...?

• I've got six counters, but I need double that number. How many counters would that be?

Same day intervention

Enrichment

• Call out a number from 0 to 10 and see if learners can answer you by quickly calling out the double for that number.

Number – Integers and powers

Number – Integers and powers

Lesson 3: **Doubling on a number line**

Learning objectives

Code	Learning objective
1Ni.06	Know doubles up to double 10.

Resources

Resource sheet 12: 0–20 number line (per pair); 20 counters (per pair); mini whiteboard and pen (per learner)

Revise

Use the activity *Number track doubles* from Unit 14: *Doubling* in the Revise activities.

Teach 🆂🅱 📊

- Direct learners to the first picture of the frog and lily pads in the Student's Book. Say: **The frog started on 3 and jumped on three more numbers. What has he done to his starting number?** Elicit that he has doubled it.
- Explain that a number line or track can be used to double numbers. Discuss the second picture of the frog and lily pads in the Student's Book.
- Display the **Number line tool** from 0–10 and ask a learner to pick a starting number from 1 to 5 (such as 2). Highlight this number, then say: **When you double a number, it means you add the same number again.** Draw two jumps, from 2 to 3 and 3 to 4. Highlight 4 and say: **Double 2 is 4.**
- Repeat with other numbers to 10 before extending the number line to 20 and doubling numbers to 10. Invite learners to come to the front to draw the jumps.
- Address any misconceptions, for example some learners may want to draw the jumps from 0 rather than remembering to start from the number that you are doubling. Others may forget where to stop when drawing the jumps. Rectify this by counting the jumps with learners and ensuring that the number of jumps matches the starting number.
- Discuss the Guided practice example in the Student's Book.

Practise 🆆🅱

- Workbook

Title: Doubling on a number line

Page: 60

- Refer to Activity 2 from the Additional practice activities.

Apply 👥 🖥 [TWM.07]

- Display **Slide 1**.
- Give pairs of learners a 0–20 number line, 20 counters and a mini whiteboard and pen.
- Learners double the two red numbers on the slide. Learner A uses the number line to do this. Learner B uses the counters. They write their answers on mini whiteboards. Encourage learners to compare the outcomes of their doubling with each other. Did both methods achieve the same answer? Which method was quicker? Say: **You are *critiquing* the methods to find the most suitable one for the job.**
- Learners then swap roles and double the two blue numbers on the slide.

Review [TWM.07]

- Referring to the **Apply** activity, 🗣 ask: **Did you prefer using the counters or the number line to find the doubles?** Discuss this with learners, talking about which method they found easier or harder, which method was quicker and which method you need more equipment for.
- [TWM.07] Draw conclusions about which method would be better to use in different circumstances (for example, it is quicker to use the number line so that would be a better method if you had to find a double quickly).

Assessment for learning

- What is double 7?
- How would you use a number track or line to find double 7?

Same day intervention
Support

- Give learners 0–10 number lines and focus on doubling numbers to 5 before moving on to doubling numbers to 10.

Lesson 4: **Doubling facts to 10**

Learning objectives

Code	Learning objective
1Ni.06	Know doubles up to double 10.

Resources

Resource sheet 1: 0–20 number cards (per pair); 0–10 number cards from Resource sheet 1: 0–20 number cards (per pair); Resource sheet 12: 0–20 number line (per pair); random card from Resource sheet 1: 0–20 number cards (per learner)

Revise

Choose an activity from Unit 14: *Doubling* in the Revise activities.

Teach 🆂🅱 📊

- Direct learners to the picture in the Student's Book. **[T&T]** Ask: **What do you think will be the last number to come out of the doubling machine? Why?** Take suggestions then use the **Number line tool** to check double 4 and agree that it would be 8.
- Remind learners that they have learned how to double numbers to 10 using countable objects and number lines. Tell them that they will now be trying to learn doubles to 10 by heart – remembering the answer without having to work it out or without using equipment such as counters or a number line.
- Write the numbers 0–10 on the board and ask learners to help you to recall as many doubling facts for them as possible. If appropriate, ask less able learners to check these by using counters or a number line.
- Give pairs of learners a set of 0–20 number cards. Call out numbers from 0 to 10 and ask learners to find the number card that matches double that number and hold it up. Every time they do this, say: **That's right – double X = Y** to help learners to recall this fact in the future.
- Discuss the Guided practice example in the Student's Book.

Practise 🆆🅱

- Workbook

Title: Doubling facts to 10

Page: 61

- Refer to Activity 2 from the Additional practice activities.

Apply 👥 🖥

- Display **Slide 1**.
- Give each pair a set of shuffled 0–10 number cards and a 0–20 number line.
- They pick a number card and compete to say the number that is double that number first. They check their answers, using the number line, then repeat for all the number cards.

Review

- Give each learner a 0–20 number card at random. 🅿 Ask questions such as: **Who has double 7?** Any learners with that card must hold it up.

Assessment for learning

- What is double 4?
- How many doubles to 10 can you tell me?

Same day intervention

Support

- Encourage learners to work on memorising doubles to 5 first.

Number – Integers and powers

177

Additional practice activities

Activity 1

Learning objective
• Find and make doubles of amounts for the numbers 0–5 (or 0–10, variation only)

Resources
set of 0–5 (or 0–10 for variation) number cards from Resource sheet 1: 0–20 number cards (per learner); modelling clay (per learner); two stick-on eyes (per learner); ten toothpicks or lolly sticks (per learner)

What to do
• Learners roll the clay into a ball and stick the eyes on to make it into a creature.
• They pick a number card and give their creature that number of legs on one side of its body. They double the number of legs by putting the same number on the other side.
• They can repeat this, removing the legs then picking a different number card.

Variations
1 Give Challenge 1 learners a creature with an amount of legs to 5 (10 for the lesson on doubles to 10) already stuck to one side of its body. These learners must count out a matching amount of legs and add them to the other side of the body.

2 Give learners 0–10 number cards for Lesson 2 so that they can find doubles to 10.

3 Learners make a doubles creature for each number to 5 (10 for the lesson on doubles to 10).

Activity 2 or

Learning objective
• Recall doubles to 10

Resources
set of shuffled 1–10 number cards from Resource sheet 1: 0–20 number cards (per group/pair); set of shuffled 2, 4, 6, 8, 10, 12, 14, 16, 18 and 20 number cards from Resource sheet 1: 0–20 number cards (per group/pair); 0–20 number line (per group/pair) (variation only)

What to do
• Learners work in small groups or pairs.
• Spread the 1–10 cards face down on one side of the table, then spread the 2, 4, 6, 8, 10, 12, 14, 16, 18 and 20 number cards face down on the other side of the table.

• Learners take turns to pick one card from each of the two groups. If one card that they pick is the double of the other card that they pick, they can keep the cards and have another go. If the cards are not a double match, they must put the cards back where they found them and it is then the next learner's turn.
• The winner is the learner with the most pairs of cards when there are no cards left to take.

Variations
1 Limit Challenge 1 learners to number cards from 1–5, then 2, 4, 6, 8 and 10 number cards only until they are secure in recalling doubles to 5.

2 Check the pairs of cards with a 0–20 number line.

Unit 15: Money

Collins International Primary Maths
Recommended Teaching and
Learning Sequence: Term 3, Week 6

Learning objectives

Code	Learning objective
1Nm.01	Recognise money used in local currency.

Unit overview

This unit introduces learners to the concept of money. They learn about what money is used for and how to differentiate coins and notes from other similarly shaped items. Learners learn to recognise their local currency, learning that different countries use different types of coins and notes and that coins have been used for thousands of years. They handle coins and notes during observational and role-play activities and sort coins and notes according to various criteria.

Prerequisites for learning

Learners need to:
• recognise simple 2D shapes
• recognise a variety of colours
• have some experience of how money is used in various real-life situations (e.g. being present while a parent pays for groceries).

Vocabulary

money, coins, notes, sort

Common difficulties and remediation

Some learners may have trouble differentiating between coins and other shiny objects such as buttons or jewellery. Giving them activities in which they have to identify coins and remove them from a set of miscellaneous objects will help them to form an idea of what is a coin and what is not.

Different countries have different currencies. If you do not see your currency represented in this unit, substitute your own throughout the lessons. Learners need to be able to identify their own local currency so it is counter-productive to teach this unit with currency from a different country, except for when introducing them to the differences between coins and notes across the world.

Supporting language awareness

Display the key words for each lesson and discuss them with learners. Use and refer to these words throughout the lesson and encourage learners to use the words too, prompting them when necessary.

Teach learners the correct name for their country's currency – such as dollars, euros, rupees and paise/dirhams and fils.

Promoting Thinking and Working Mathematically

TWM.06 Classifying

Learners sort coins and notes according to different criteria.

Success criteria

Learners can:
• identify the coins and notes of their local currency
• understand that money is used to pay for things
• sort coins and notes according to different criteria.

Number – Money

Lesson 1: **What is money?**

Learning objectives

Code	Learning objective
1Nm.01	Recognise money used in local currency.

Resources

small selection of local currency coins and notes (per pair); small selection of coins and notes from different currencies (per pair); a few objects that could be mistaken for money, e.g. bottle tops, jewellery, note paper and buttons (per class)

Revise

Use the activity *I went shopping...* from Unit 15: *Money* in the Revise activities.

Teach 🆂🅱 📊 🖥

- Direct learners to the picture in the Student's Book. Ask: **How will the children pay for their ice creams? What will they use to pay for their ice creams?**
- Show learners some coins and notes. **[T&T]** Ask: **What is this? What do we use it for?** Elicit that it is money and that we use money to pay for things.
- Display local currency on the **Money tool**. If your local currency is not one of the pre-populated currencies in the Money tool then display images of local currency coins and notes. Say: **In this country, our money is called [insert name of local currency e.g. dollars).** Explain that coins are made out of metal, and notes are made from paper (or plastic). **[T&T]** Ask: **Talk to your partner about a time that you have spent some money or seen your parents pay for something.**
- Hold up a button and ask: **Is this money? Why/why not?** Do the same with other small objects that could be mistaken for money, such as jewellery and bottle tops. Do the same for notes. Discuss the features of money with learners.
- Display **Slide 1**. Explain that money has different features in different countries. Point out some features that your local currency always has and note how those features are absent on the coins and notes of other countries.
- Discuss the Guided practice example in the Student's Book.

Practise 🆆🅱

- Workbook

Title: What is money?

Page: 62

- Refer to Activity 1 from the Additional practice activities.

Apply 👥 🖥 [TWM.06]

- Display **Slide 2**.
- Give pairs of learners a small selection of local currency coins and notes in a dish. Also add some coins and notes from other countries. Learners examine the coins and notes to work out whether they are local currency or not. They separate the money into local currency and not local currency.

Review

- Ask learners to show the class anything from their dish in the **Apply** activity that they were unsure about. Look at the items together and decide whether they are local currency, or if they are from another country. 🔁 Ask: **How can we tell if the item is money? What do we need to look for?**

Assessment for learning

- Could you use this object to pay for something in a shop?
- Is this coin/note local currency?
- What do we use money for?
- What are some of the features of our local currency?

Same day intervention

Support

- For the **Apply** activity, give learners some local currency and some small objects that could be mistaken for money, such as bottle tops, buttons, jewellery and notepaper. They differentiate between the objects that are money and the objects that are not money.

Lesson 2: **Recognising coins**

Learning objectives

Code	Learning objective
1Nm.01	Recognise money used in local currency.

Resources

set of coins from your local currency (per learner); coloured pencils (per learner) (for the Workbook); book (per learner)

Revise

Use the activity *Describing coins and notes* from Unit 15: *Money* in the Revise activities.

Teach [SB] 🖥

- Direct learners to the picture in the Student's Book. Ask: **Is this a coin from our currency? How could you describe this coin?**
- Pass around coins from your currency for learners to handle. Tell learners what your currency is called (e.g. dirhams and fils). Ask learners if they recognise any of the coins and ask volunteers to describe some of the features of each coin (e.g. colour, shape, sides, images on either side, numbers). Allow this to lead into a general discussion about coins and learners' experience of money.
- 🗣 Ask: **Is there anything that is always the same about a coin from our currency?**
- Discuss the Guided practice example in the Student's Book, including the various features of the three coins.

Practise [WB]

- Workbook

Title: Recognising coins

Page: 63

- Refer to Activity 2 from the Additional practice activities.

Apply 👥 🖥

- Display **Slide 1**.
- Give each learner a set of coins from your currency. They work in pairs.

- Learner A secretly chooses a coin and hides it behind a book while they look at it. They describe the coin to learner B who uses the description to identify the coin from their set. Learner B holds up the coin that matches the description.
- Learners swap roles and repeat.

Review

- Give each learner one coin from your local currency. Make statements such as: **Everyone with a silver coin clap three times** or **Everyone with a round coin swap places.**
- Learners can swap their coins after a couple of minutes and the game can begin again.

Assessment for learning

- Describe this coin to me.
- I'm thinking of a coin and it looks like this (describe coin). Can you find the coin that I am describing?
- What are some features of coins we use?
- What shape/colour is this coin?

Same day intervention

Support

- Ask learners to focus on one feature of the coins when describing them (e.g. colour).

Enrichment

- Give learners two coins from your local currency. Ask learners to point out all the differences they can see between the two coins.

Number – Money

181

Lesson 3: **Recognising notes**

Number – Money

Learning objectives

Code	Learning objective
1Nm.01	Recognise money used in local currency.

Resources

set of banknotes or replica banknotes to be used in schools from your currency (per learner); coloured pencils (per learner) (for the Workbook); ten counters (per pair); book (per learner)

Revise

Use the activity *Describing coins and notes* from Unit 15: *Money* in the Revise activities.

Teach 🆂🖥

- Direct learners to the picture in the Student's Book. Ask: **What do you think this is?** Discuss whether learners think it is money, allowing you to explain that money comes as coins and notes in most countries and that notes are generally worth more (they have a higher value) than coins. Ask: **Do you think this note is from our local currency?** Look for features of your local currency on the note to decide on the answer.

- Pass around notes from your currency for learners to handle. Remind learners what your currency is called. Discuss the different colours used in your currency's banknotes. **[T&T]** Ask: **Why do you think the notes are different colours?** Take suggestions, then explain that it helps people to tell the difference between the different notes without having to look closely to find the numbers. Tell learners that this is important because some notes are worth more than others.

- Look for things that the notes have in common with the coins from your local currency that you looked at in Lesson 2. What features do they have in common? What is different?

- Discuss the Guided practice example in the Student's Book, including the various features of the note.

Practise 🆆🅱

- Workbook

Title: Recognising notes

Page: 64

- Refer to Activity 2 (variation) from the Additional practice activities.

Apply 👥 🖥

- Display **Slide 1**.

- Give each learner a set of banknotes from your local currency and a book, and each pair ten counters.

- Each learner hides their banknotes behind a book and faces their partner across the table. They count to three and each holds up a note. If the notes match they pick up a counter. The aim is to collect as many counters as possible.

- Encourage the pairs of learners to look closely at the notes that they are holding up to check for common features to ensure that they match.

Review

- Ask a volunteer learner to describe a banknote to the class. Pairs of learners look through their notes and hold up the one they think matches.

- Repeat with different volunteers and notes.

Assessment for learning

- Describe this note to me.
- I'm thinking of a note and it looks like this (describe note). Can you find the note that I am describing?
- What are some features of notes we use?
- What shape/colour is this note?

Same day intervention

Support

- Ask learners to focus on one feature of the notes when describing them (e.g. colour).

Enrichment

- Give learners two notes from your local currency. Ask learners to point out all the differences they can see between the two notes.

Lesson 4: **Sorting coins and notes**

Learning objectives

Code	Learning objective
1Nm.01	Recognise money used in local currency.

Resources

coloured pencils (per learner) (for the Workbook); variety of coins and notes in your local currency (per pair); small pieces of paper or labels (per pair)

Revise

Choose an activity from Unit 15: *Money* in the Revise activities.

Teach [SB] [📊] [TWM.06]

- Direct learners to the picture in the Student's Book. Explain that some money has been sorted in the picture. **[TWM.06]** Ask: **What do you think the sorting rule is?** Agree that the money has been sorted into coins and notes. Say: **When we sort objects or numbers, we are** *classifying* **them.**
- Tell learners that coins and notes can be sorted in lots of different ways. Use the **Money tool** to show learners a set of coins from your currency. If your local currency is not one of the pre-populated currencies in the Money tool, then display images of local currency coins. **[T&T]** Ask: **How could we sort these coins?** Take suggestions (e.g. by size, shape or colour) and write them on the board. Choose one of the ways to sort the coins and do this with learners' help.
- Now show learners a set of notes from your currency and make a list of ways in which they could be sorted.
- Remind learners to look closely at the coins and notes to help them to decide how they will sort them and which set each coin or note should go into when they are sorting. Remind them to look for what is the same and what is different.
- Discuss the Guided practice example in the Student's Book.

Practise [WB]

- Workbook

Title: Sorting coins and notes

Page: 65

- Refer to Activity 1 (variation) from the Additional practice activities.

Apply 👥 🖥 [TWM.06]

- Display **Slide 1**.
- Ask pairs of learners to decide whether they are going to sort coins or notes or both. Ensure that learners have access to any of the coins and notes that they want to sort.
- Learners pick a sorting rule to sort by and sort their coins or notes into two sorting hoops. They write a label for each hoop to show the sorting rule.

Review

- Look at some of the ways that learners sorted the money in the **Apply** activity. Can learners spot any coins or notes that have not been sorted correctly? Ask: **Can you think of a way to sort the money that nobody has tried yet?**

Assessment for learning

- What is the sorting rule for these coins?
- Show me two different ways to sort these coins/notes.
- What is the same about these coins/notes? What is different?

Same day intervention

Support

- Give learners only coins or only notes and encourage them to sort them by the more obvious characteristics such as colour.

Number – Money

Number – Money

Additional practice activities

Activity 1

Learning objectives
- Become familiar with local currency
- Understand what money is used for
- Sort coins and notes (variation)

Resources
selection of coins and notes from local currency (per group); shopping bags, purses and wallets (per group); items 'for sale' – dependent on what sort of role-play shop you would like, e.g. toys or plastic fruit and vegetables (per group)

What to do
- Set up a role-play shop in a corner of the classroom. Give learners shopping bags and purses or wallets containing local currency. Provide items 'for sale' (e.g. plastic food or toys) and something to use as a till.
- Learners play shops, using their money to buy items from the role-play shop. As learners are not yet learning the value of various coins it is not necessary to explore this, but learners could make labels for the items for sale that include the unit of local currency (e.g. $4).
- The idea is to allow learners to get used to handling local currency in an informal role-play situation. They should be encouraged to hand over money

to pay for items and use the name of the currency throughout. Allow learners to add to the role play, for example giving change, if they have prior knowledge of this, adding price labels or talking about which items are cheap or expensive.
- Encourage learners to look carefully at the money as they handle it and to describe it to you.

Variations [TWM.06]
1 Ask learners to sort currency from other objects that may be mistaken for this (e.g. counters, bottle tops, jewellery, metal buttons) to put in their purses to use in the shop.

2 Ask learners to sort the coins and notes before beginning the role play (e.g. into coins and notes/into small coins and large coins/into gold coins and silver coins/according to size/according to shape). Tell them that they can only use one of the categories during their time playing with the shop (e.g. silver coins only).

Activity 2

Learning objectives
- Observe and identify the features of local currency

Resources
selection of coins in your local currency (per learner); wax crayon (per learner); paper (per learner); selection of coins in your local currency (for the second variation) (per learner)

What to do
- Give each learner a wax crayon, some paper and a selection of coins.
- Learners put the paper over the top of the coins and rub them with the wax crayon on its side to reveal the pattern and details on the coin beneath.
- Use this activity as a prompt for discussion about the pictures, numbers, words and patterns that appear on your local currency.

Variations
1 Include coins from other currencies to prompt discussion about the differences from your own currency.

2 Ask learners to draw simple pictures of notes in your local currency. Encourage them to focus on the most distinguishing features of the notes such as the most predominant image, currency symbol and number.

3 Ask learners to look carefully at their rubbings and find two things that are the same about each coin and at least one thing that is different.

Unit 16: Place value and ordering to 10

Learning objectives

Collins International Primary Maths
Recommended Teaching and
Learning Sequence: Term 1, Week 3

Code	Learning objective
1Np.01	Understand that zero represents none of something.
1Np.03	Understand the relative size of quantities to compare and order numbers from 0 to 10 [20].
1Np.04	Recognise and use the ordinal numbers from 1st to 10th.

Unit overview

In this unit, learners begin to compare and order numbers and quantities. They compare different numbers of objects and order them, then count the objects to check that they were correct. They also begin to understand the magnitude of numbers by placing a given number between two other numbers on a number line (for example, learners know that 7 falls somewhere between 5 and 10 on a number line). This work on ordering numbers develops further as learners explore ordinal numbers.

Learners also discover the concept of zero and, through various practical tasks, begin to understand that it means 'none of something'.

Prerequisites for learning

Learners need to:
- be able to count to 10
- be able to recognise, read and write numbers to 10
- be familiar with a number track
- be able to count up to ten objects accurately.

Vocabulary

zero, nothing, nought, none, compare, more, less, greater, smaller, order, between, smallest, largest, most, least, greater than, less than, first, second, third, fourth, fifth, sixth, seventh, eighth, ninth, tenth, last

Common difficulties and remediation

Some learners may not be able to grasp straight away that zero means nothing, as all other numbers can be represented with quantities of objects. It is essential to provide a range of practical activities regarding this to ensure that learners understand the concept of zero.

Learners have used number tracks before but this is the first time that they will have used a number line. They may be confused about where the numbers are placed on a number line as on a track they are within each segment, whereas on a number line they are not. Give learners plenty of practice at counting along a number line until they become familiar with this.

When learning ordinal numbers, some learners may be confused as not all of the numbers follow the same pattern: fourth, fifth, sixth, seventh, eighth, ninth and tenth are easy to remember, but first, second and third are different. To remind learners of the names for the first three positions, play lots of games where only the first three positions are used. Introduce fourth and beyond when they are secure with these words.

Supporting language awareness

Display the key words for each lesson and discuss them with learners. Use and refer to these words throughout the lesson and encourage learners to use the words too, prompting them when necessary.

Display number names and the numerals that match them together, and ensure that these are displayed in various places around your classroom, as number lines and also as labels for objects (for example, 3 windows). Use signs for ordinal numbers around your classroom where appropriate.

Use every available opportunity to count with learners, for example using ordinal numbers when learners are lined up.

Success criteria

Learners can:
- understand that zero means none of something
- compare and order amounts of objects to 10
- compare and order numbers to 10
- use ordinal numbers in context.

Lesson 1: **Zero**

Learning objectives

Code	Learning objective
1Np.01	Understand that zero represents none of something.

Resources

dish (per learner); two sets of 0–5 number cards from Resource sheet 1: 0–20 number cards, shuffled together (per pair); approximately 30 counters or other countable objects (per pair); eight large building blocks (per class)

Revise

Use the activity *Zero!* from Unit 16: *Place value and ordering to 10* in the Revise activities.

Teach 🆂🅱 📊

- Direct learners to the picture in the Student's Book. Ask: **How many apples are in the first bowl?** Elicit that there are three apples. Now ask: **How many apples are in the other bowl?** Discuss the fact that there is nothing in the bowl – no apples at all. Explain that there are zero apples and that zero means 'nothing', 'none' and 'nought'. Write 0 on the board.

- Rearrange learners so that there are different numbers of learners at each table, leaving one table empty. Ask each group: **How many learners are at your table?** Now point to the empty table and ask: **How many learners are at this table?** Agree that, again, there are zero because there is nobody there.

- Now tell learners that you want to find out what happens when you add zero to a group of objects. Display the **Tree tool**, place four apples on the tree and count them with learners. Ask a learner volunteer to come to the front and add zero apples to the tree. If the learner attempts to add some apples, remind them of zero's value by asking: 🗣 **What does zero mean?** Establish that they can't add any apples to the tree because adding zero apples means adding no apples. Repeat this twice with different objects.

- Discuss the Guided practice example in the Student's Book.

Practise 🆆🅱

- Workbook

Title: Zero

Page: 66

- Refer to Activity 1 from the Additional practice activities.

Apply 👥 🖥

- Display **Slide 1**.
- Give pairs of learners a dish each, two sets of 0–5 number cards (shuffled together) and about 30 counters or other countable objects.

- Learners take turns to pick a number card and count that number of counters into their dish.
- When a learner picks a 0 card, they empty their dish of counters so that there are zero counters in it and start again. Their partner does not empty their dish.
- The winner is the learner with more counters in their dish when you stop the game.

Review

- Put eight large building blocks in front of learners and count them together.
- 🗣 Ask: **What would happen if we put zero more blocks in this pile?** Elicit that the number of blocks wouldn't change because you wouldn't have put any more blocks in the pile.
- Now ask: **What if I take zero blocks away from this pile? How many blocks would be left?** Elicit that there would, again, be no change to the number because you wouldn't have removed any blocks.
- Give pairs of learners an amount of blocks to 10 and place a dish of extra blocks on their table. Ask them to investigate what happens to their set of blocks when they add zero more blocks to it and when they take zero blocks away from it. Learners then report back to the class with their findings.

Assessment for learning

- Which of these trees has zero apples on it?
- What does 'zero' mean?
- If you have two marbles and I give you zero more, how many marbles will you have now?
- If you have two marbles and I take away zero of them, how many marbles will you have left?

Same day intervention
Support

- Give learners lots of practice at recognising zero in practical contexts (for example, identifying which bowls contain zero of something, asking them to take zero steps forward, and so on).

Lesson 2: **Comparing numbers to 10**

Learning objectives

Code	Learning objective
1Np.03	Understand the relative size of quantities to compare [and order] numbers from 0 to 10 [20].

Resources

11 interlocking cubes (per class); Resource sheet 4: 0–10 number track (per pair); 20 counters (per pair); set of 0–10 number cards from Resource sheet 1: 0–20 number cards (per pair)

Revise

Use the activity *More or less?* from Unit 16: *Place value and ordering to 10* in the Revise activities.

Teach 📘 📊

• Direct learners to the picture in the Student's Book. Ask: **Which child has more scoops of ice-cream on their cone?** Count the scoops on the cones to check and say: **Abdul has two scoops and Karim has three scoops so Karim has more.**

• Now ask two learners to come to the front of the class. Give one learner seven interlocking cubes and the other learner four interlocking cubes. Tell learners that there are different ways to work out who has more and who has less. First, ask them to look at the cubes. Ask: **Can you tell if one learner has more than the other by looking?** Some learners will be able to tell but others may not.

• Display the **Number track tool**. Count the first learner's cubes with the other learners and highlight the number of cubes (7) on the number track. Now count the second learner's cubes and highlight their number (4) too. Point to both numbers and explain that we can tell that 7 is more than 4 because it comes after 4 on the number track.

• Explain that on the number track, 0 must be the smallest amount because it means 'nothing'. Show learners how numbers and amounts get greater the further you progress up the number line and smaller the further you progress down the number line. Use both the number track and countable objects to demonstrate this for every number up to 10, making sure to emphasise the words: more, greater, less and smaller.

• Discuss the Guided practice example in the Student's Book. Ask: **Could you draw any other number of dots in the 'less' box or does it have to be 5? Why?**

Practise 📒

• Workbook

Title: Comparing numbers to 10

Page: 67

• Refer to Activity 1 (variation) from the Additional practice activities.

Apply 👥 🖥️

• Display **Slide 1**.

• Give pairs of learners a set of 0–10 number cards, a 0–10 number track and 20 counters.

• Each learner picks a number card then counts out that number of counters.

• Learners compare the numbers to work out whose number is larger. They do this by comparing the numbers of counters visually and by finding both numbers on the number track, then working out whose number comes first and whose comes after that.

Review

• Ask two learners to come to the front. Give each of them a 0–10 number card assigned at random. They hold up their cards to show the other learners.

• 🗣 Ask: **Whose number is larger? Whose number is smaller?**

• The other learners use their number tracks to work out the answer. Challenge 1 learners would benefit from using counters to support them with this.

Assessment for learning

• Which number is larger, 6 or 2?

• Which number is smaller, 8 or 3?

• Which of these sets of objects contains more/ less?

• Show me how you would use a number track to work out which of these numbers is the larger/ smaller.

Same day intervention

Support

• Give learners lots of practice at differentiating between amounts of objects before moving on to comparing numbers on a number track.

Enrichment

• Ask learners to describe the difference between the two amounts that they are comparing (e.g. 'There is not much difference between 8 and 9 – they are next to each other when you count' or 'There is a difference of 2 between 6 and 8.'

Number – Place value, ordering and rounding

Unit **16** Place value and ordering to 10

Lesson 3: **Ordering numbers to 10**

Learning objectives

Code	Learning objective
1Np.03	Understand the relative size of quantities to [compare and] order numbers from 0 to 10 [20].

Resources

0–10 number cards from Resource sheet 1: 0–20 number cards (per learner); 0–10 number line from Resource sheet 11: 0–10 number line (per learner); mini whiteboard and pen (per learner)

Revise

Use the activity *Order the numbers* from Unit 16: *Place value and ordering to 10* in the Revise activities.

Teach 🔲 📊

- Direct learners to the picture in the Student's Book. Read the numbers on the racing cars to learners and point out that a number is missing. Some Challenge 3 learners may be able to work out the missing number straight away.
- Display the **Number line tool**. Remind learners that a number line is similar to a number track and helps them to count and order numbers. Discuss the difference between a number track and a number line, pointing out that each number is joined by a very small line to the number line rather than having its own segment on a number track. Count along the number line with learners.
- Tell learners that you are going to use the number line to find out the missing car number. Circle the numbers 4, 5, 7 and 8. Ask: **Can you see a number that comes between those numbers?** Elicit that the missing number is 6.
- Display the **Number line tool** again, hiding the number 7. 🔲 Ask: **How can we work out which number is missing?** Count along the number line with learners until you reach the gap. Learners will automatically say the missing number as they continue to count. Fill in 7 on the line and say: **7 comes between 6 and 8.** Repeat this with different missing numbers.
- Finally, use the number line to order sets of three numbers. Write the numbers 4, 9 and 2 on the board and show learners how to start at the beginning of the number line to look for the numbers in order. Circle 2 first, then 4, then 9. Then write the three numbers on the board in order. Repeat for different sets of three numbers emphasising the words 'smallest' and 'largest'.
- Discuss the Guided practice example in the Student's Book.

Practise 📘

- Workbook

Title: Ordering numbers to 10

Page: 68
- Refer to Activity 2 (2nd variation) from the Additional practice activities.

Apply 🖥️ 👤

- Display **Slide 1**.
- Give each learner a set of 0–10 number cards, a 0–10 number line and a mini whiteboard and pen.
- Learners pick three number cards and use the number line to order them. They write the numbers in order on their whiteboards.
- Learners repeat several times.

Review

- Draw a blank 0–10 number line on the board.
- Give each of 11 learners a number card picked randomly from a 0–10 set.
- Ask probing questions to encourage learners to think about where to place their number cards on the number line, for example: **What number must go at the beginning/end of the number line? We have filled in 0 and 10, so which numbers go between them? Which number goes between 8 and 10? Which number comes after 4?**

Assessment for learning

- Can you put these 0–10 number cards in order?
- What number comes between 6 and 8?
- What is the missing number on this number line?
- Order these three numbers. Which is the largest? Which is the smallest?

Same day intervention
Support

- Give learners more practice in ordering the numbers 0 to 10 before they attempt to find missing numbers or say which number comes between two other given numbers.

Enrichment

- Challenge learners to tell you two or three numbers that fall between two numbers that are further apart (for example, 4 and 10).

Number – Place value, ordering and rounding

Lesson 4: **Ordinal numbers**

Learning objectives

Code	Learning objective
1Np.04	Recognise and use the ordinal numbers from 1st to 10th.

Resources

ten toy animals (per class); Resource sheet 14: Ordinal number cards (one set per class); toy cars (one per learner); mini whiteboard and pen (per learner); chalk or tape (per class)

Revise

Use the activity *Order the numbers* (variation) from Unit 16: *Place value and ordering to 10* in the Revise activities.

Teach [SB]

- Look at the picture in the Student's Book. Ask: **Have you ever run in a race before? Where did you come?** Take feedback from the class. Ask: **Who is the winner in this race? Who is last? Who came second?**

- Write several ordinal numbers on the board, reading them out as you do so. (Decide prior to the lesson the format in which you want to present ordinal numbers, that is, either as numerals and abbreviations: 1st, 2nd, ... or as words: first, second, ... or both: first – 1st.) **[T&T]** Ask: **"Do you know what these numbers are used for? Can you think of a time when you or someone you know used them?** Establish that ordinal numbers are used to help put things in order.

- Line up ten toy animals in a race and ask for volunteers to help you label them by using the ordinal number cards. Remind learners that it is just like counting in ones, but the number words are slightly different. Point out that the last animal in the race could be referred to as either last or tenth.

- Repeat this, moving the animals into different positions in the race and asking different volunteers to label them.

- Discuss the Guided practice example in the Student's Book.

Practise [WB]

- Workbook

Title: Ordinal numbers

Page: 69

- Refer to Activity 2 from the Additional practice activities.

Apply 👥 🖥

- Display **Slide 1**.
- Split learners into groups of ten or fewer.
- Give each learner a toy car and a mini whiteboard and pen. Mark a starting line on the floor with chalk or tape.
- Learners place their cars on the starting line and each give their car one strong push to make it go as far as possible across the floor.
- Each learner places their mini whiteboard next to their car and writes on it the ordinal number to show its position among all the cars.
- Repeat this activity as many times as necessary.

Review

- Discuss ordinal numbers as a way of organising events. Ask learners to use ordinal numbers to describe to a partner what they do when they get up in the morning, for example: 'First I wash my face. The second thing I do is eat breakfast.'

Assessment for learning

- Where did this toy come in the race?
- Who is seventh?
- Who is first/last?
- Use ordinal numbers to tell me how to make a sandwich.

Same day intervention
Support

- Play games with learners in which there are only first, second and third places to reinforce these words as they are harder to remember than the other ordinal numbers because they don't follow the same pattern.

Number – Place value, ordering and rounding

189

Additional practice activities

Activity 1 👥 ⚛

Learning objectives
- Recognise that zero means none of something
- Compare numbers and amounts (variation only)

Resources

Resource sheet 15: Tree (per learner); piles of counters (per learner); shuffled 0–10 number cards from Resource sheet 1: 0–20 number cards – with four extra zeros added to the pack (per pair)

What to do

- Learners pick a number card and place that many counters on their tree, to look like apples.
- They swap places with their partner and count the apples on each other's trees to check the amounts are correct.
- They then repeat the activity with a different number card.
- There are four zero number cards in the pack to give learners plenty of practice at adding nothing to the trees when they pick a zero card.

- At the end of the activity, ask: **What happened to the amount of apples on your trees every time you had to add zero apples to it?** Elicit that the amount of apples stayed the same. **[T&T]** Ask: **Why did the amount of apples not change?** Ask learners for their explanations, agreeing that the amount can't change if you add or subtract zero as zero means 'nothing'.

Variation 🧍

2 Take the extra zeros out of the number card pack. Each learner picks a card and places that number of counters on their tree. They then compare both trees pointing and saying which tree has more and which has less.

Activity 2 👥 ⚛

Learning objectives
- Use ordinal numbers
- Order numbers from 1 to 10 (variation only)

Resources

large cut-out paper fish (per learner); newspapers (six per class)

What to do

- Cut out large fish shapes from paper. Give one to each learner.
- Divide learners into groups of six. Groups take turns to race by flapping their fish towards the finish line with their newspapers.
- When all fish are over the finish line, learners write the ordinal number to indicate the position of the fish onto the fish.

Variations

1 Encourage these learners to watch the finish line carefully when it is not their turn to race. Ask them which position each fish came in to prompt their use of ordinal number language.

2 Adapt this activity for ordering numbers by writing the numbers 1–9 on nine paper fish and replicating a basic number line by labelling one side of the finish line with '0' and the other side with '10'. Learners work together as a group to flap the nine fish in order from 1 to 9. Encourage discussion between the learners as they do this, asking questions to support this such as: **Is your number closer to 10 or 0?**

3 Give Challenge 3 learners all of the fish when the race is over and the ordinal numbers have been written on them and ask them to order them from first to sixth.

Unit 17: Place value and ordering to 20

Learning objectives

Collins International Primary Maths Recommended Teaching and Learning Sequence: Term 2, Week 8

Code	Learning objective
1Np.02	Compose, decompose and regroup numbers from 10 to 20.
1Np.03	Understand the relative size of quantities to compare and order numbers from 0 to 20.

Unit overview

In this unit, learners are introduced to place value, learning how to decompose and compose teen numbers by breaking them down into a ten and some ones or putting a ten and some ones together to make a number. They use Base 10 equipment to boost their comprehension of this concept.

Learners also discover that a number can be made up in different ways, for example, 13 = 10 + 3, 13 = 10 + 2 + 1, and so on.

The unit culminates in learners comparing and ordering quantities and numbers to 20, using both concrete objects and numbers to do this.

Prerequisites for learning

Learners need to:
- recognise the numbers 0–20
- be able to count up to 20 objects accurately
- be familiar with various addition strategies
- have experience of comparing and ordering quantities and numbers to 10.

Vocabulary

numbers, amounts, tens, ones, partition, compose, decompose, groups, regrouping, compare, order, more, most, larger, largest, less, least, smaller, smallest

Common difficulties and remediation

Partitioning can be difficult for learners to understand as they are just beginning to find out what each digit in a number represents. Use lots of visual representations of numbers as tens and ones (Base 10 equipment, digital representations, and so on) to promote deeper understanding.

Some learners may find it difficult to order teen numbers as they all start with 1. Encourage learners to look at the second digit, reminding them that the pattern is the same as for the numbers 1 to 9.

Supporting language awareness

Display the key words for each lesson and discuss them with learners. Use and refer to these words throughout the lesson and encourage learners to use the words too, prompting them when necessary.

Displaying the number symbols along with their names in the classroom is good practice and helps learners to read and write the number names.

Promoting Thinking and Working Mathematically

TWM.05 Characterising
Learners investigate numbers that are made up of a ten and some ones and compose and decompose these to demonstrate this property.

Success criteria

Learners can:
- compose and decompose numbers from 10 to 20
- regroup numbers from 10 to 20
- compare and order numbers from 0 to 20.

Number – Place value, ordering and rounding

Unit 17 Place value and ordering to 20

Lesson 1: **Partitioning numbers into tens and ones**

Number – Place value, ordering and rounding

Learning objectives

Code	Learning objective
1Np.02	Compose, decompose [and regroup] numbers from 10 to 20.

Resources

Base 10 equipment – tens and ones (per pair); coloured pencils (per learner) (for the Workbook); mini whiteboard and pen (per pair); 10–20 number cards from Resource sheet 1: 0–20 number cards (per pair)

Revise

Use the activity *Tens and ones* from Unit 17: *Place value and ordering to 20* in the Revise activities.

Teach 🔲 📊 [TWM.05]

- Direct learners to the picture in the Student's Book. Explain that the picture shows what the number 13 is made of. Say: **13 is made of one ten and three ones.**
- Display the **Base 10 tool** and give pairs of learners some Base 10 equipment tens and ones to manipulate. Ask learners to make 13 from their equipment, using a ten and three ones. Demonstrate this with the **Base 10 tool**. Then show them how to separate these into tens and ones again. Explain that this is called partitioning.
- Demonstrate how to partition 13 on the board.

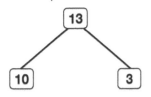

- Now write 18 on the board. **[TWM.05]** Ask: **How do you think you could make 18 from your Base 10 equipment?** Ask learners to demonstrate, referring to how many tens and ones there are in 18 and which Base 10 blocks represent the tens and ones. Say: **By identifying how many tens and ones a number is made up of, you are *characterising*.** Invite a learner to partition 18 on the board as above.
- Address any misconceptions (such as that one block of 10 is worth 10 tens rather than 1 ten) while you partition another teen number.
- 📢 Ask: **What do you think 20 is made of?** Elicit that it is made of 2 tens and 0 ones. Represent this with Base 10 equipment.
- Discuss the Guided practice example in the Student's Book.

Practise 🔲

- Workbook

Title: Partitioning numbers into tens and ones

Page: 70

- Refer to Activity 1 from the Additional practice activities

Apply 👥 🖥 [TWM.05]

- Display **Slide 1**.
- Give pairs of learners a mini whiteboard and pen, Base 10 tens and ones equipment and a set of 10–20 number cards.
- Learners pick a number card. Learner A partitions the number into tens and ones on the whiteboard. Learner B makes it with Base 10 equipment.
- Learners swap roles and repeat.

Review 📊

- Invite learners to partition numbers to 20 on the **Base 10 tool**, addressing any issues or misconceptions that arose during the lesson as they do so.
- 📢 Ask: **Can we partition one-digit numbers? Why/ why not?** Discuss this with reference to the fact that there are no tens in a one-digit number so the number 5 (for example) is made up of 0 tens and 5 ones.

Assessment for learning

- Partition 14 on your whiteboard.
- How many tens and ones is 14 made of?
- Show me 14 with Base 10 equipment.

Same day intervention
Support

- Label each Base 10 equipment tens with **1 ten** to remind learners that they are worth 1 ten, not 10 tens.

Lesson 2: **Combining tens and ones**

Learning objectives

Code	Learning objective
1Np.02	Compose, decompose [and regroup] numbers from 10 to 20.

Resources

10 number card from Resource sheet 1: 0–20 number cards (per learner); shuffled 1–9 number cards from Resource sheet 1: 0–20 number cards (per learner); mini whiteboard and pen (per learner); Base 10 equipment – tens and ones (per learner) (optional)

Revise

Use the activity *Tens and ones* from Unit 17: *Place value and ordering to 20* in the Revise activities.

Teach 🟦 📊

• Direct learners to the picture in the Student's Book. Ask: **What number is the Base 10 equipment making?** Point out that there is 1 ten and 6 ones, so it must make 16.

• Tell learners that we can make numbers from tens and ones. Display the **Base 10 tool** and put 1 ten and 9 ones on the screen. Ask: **What number have I composed?** Remind learners that 1 ten is worth 10 and each one is worth 1. Agree that there is 1 ten then count the ones together. Write 19 on the board when you have agreed that you have composed this number.

• Now show learners how to compose a teen number from two numbers. Write 10 and 2 on the board. To help learners to make this connection, explain that this is the same as showing 1 ten and 2 ones with Base 10 equipment. Agree that you have made 12 and write this as 1 ten + 2 ones = 12, then 10 + 2 = 12.

• Explore both methods with other teen numbers.

• Discuss the Guided practice example in the Student's Book.

Practise 📒

• Workbook

Title: Combining tens and ones

Page: 71

• Refer to Activity 1 from the Additional practice activities.

Apply 👤💻

• Display **Slide 1**.

• Give each learner a 10 number card, a set of shuffled 1–9 number cards and a mini whiteboard and pen.

• Learners pick a 1–9 number card to combine with the 10 to compose a number. They write the two numbers on their whiteboards as an addition followed by the number that they compose.

• Provide Base 10 equipment for learners who need extra support during this activity.

Review

• Tell learners that stamps of the feet are tens and claps of the hands are ones.

• Stamp once, then clap six times and ask: **What number have I made?**

• Invite learners to the front of the class to try composing numbers in this way.

Assessment for learning

• What number have I made with the Base 10 equipment?

• What number is made from 1 ten and 8 ones?

• Write an addition number sentence to show what number is made from 1 ten and 8 ones.

Same day intervention

Support

• Ensure that learners use Base 10 equipment throughout to support them with this concept.

Number – Place value, ordering and rounding

Lesson 3: **Representing numbers in different ways**

Number – Place value, ordering and rounding

Learning objectives

Code	Learning objective
1Np.02	Compose, decompose and regroup numbers from 10 to 20.

Resources

Base 10 equipment – tens and ones (per pair); set of shuffled 10–20 number cards from Resource sheet 1: 0–20 number cards (per pair); mini whiteboard and pen (per pair); ten interlocking cubes (per class)

Revise

Use the activity *Number race* from Unit 17: *Place value and ordering to 20* in the Revise activities.

Teach 🔲 📊

- Direct learners to the picture in the Student's Book. Explain that the picture shows some different ways to make 15. Use the **Base 10 tool** to make 15, then show learners how to regroup the ones to match the calculations in the Student's Book.
- Now write 14 on the board and ask learners to move the ones on the **Base 10 tool** to show different ways of regrouping the ones to express the number in different ways. Keep the ten intact. Remind learners that they know lots of number facts to 10 that can help them with this as they know that 3 + 1 = 4, 1 + 3 = 4 and 2 + 2 = 4.
- Give pairs of learners Base 10 equipment and ask them to explore different ways of regrouping the ones to make 16. Make a list of these on the board and use the **Base 10 tool** to match their suggestions.
- Discuss the Guided practice example in the Student's Book.

Practise 📕

- Workbook

Title: Representing numbers in different ways

Page: 72

- Refer to Activity 1 (variation) from the Additional practice activities.

Apply 👥 🖥

- Display **Slide 1**.
- Give pairs of learners Base 10 equipment, a set of shuffled 10–20 number cards and a mini whiteboard and pen.

- Learners pick a number card and use their Base 10 equipment to find different ways of regrouping the ones to represent the number. They write their calculations on their whiteboards (for example, 10 + 8 = 18, 10 + 4 + 4 = 18, 10 + 7 + 1 = 18).

Review [TWM.03]

- Write 13 on the board and these number sentences: 10 + 3 = 13, 10 + 1 + 2 = 13, 10 + 2 + 1 = 13.
- **[TWM.03]** Ask: **How could you regroup the ones that make the ten?** Remind learners that they know number facts for 10 to help them with this.
- Make a list of ways to regroup the ten, using ten interlocking cubes to demonstrate this.

Assessment for learning

- How many ways can you regroup the ones in 16?
- Write a number sentence that regroups 16.
- 10 + 4 + something = 14. What is the missing number?
- Use Base 10 equipment to regroup 16.

Same day intervention
Enrichment

- Encourage learners to regroup the numbers further by regrouping the ones that make up the ten in each number too.

Lesson 4: **Comparing and ordering numbers to 20**

Learning objectives

Code	Learning objective
1Np.03	Understand the relative size of quantities to compare and order numbers from 0 to 20.

Resources

60 counters or other countable objects (per pair); Resource sheet 1: 0–20 number cards (shuffled) (per pair); Resource sheet 12: 0–20 number line (per pair); 35 counters (per learner) (for the Workbook)

Revise

Use the activity *Order the numbers* from Unit 17: *Place value and ordering to 20* in the Revise activities.

Teach 🔲 📊

- Direct learners to the picture in the Student's Book. Ask: **How many red fish are there? How many blue fish are there?** Count the fish with learners and write the numbers 13 and 18 on the board. Say: **We can see that there are more blue fish, so 18 must be greater than 13.**
- Use the **Number line tool** to demonstrate to learners that 18 is further along the number line than 13.
- Give pairs of learners 40 counters and ask them to count out 16 counters and 11 counters. Ask: **Which set has more counters?** Agree that it is the group with 16 counters. 🔄 Ask: **How else could we check to see which is the greater number?** Elicit that you could use a number line, then ask a learner to demonstrate how to do this.
- Write the numbers 12, 5, 20 and 9 on the board. Ask less able learners to count out groups of counters to match these numbers. Ask the rest of the class to look at the number line to work out which order the numbers come in from smallest to greatest.
- Make statements about each of these numbers, for example: **12 is greater than 5 but smaller than 20.** Order the numbers with learners, then ask the less able learners to look at the size of their groups of counters to check the answer. Repeat this with four different numbers to 20.
- Discuss the Guided practice example in the Student's Book.

Practise 📓

- Workbook

Title: Comparing and ordering numbers to 20

Page: 73

- Refer to Activity 2 from the Additional practice activities.

Apply 👥 💻

- Display **Slide 1**.
- Give each pair a shuffled set of 0–20 number cards, about 60 countable objects (e.g. cubes, counters, paper clips) and a 0–20 number line.
- They pick three number cards and count out groups to match each number then place the groups in order, smallest amount to greatest amount based on the size of each group.
- They then use their number line to check which order the numbers come in.
- Repeat with three different numbers.

Review

- Ask learners to secretly pick a number between 0 and 20. Ask volunteer learners to make a greater than/less than statement about their number (for example, 'It's greater than 10 but less than 14') and make a list of all the numbers the secret number could be on the board (such as 11, 12, 13).

Assessment for learning

- Put the numbers 18, 6 and 10 in order.
- How would you use a number line to order these numbers?
- Put these groups of objects in order, from least to most.
- Is 16 more or less than 19?
- Make a greater than/less than statement about 15.

Same day intervention

Support

- Let learners use counters or cubes to compare and order quantities that match the numbers you want them to compare in order to boost understanding of the relationship between numbers and amounts.

Enrichment

- Ask learners to order three numbers by counting and working out which order they come in. They can check this using a number line afterwards.

Number – Place value, ordering and rounding

Additional practice activities

Activity 1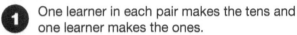

Learning objective
• Partition numbers from 10 to 20, composing, decomposing and regrouping them

Resources
10–20 number card from Resource sheet 1: 0–20 number cards chosen at random (per pair); natural objects from the school grounds such as stones, sticks and leaves (per pair); + and = cards from Resource sheet 13: Operations cards (per pair) (for variation)

What to do
• Give each pair a 10–20 number card.
• Learners work together to 'partition' this number, using natural resources in the style of Base 10 equipment, for example stones represent the ones and long sticks represent the tens.
• Check learners' work then give them another number to partition.

Variations

1 One learner in each pair makes the tens and one learner makes the ones.

2 Learner A makes a number from natural resources in the style of Base 10 equipment. Learner B must work out which number has been made. Learners swap roles and repeat.

2 Learners use natural resources to regroup numbers and represent them in different ways. Give learners + and = cards to use with the sticks and stones to represent calculations.

3 Learner A partitions two different numbers using the natural materials – one to match the number card they have chosen and one to match any other number to 20. Learner B must work out which represents the number on the card.

Activity 2

Learning objective
• Compare and order numbers to 20

Resources
sticky notes on which you have written a number from 1–19 (per learner)

What to do
• Put a sticky note on which you have written a number from 1 to 19 on each learner's back. Ensure that they do not see their number.
• Draw a number line on the wall or floor with chalk and write 0 at one end and 20 at the other end.
• Learners must put themselves in ascending numerical order along the number line by making statements about the other learners' numbers without revealing the numbers, for example: 'Your number is greater than 5 but less than 10.'
• Check the human number lines. If any numbers need to be moved, make a statement about the number to help learners to reposition it.

Variation

1 Do this as a adult-led group activity by asking each learner to come to the front in turn so that the other learners can make statements about their numbers.

Unit 18: Half (A)

Collins International Primary Maths
Recommended Teaching and
Learning Sequence: Term 3, Week 4

Learning objectives

Code	Learning objective
1Nf.01	Understand that an object or shape can be split into two equal parts or two unequal parts.
1Nf.02	Understand that a half can describe one of two equal parts of a quantity or set of objects.

Unit overview

In this unit learners are introduced to the concept of fractions, finding and identifying halves of objects, shapes and amounts. They identify objects and shapes that are split into two unequal parts, distinguishing these from objects and shapes that are divided into two halves. They use halved shapes to help them find halves of amounts by sharing the objects equally between the two halves of a shape.

Prerequisites for learning

Learners need to:
• recognise simple shapes
• be able to count up to 20 objects.

Vocabulary

whole, half, halves

Common difficulties and remediation

Some learners may believe that dividing a shape or object into two unequal pieces means that it is 'in half'. Giving learners foldable shapes helps to remedy this, as they can see that both sides must match up and be equal in order for the shape to be divided into halves.

Likewise, learners may not understand that 'half' means 'two equal pieces'. Remind learners that there must be two pieces and that those pieces must be the same. Teaching learners to identify unequal pieces as well as halves helps to remedy this.

At this stage, learners may not understand that concrete objects can be halved. Giving them objects made from modelling clay, which they can cut in half themselves, can help them to understand this.

Halving numbers is an abstract concept, so it is introduced in this unit as halving amounts of concrete objects by sharing them into two equal groups. It is essential that learners have plenty of opportunities to practise this skill, to allow them to understand halving numbers in the next unit.

Supporting language awareness

Display the key words for each lesson and discuss them with learners. Use and refer to these words throughout the lesson and encourage learners to use the words too, prompting them when necessary.

The words 'half', 'halves' and 'whole' are very important in this unit. Display these words next to physical representations such as halved shapes, halved objects and halved amounts so that learners get used to seeing what they mean.

Use the words 'half', 'halves' and 'whole' regularly in the classroom. Say, for example: **Please will half of the class line up over here?** and: **I'm going to cut these whole apples into halves for our snack time.**

Promoting Thinking and Working Mathematically

TWM.01 Specialising
Learners choose a shape that has been divided into two parts and examine it to determine whether it is in halves or two unequal parts.

TWM.04 Convincing
Learners check if a shape has been divided into two halves by comparing the pieces to establish whether they are the same size and shape and counting them to ensure that there are two pieces. They use this information to explain their reasoning.

Success criteria

Learners can:
• identify and make halves of shapes
• identify and make halves of objects
• halve amounts of objects to 20.

Number – Fractions, decimals, percentages, ratio and proportion

Unit **18** Half (A)

Number – Fractions, decimals, percentages, ratio and proportion

Lesson 1: **Halving objects**

Learning objectives

Code	Learning objective
1Nf.01	Understand that an object [or shape] can be split into two equal parts or two unequal parts.

Resources

modelling clay (per learner/pair); blunt or plastic knife (per learner/pair); 12 interlocking cubes (per pair)

Revise

Use the activity *Match the halves* (variation) from Unit 18: *Half (A)* in the Revise activities.

Teach [SB] [TWM.01]

- Direct learners to the picture in the Student's Book. Ask: **What has happened to the apple? If two people had one piece each, would they get the same amount?** Elicit that the apple has been cut in half.
- Ask: **What does it mean if you cut something in half?** Elicit that when you cut something in half, you have two pieces of exactly the same size. If the two pieces are not the same size, they are not halves.
- [TWM.01] Show learners a representation of an item of food, such as a modelling clay cake, cut into two unequal pieces. Ask: **If you had one piece and your friend had the other piece, would that be fair?** Discuss this, explaining that it would not be fair as one part is bigger than the other and that the two parts are not equal–the same. To make it fair, it would have to be cut into halves–into two equal pieces. Recombine the clay pieces, ask two learners to come to the front, and cut the clay 'cake' again, giving them one half each. Ask again: **Is it fair?** Repeat this, giving them two equal or unequal parts each time and asking learners to explain why these are or are not halves. Say: **You are checking whether an object is in halves or two unequal parts – you are *specialising*.**
- Now cut an identical item into halves. Model how to make a line down the middle so that both sides look the same. Explain that when a whole object is cut in half, the pieces must be the same size.
- Discuss the Guided practice example in the Student's Book.

Practise [WB]

- Workbook

Title: Halving objects

Page: 74

- Refer to Activity 1 from the Additional practice activities.

Apply 👤 or 👥👥 🖥

- Display **Slide 1**.
- Give each learner modelling clay and a plastic knife. From the clay, they make a cake, an orange and a biscuit. They use the knife to cut their objects in half.
- Learners can work in pairs if you wish.

Review

- Give pairs of learners 12 interlocking cubes.
- Say an even number from 2 to 12, and ask pairs of learners to use their cubes to make a 'bar of chocolate' consisting of that many pieces.
- Learners then then break their chocolate bar in half so that both learners have the same amount. Ask: **How many pieces of chocolate do you each have?**
- Repeat for other even numbers to 12. If appropriate, say an odd number and discuss with the class how the chocolate bar has been split into two unequal parts.

Assessment for learning

- Is this cake cut into halves? How can you tell?
- If I had this part of the apple and you had the other part, would it be fair? Why/why not?
- How many pieces must I cut this into for it to be in halves?
- Where would you draw a line to cut this in half?

Same day intervention
Support

- Give learners objects that have been cut into halves and objects that have been cut into two extremely unequal pieces, so that the difference is obvious to start with. Then progress into objects where the unequal pieces look a little closer to halves.

Lesson 2: **Halves of shapes**

Learning objectives

Code	Learning objective
1Nf.01	Understand that a [an object or] shape can be split into two equal parts or two unequal parts.

Resources

large paper triangle, square, circle and rectangle (per class); small paper triangle, square, circle and rectangle (per learner); scissors (per learner); glue stick (per learner); paper shapes (triangle, square, circle and rectangle) cut into halves (half a shape per learner)

Revise

Use the activity *Match the halves* from Unit 18: *Half (A)* in the Revise activities.

Teach [SB] [📊] [TWM.01/04]

- Direct learners to the picture in the Student's Book. Ask: **What has happened to these shapes?** Elicit that they have been split (divided) into halves. **[TWM.04]** 🖱 Ask: **How do you know that the shapes are in halves?** Discuss and agree that they must be in halves because both pieces are the same size for both shapes.
- Display the **Symmetry tool**, bringing up different shapes and asking learners where you should draw a line to divide the shape into halves.
- **[TWM.01]** Draw the line incorrectly a few times, making the two parts unequal. Ask the learners to correct you and put the line in the correct place.
- Now demonstrate how to divide a shape in half, folding large paper shapes so that the sides fit together perfectly. Cut along the crease and show learners the two halves taken apart and put back together again to make a whole shape.
- Discuss the Guided practice example in the Student's Book.

Practise [WB]

- Workbook

Title: Halves of shapes

Page: 75

- **[TWM.01]** Refer to Activity 1 from the Additional practice activities.

Apply 👤🖥

- Display **Slide 1**.
- Give each learner a set of paper shapes. They fold each shape in half, then cut the shape along the crease and, if appropriate, stick both halves of each shape into their exercise books.

Review

- Give each learner half of a paper shape.
- Each learner must find somebody with the other half of their shape.
- When they find each other, they put their shape together to create a whole shape.

Assessment for learning

- Where would you draw the line to divide the shape into halves?
- If I draw the line here (incorrectly), why is this shape not divided into halves?
- Match two halves to make one whole shape.

Same day intervention

Enrichment

- Ask learners to fold the shapes in different ways to see if they can find more than one way of dividing them into halves.

Number – Fractions, decimals, percentages, ratio and proportion

199

Unit **18** Half (A)

Lesson 3: **Halves of amounts (1)**

Learning objectives

Code	Learning objective
1Nf.02	Understand that a half can describe one of two equal parts of a quantity or set of objects.

Resources

ten marbles (per class); dish (per learner); dish containing an even number of small objects to share (per pair)

Revise

Use the activity *Tree tool halving* from Unit 18: *Half (A)* in the Revise activities.

Teach SB [TWM.04]

- Direct learners to the picture in the Student's Book. Ask: **How many cherries are there? How can you make sure that both children get the same number of cherries?** Take suggestions and tell learners that you can halve amounts as well as shapes and objects, and if both children get half of the cherries each, it means that they will each get the same number of cherries, and that it will be fair.
- Tell learners that you want to share ten marbles with a volunteer. Give the volunteer two marbles and keep eight for yourself. Ask: **Is that fair? What must I do to make it fair?** Elicit that you must have half of the marbles each, then share the marbles between yourself and the volunteer, one by one, saying: **One for you, one for me.**
- [TWM.04] Remind learners that both halves must be the same. Ask another learner to count how many marbles you each have to ensure that you have half the amount each. Remind the learners that by counting to check that both learners have equal amounts, they are *convincing* – giving evidence that the amount has been halved.
- Repeat for other amounts of marbles.
- Discuss the Guided practice example in the Student's Book.

Practise WB

- Workbook

Title: Halves of amounts (1)

Page: 76

- Refer to Activity 2 from the Additional practice activities.

Apply 👥💻 [TWM.04]

- Display **Slide 1**.
- Give each learner a dish and each pair of learners a dish containing an even number of objects to share between them. Learners share the objects as demonstrated in **Teach**. They then count to ensure that they each have half of the objects.
- They can then swap their dish of objects with another pair and repeat the exercise.
- Ask pairs: **Did you both have the same amount? Does that mean that you had half each? Why?**

Review

- Divide the class into halves by sending alternate learners to opposite sides of the classroom. If there is an odd number of learners in the class, ask one learner to be your 'helper' and assist with organising the class and counting. Count the learners on each side to check that the class is in halves.
- Now put the learners into small groups of even numbers and challenge the groups to divide themselves into halves.

Assessment for learning

- Show me how you would share these marbles so that we would have half each.
- I've got seven marbles and you've got three marbles. Is that half each? Why/why not?
- How can you check to make sure that we have half each?

Same day intervention
Support

- Give learners small numbers of large objects (such as six building blocks) to divide into halves. Then they shouldn't be overwhelmed by having to share too many objects and the objects won't be lost or roll away.

Number – Fractions, decimals, percentages, ratio and proportion

Lesson 4: **Halves of amounts (2)**

Learning objectives

Code	Learning objective
1Nf.01	Understand that an object or shape can be split into two equal parts or two unequal parts.
1Nf.02	Understand that a half can describe one of two equal parts of a quantity or set of objects.

Resources

large rectangular sheet of paper (per class); two large paper shapes (per class); about 20 stickers (per class); red card oval 'ladybird' shape (per learner); any even number of counters from 2 to 20 (per learner)

Revise

Use the activity *Tree tool halving* from Unit 18: *Half (A)* in the Revise activities.

Teach [SB]

- Direct learners to the picture in the Student's Book. **[T&T]** Ask: **How many spots does each ladybird have? What do you notice about the way the spots are arranged on each of these ladybirds?** Elicit that they are all arranged in two halves.
- Remind learners that you can find half of a paper shape by folding it so that the sides match. Do this with a large rectangular sheet of paper, then tape it to the board.
- Explain that you can use halved shapes to help you to find halves of amounts. Show the learners eight stickers, then invite a learner to find half of that amount by sharing the stickers equally between the two halves of a paper shape. Ensure that they place one sticker at a time on each side.
- Invite a learner to count the stickers on both sides to check that each side contains half of the stickers.
- Repeat this with another folded paper shape and a different starting number of stickers.
- Discuss the Guided practice example in the Student's Book.

Practise [WB]

- Workbook

Title: Halves of amounts (2)

Page: 77

- Refer to Activity 2 from the Additional practice activities.

Apply 👤 🖥 [TWM.04]

- Display **Slide 1**.
- Give each learner a red oval ladybird shape and an even number of counters.
- Learners fold the oval in half vertically then draw a line down the crease.

- They arrange their counters as 'ladybird spots', putting half on each side. They count how many counters on each side to check that each side has half each.
- **[TWM.04]** Ask: **How do you know that your ladybird's spots are shared into halves?** Elicit that the ladybirds have the same amount of spots on either side of their body which means that they must have been divided exactly in half.
- If you want to use the ladybirds in a display, ask learners to draw around each counter and colour in the spots.

Review 📊

- Display the **Tree tool** showing two trees.
- Ask a learner to put an even number of birds in the sky.
- Ask a volunteer to share the birds between the trees so that each tree contains half of the amount of the birds. Count the birds on each tree to find half of the starting amount.

Assessment for learning

- Show me how to put half of the counters on each side of the ladybird.
- There are four counters on this side and six counters on the other side – are half the counters on each side? Why/why not?
- How can you check that each side has half of the starting amount?

Same day intervention

Enrichment

- Find various objects around the classroom for learners to divide into halves (e.g. books in the book corner, pencils in pots). Ask them to keep a list of what they have halved, how many there were to start with and what half of that amount is.

Number – Fractions, decimals, percentages, ratio and proportion

Additional practice activities

Activity 1

Learning objectives
- Determine whether shapes have been divided into halves
- Divide representations of objects into halves (variation)

Resources

various shapes, some of which have been divided into halves with a line and some of which have been divided into two unequal parts with a line (per pair); two sorting hoops labelled 'halves' and 'not halves' (per pair)

What to do [TWM.01]
- Pairs of learners pick a shape.
- They decide together whether it has been divided into halves or two unequal parts. They can check this by folding the shape.

- They put the shape into the correct sorting hoop then pick another shape to sort.

Variation

 Use paper cut-out representations of objects instead of shapes.

Activity 2

Learning objective
- Find half of an amount of objects

Resources

two paper plates (per learner); cut-out shapes to represent food items, e.g. sandwiches, fruit (per learner); glue stick (per learner)

What to do
- Learners count how many food cut-outs they have, then share them between the paper plates so that there is half of the whole amount on each plate.
- They stick the cut-out food onto the plates.

- [TWM.04] Ask learners to explain how they could check that they have shared the amount of food into halves.

Variations

1 Challenge 1 learners should do this as a paired activity, sharing the food between one paper plate each.

2 Give learners one paper plate each, divided into halves with a line down the middle. Learners stick half of the food on one half and the other half of the food on the other half of the plate.

Number – Fractions, decimals, percentages, ratio and proportion

Unit 19: Half (B)

Collins International Primary Maths
Recommended Teaching and
Learning Sequence: Term 3, Week 5

Learning objectives

Code	Learning objective
1Nf.03	Understand that a half can act as an operator (whole number answers).
1Nf.04	Understand and visualise that halves can be combined to make wholes.

Unit overview

In this unit, learners continue to consolidate and develop their understanding of halving, this time discovering that two halves make one whole and using this understanding to make whole shapes and objects from halves. They also use their prior knowledge of halving amounts to halve numbers to 20, building the understanding that a half can act as an operator, subtracting (taking away) from a number and making it smaller.

Prerequisites for learning

Learners need to:
- have experience of halving amounts to 20
- have experience of halving shapes and objects
- understand that if an object or shape is divided into halves, it will be in two equally sized parts.

Vocabulary

whole, half, halves, operator

Common difficulties and remediation

Some learners may not understand that when an object or number is halved, it can be made whole again by putting the halves back together, therefore a lot of practical, hands-on activities are required for learners to gain a full understanding of this.

While learners may grasp that two halves make one whole, they may not understand that four halves make two wholes or six halves make three wholes, and so on. Again, provide practical activities involving making wholes from halves to remedy this.

The concept of half as an operator can be difficult to understand. However, ensuring that learners can see that halving an object or number gives a smaller object or number, helps them to relate this to subtraction.

Supporting language awareness

Display the key words for each lesson and discuss them with learners. Use and refer to these words throughout the lesson and encourage learners to use the words too, prompting them when necessary.

The words 'half', 'halves' and 'whole' are very important in this unit. Display these words next to physical representations such as halved shapes, halved objects and halved amounts so that learners get used to seeing what they mean.

Use the words 'half', 'halves' and 'whole' regularly in the classroom. Say, for example: **Please will half of the class line up over here?** And **I'm going to cut these whole apples into halves for our snack time.**

Promoting Thinking and Working Mathematically

TWM.03 Conjecturing
Learners form the understanding that a half can act as an operator by investigating what happens to a number or shape when it is halved.

TWM.05 Characterising
Learners identify halves and put them together to form wholes. They identify whole shapes and objects, recognising that they have no part missing.

Success criteria

Learners can:
- make one whole from two halves
- identify when an object or shape is whole
- understand that half can act as an operator to make numbers and amounts smaller
- halve numbers to 20.

Number – Fractions, decimals, percentages, ratio and proportion

203

 Unit **19** Half (B)

Lesson 1: **Halving numbers to 10**

Learning objectives

Code	Learning objective
1Nf.03	Understand that a half can act as an operator (whole number answers).

Resources

ten interlocking cubes (per learner); ten counters (per learner); paper rectangle divided into halves (per learner)

Revise

Use the activity *Halve the towers* from Unit 19: *Half (B)* in the Revise activities.

Teach [SB] [TWM.03]

• Direct learners to the pictures in the Student's Book. Ask: **Look at each pair of pictures. What has happened to the second object in each pair?** Elicit that they have all been halved. Say: **Lets** *conjecture*: **Has halving them made them bigger or smaller/more or less?** Agree that the objects, shapes, amounts and numbers have been made smaller.

• **[TWM.03]** Write 8 + 2 = 10 and 8 – 2 = 6 on the board. Ask: **Which operation gives numbers that are smaller – add or subtract?** Agree that subtracting gives smaller numbers. Explain that when you halve a number, it works as an operator just like subtraction, and that you are subtracting half of the number from itself.

• Write 10 – 5 = ? on the board and ask learners to make a tower of ten interlocking cubes. Ask them to subtract 5 of the cubes. Ask: **What has happened to the tower?** Discuss how it has become smaller and is now in two halves. Count the cubes in each half to demonstrate that the starting number (10) has been halved.

• Repeat this with 8 as a starting number.

• Explain to the learners that you don't have to write subtractions to find half of a number: instead of '10 – 5', you can say 'half of 10' because half is the operator and does the same job as the – sign.

• Remind learners of how they have been using objects to find halves of amounts to assist them with the activities in this lesson.

• Discuss the Guided practice example in the Student's Book.

Practise [WB]

• Workbook

Title: Halving numbers to 10

Page: 78

• Refer to Activity 1 from the Additional practice activities.

Apply 👤🖥

• Display **Slide 1**.

• Give each learner ten counters and a paper rectangle divided into halves.

• They use the counters and rectangle to find half of the numbers listed on the slide.

Review

• Ask quickfire questions: **What is half of (any even number to 10)?** Learners show the answers on their fingers.

Assessment for learning

• Is the answer bigger or smaller when you halve a number?

• If you halve a number, are you adding or subtracting?

• What is half of (any even number to 10)?

Same day intervention

Support

• Give learners paper shapes to halve before halving numbers, to boost their understanding of something getting smaller when it is halved.

Enrichment

• Ask learners to write a subtraction number sentence to match each number that they halve.

Lesson 2: **Halving numbers to 20**

Learning objectives

Code	Learning objective
1Nf.03	Understand that a half can act as an operator (whole number answers).

Resources

tower of 16 interlocking cubes (per class); large paper rectangle (per class); 20 counters or cubes (optional) (per learner); mini whiteboard and pen (per learner)

Revise

Use the activity *Halve the towers* from Unit 19: *Half (B)* in the Revise activities.

Teach [SB]

- Direct learners to the picture in the Student's Book. Ask: **What do you notice about the children's ages?** Elicit that they have been halved each time. Check this by using a tower of 16 interlocking cubes and halving it twice. Ask: **If the four year old had a younger brother who was half of his age, how old would the younger brother be?** Agree that he would be two years old.

- Draw attention to the $\frac{1}{2}$ symbol beside the apple in the Student's Book and ask: **Does anybody know what this is?** Take suggestions then explain that it is the sign for 'one half'.

- Fold a rectangle of paper in half, draw a line down the crease and write $\frac{1}{2}$ on each side. Tell learners that this shows them that each side is one half.

- Now write $\frac{1}{2}$ of 6 = on the board and explain that this is asking us to find half of 6. Ask: **How can we find half of 6?** Elicit that we could use counters or cubes shared into two equal groups or draw dots to remember the halving fact. Write the answer (3) at the end of the number sentence.

- Recap halves of numbers to 10, then write $\frac{1}{2}$ of 14 = on the board and ask learners to find the answer, reminding them that $\frac{1}{2}$ acts as an operation to make the number smaller. Learners can use cubes, counters or known facts to do this. Write the answer on the board and repeat this for other even numbers from 10 to 20.

- Discuss the Guided practice example in the Student's Book.

Practise [WB]

- Workbook

Title: Halving numbers to 20

Page: 79

- Refer to Activity 1 from the Additional practice activities.

Apply 👤🖥

- Display **Slide 1**.

- Learners write on their mini whiteboards as many halving facts for the numbers to 20 as they can remember. They check the facts they can't remember by finding half of those numbers using counters or interlocking cubes.

Review

- Choose two learners to play against each other. They sit at the front of the classroom and you ask them a halving question, such as: **What is half of 4?** The first learner to answer correctly stays where they are and the other learner goes back to their seat. Pick another learner to compete.

- Continue until everyone has had a turn. Start with your less able learners and use larger starting numbers for your more able learners.

Assessment for learning

- Is the answer getting larger or smaller when a number is halved?
- What is half of (any even number to 20)?
- How many halving facts can you remember for the numbers to 20?
- What does $\frac{1}{2}$ mean?
- How do you write one half in numbers?

Same day intervention

Enrichment

- Ask learners to make a halving facts display for the wall using the facts that they have memorised so far. Encourage them to use simple illustrations (pictorial representations) to demonstrate each fact.

Number – Fractions, decimals, percentages, ratio and proportion

Unit 19 Half (B)

Lesson 3: Combining halves (1)

Number – Fractions, decimals, percentages, ratio and proportion

Learning objectives

Code	Learning objective
1Nf.04	Understand and visualise that halves can be combined to make wholes.

Resources

eight oranges cut into halves and one orange cut into a half and two quarters (per class); set of paper shapes – circle, triangle, square, rectangle – cut into halves plus one shape cut into two unequal parts (per learner)

Revise

Use the activity *How many wholes?* from Unit 19: *Half (B)* in the Revise activities.

Teach [SB] 🖵

- Direct learners to the picture in the Student's Book. Ask: **How many grapefruit halves are in the picture? How many whole grapefruits do you think that would be?**
- Explain that if you put two halves together, they make one whole. Show learners two orange halves. Put them together to make one whole. Display **Slide 1** to reinforce the concept.
- Now show learners one orange half and one orange quarter. Put them together and ask: **Have I made one whole? Why not?** Explain that the orange was not cut into two equal halves so therefore the two pieces can't make one whole when you put them together.
- Now show learners four orange halves. Agree that they are all the same size. Ask a learner to put the halves together to make wholes. Count the whole oranges together and say: **Four halves make two wholes.** Display **Slide 2** to reinforce the concept.
- Show learners five orange halves and put them together to makes wholes. Point out that there is half an orange left over. Say: **Five halves make two wholes and one half left over.**
- Discuss the Guided practice example in the Student's Book.

Practise [WB]

- Workbook

Title: Combining halves (1)

Page: 80

- Refer to Activity 2 from the Additional practice activities.

Apply 🧑 🖵 [TWM.05]

- Display **Slide 3**.
- Give each learner a set of paper shapes (circle, square, triangle and rectangle) that have been cut into halves. Also include one shape that has been cut into two unequal parts.
- Learners put the halves of each shape together to make whole shapes. They identify the two unequal pieces and leave them unmatched.
- Ask learners to show the class their unequal pieces and to demonstrate why these would not fit together with any of the other pieces to make one whole. Say: **You are *characterising* – describing why a piece of shape cannot be one half.**

Review

- Tell learners that you have made eight whole oranges by putting halves together. 🖐 Ask: **How many orange halves do you think I had to start with?** Take suggestions then work out the answer as a class, to check.

Assessment for learning

- Will these two parts make a whole if you put them together? Why/why not?
- How many whole lemons will eight half lemons make?
- Show me how to combine two halves of a shape to make a whole shape.

Same day intervention

Support

- Give learners lots of practice at combining and breaking apart halves of concrete objects. Modelling clay is useful for this.

Lesson 4: **Combining halves (2)**

Learning objectives

Code	Learning objective
1Nf.04	Understand and visualise that halves can be combined to make wholes.

Resources

Resource sheet 10: Part–whole diagram (per learner); 20 counters (per learner); set of even number cards from 2 to 20 from Resource sheet 1: 0–20 number cards (per learner)

Revise

Use the activity *How many wholes?* from Unit 19: *Half (B)* in the Revise activities.

Teach [SB]

- Direct learners to the picture in the Student's Book. **[T&T]** Ask: **How many cakes is each child holding? How can we find how many cakes they made altogether?** Discuss this, with reference to the fact that they are each holding half the amount of cakes that were baked, so if you put all the cakes together again you could find out the whole amount.

- Remind learners that they have been combining halves to make wholes. Explain that this applies to amounts and numbers too – if a number of objects is halved and you then count both halves together, you will get the whole amount that you started with.

- Demonstrate this with groups of learners. Ask eight learners to come to the front. Ask: **If I split the group of learners in half, how many learners will be in each half?** Elicit that there will be 4 in each half and do this, counting each half to check. Now count all the learners together, moving the groups back together as you do so, saying: **Half of eight is four. Two lots of four makes eight.**

- Repeat this with a larger group of learners.

- Some learners may notice that combining two halves is the same as doubling – it can be useful to discuss this, explaining that doubling is the opposite of halving, and that is what we are doing when we put two halves of an amount or number together.

- Discuss the Guided practice example in the Student's Book.

Practise [WB]

- Workbook

Title: Combining halves (2)

Page: 81

- Refer to Activity 2 (variation) from the Additional practice activities.

Apply 👤💻

- Display **Slide 1**.

- Give each learner a part–whole diagram resource sheet, a set of even number cards from 2 to 20 and 20 counters.

- Learners pick a number card and count out that many counters into the 'whole' section of the diagram.

- They move half of the counters into one 'part' section and the other half into the other 'part' section. They count how many counters there are in each half.

- Learners then recombine the counters into the whole section and re-count them to check that they have remade their starting number.

Review 📊

- Display the **Tree tool** showing two trees.

- Ask a learner to put an even number of apples in the sky between the trees. Count the apples.

- Ask a learner to share the apples between the two trees to find half the amount, then recombine them between the trees to find the whole starting amount.

- Ask: **If I split this number of apples into halves again and then put the whole amount back together again, will either of those amounts change?** Give pairs of learners ten interlocking cubes. Ask them to use these to test what happens if you keep halving and recombining to prove that the whole starting amount and the two halves will never change unless you add or take away more apples to or from the whole starting amount.

Assessment for learning

- If half of 6 counters is 3, how many counters will there be if I put the two halves back together again?

- Show me how to split these counters into halves and put them back together as one whole amount using the part–whole diagram.

Same day intervention

Support

- Keep amounts to 10 or less and use large countable objects to split amounts into halves and recombine them.

Number – Fractions, decimals, percentages, ratio and proportion

Additional practice activities

Activity 1

Learning objective
- Find half of numbers to 20

Resources
squares of card in different colours with a hole punched in the top and bottom of each (approx. four per learner); counters or cubes (optional) (per learner); squares of card on which $\frac{1}{2}$ has been written (approx. four per learner); piece of string approximately 50 cm long (per learner)

What to do [TWM.01]
- Learners choose an even number from 10 to 20.
- They write this number on a piece of card. Then they halve the number (using counters or cubes if required) and write the resulting number on another piece of card. They continue to halve each resulting number until they reach an odd number that cannot be halved (e.g. 12, 6, $\frac{3}{20}$, 10, $\frac{5}{18}$, 9).

- They put their cards in order and add $\frac{1}{2}$ cards between each card to show that each number has been halved.
- They thread the cards onto string to make a halving facts mobile to hang from the ceiling.

Variations
1 Learners could use 8 as their starting number for their mobiles as this would still involve halving three numbers but avoid halving larger numbers, which they may find difficult.

3 Ask Challenge 3 learners to investigate various starting numbers to 20 by halving them (as detailed in the activity) and finding out how many numbers it takes them to reach a number which cannot be halved (an odd number).

Activity 2

Learning objective
- Combine halves to make wholes

Resources
20 paper triangles (per pair); glue stick (per pair); set of even numbered cards from 2 to 20 from Resource sheet 1: 0–20 number cards (per pair)

What to do
- Give each pair 20 paper triangles, a glue stick and a set of even numbered cards from 2 to 20.
- Explain that each triangle represents half of a butterfly.
- Learners choose a number card and count out that many triangles.

- They stick triangles together with their points touching to create 'butterflies'. They count how many butterflies they have made and write the starting number of halves and how many wholes they made.
- Learners can then decorate their butterflies.

Variation
2 Draw numbers of spots to 10 on each triangle. Learners then find the two butterfly wings with the same number of spots and stick them together to find out what the whole starting number would have been, by counting all of the spots on their butterfly.

Unit 20: Time

Collins International Primary Maths
Recommended Teaching and
Learning Sequence: Term 1, Week 6

Learning objectives

Code	Learning objective
1Gt.01	Use familiar language to describe units of time.
1Gt.02	Know the days of the week and the months of the year.
1Gt.03	Recognise time to the hour and half hour.

Unit overview

This unit focuses on the passing of time in days, weeks and months, and also in hours and half hours. Learners are taught that a week is a measure of time over seven days and nights. They learn the names of the days of the week and order them. They then learn the months of the year. They gain understanding of days, weeks and months by thinking about familiar events (e.g. **I have piano lessons on a Tuesday** or **my family goes on holiday in August every year**). They use familiar language to describe the passing of time, for example, 'today', 'yesterday' and 'tomorrow'.

Learners then move on to learning about telling the time on an analogue clock to the hour and the half hour.

Note

For those teachers/schools following the CIPM Recommended Teaching and Learning Sequence, numbers 11 and 12 are not formally introduced until Term 2, Week 6 (Unit 5). However, it is recommended that before teaching this unit, teachers spend some time ensuring learners are able to recognise numbers 11 and 12.

Prerequisites for learning

Learners need to:
- recognise the numbers 1 to 12
- be able to differentiate between two lengths (for the hands of a clock)
- have a basic understanding of the passing of time in familiar circumstances.

Vocabulary

time, clock, minute, hour, minute hand, hour hand, o'clock, half past, day, week, weekday, weekend, Monday, Tuesday, ..., month, January, February, ..., year

Common difficulties and remediation

Some learners may find it difficult to comprehend the cyclical nature of time – that at the end of every hour/day/week/month/year, another one begins. Relating this to their lives and regular familiar events can help with this.

Learners may confuse the hands of the clock, believing that when the minute hand is on 12, it is 12 o'clock, even when the hour hand is positioned on a number other than 12. Remind learners that they must check the position of both of the hands to read the time correctly.

Likewise, learners may have trouble distinguishing between the two hours that the hour hand is between for 'half past' times (for example, **Is it half past 3 or half past 4?**). Remind them that they are looking for the number that the hour hand has just passed, not the number that it hasn't yet reached.

Supporting language awareness

When reading the time, learners should say '[number] o'clock', not just the number.

Encourage learners to refer to the minute and hour hands by their names, both the long and short hand.

Refer to the passing of time relating to familiar phrases such as 'in a minute' and 'I haven't got time' to build on learners' understanding of the concept of time.

Promoting Thinking and Working Mathematically

TWM.02 Generalising
Learners form questions about the passing of time and the cyclical nature of time, leading to the understanding that as soon as one unit of time is finished, another begins.

TWM.05 Characterising
Learners learn that a week is characterised by having seven days, in a specific order, and that a year is characterised by having 12 months, also in a specific order.

TWM.06 Classifying
Learners distinguish between weekdays and weekend days. They also learn to distinguish between o'clock and half past times.

Success criteria

Learners can:
- use familiar language to describe units of time
- say the days of the week
- say the months of the year
- read and show 'o'clock' times on an analogue clock
- read and show 'half past' times on an analogue clock.

 Unit **20** Time

Lesson 1: **Days of the week**

Geometry and Measure – Time

Learning objectives

Code	Learning objective
1Gt.01	Use familiar language to describe units of time.
1Gt.02	Know the days of the week [and the months of the year].

Resources

Resource sheet 23: Days of the week cards (per pair)

Revise

Use the activity *On Monday…* from Unit 20: *Time* in the Revise activities to make an initial assessment of what learners already know about the days of the week..

Teach ⬛ [TWM.06]

- Discuss the cycle of seven pictures in the Student's Book. Ask: **What are we doing today?** (we are at school) **Were we at school yesterday? Will we be at school tomorrow? Why/why not?** (Answers may be that it was the weekend so there was no school or that we were at school because it was a weekday.) Highlight the cyclical nature of the seven pictures.
- Ask: **Can anybody name a day of the week?** Write the names on the board as they are given, grouping the weekdays together and the weekend days together.
- [TWM.06] Ask: **Why are these two groups of days different?** Establish that some of the days are weekend days and others are weekdays, as relevant to learners. Ask learners to describe the difference between the two sets of days according to their experiences. Say: **You have** *classified* **the days of the week into weekdays and weekend days.** Draw learners' attention to the cyclical nature of the seven days of the week in the Student's Book, highlighting the weekdays and weekend days. Say: **School days are weekdays. People go to work or school and then home again in the evenings. The weekend lasts for two days and two nights and it's our time to spend with friends and family and to enjoy our hobbies.**
- Ask learners to count the circles in the Student's Book image. Ask: **How long is a week?** Elicit that a week is seven days long. Explain that the weekend comes at the end of the week, and after that the cycle starts all over again. Say each day of the week together, encouraging learners to point at the correct days in the Student's Book.
- [T&T] Ask: **Do you think the week would still be made up of weekdays and weekends during the school holidays? Why/why not?**
- Discuss the Guided practice example in the Student's Book.

Practise 🔲

- Workbook
Title: Days of the week

Page: 82
- Refer to Activity 1 from the Additional practice activities.

Apply 👥 🖥 [TWM.02]

- Display **Slide 1**.
- Give pairs of learners a shuffled set of days of the week cards from Resource sheet 23.
- Learners spread the cards face up on the table.
- Learners take turns to point to a card asking their partner a question such as: 'Point to the day after this day.' 'Point to the day before this day. Is this a weekday or a day at the weekend?' 'If this is today, what was it yesterday?' 'If this is today, what day will it be tomorrow?' 'What day comes next?'
- After sufficient time, pairs order the days of the weeks cards.
- Explain to learners that by beginning to understand that the days of the week and months of the year follow the same pattern every time and always start again from the beginning when the pattern finishes, they are *generalising*.

Review [TWM.02]

- On the board, draw a circle and add seven marks to the circumference, labelling them with the days of the week.
- [TWM.02] Ask learners to tell you activities that they do regularly during the week until you have filled in an activity for each day.
- 🔁 Ask: **How many weekdays are there in a week? / How many days are there in a weekend?**

Assessment for learning

- How many days are there in a week?
- How many days are there in a weekend?
- Name the days of the week.
- What day is it today/what day was it yesterday?

Same day intervention
Support

- Find today on the diagram of the days of the week in the Student's Book and ask the learner to find what day it was yesterday and what day it will be tomorrow.

Lesson 2: **Months of the year**

Learning objectives

Code	Learning objective
1Gt.01	Use familiar language to describe units of time.
1Gt.02	Know the [days of the week and the] months of the year.

Resources

Resource sheet 16: Months of the year cards (per learner)

Revise

Use the variation activity *On Monday…* from Unit 20: *Time* in the Revise activities.

Teach [SB] [TWM.05]

- Discuss the picture in the Student's Book. Tell learners that it is a calendar and displays the months of the year.
- **[T&T]** Ask: **What do you think you would use a calendar for?** Discuss how a calendar can organise different things that happen in your life so that you know what you are doing each month.
- Ask learners if they can spot any months on the calendar that are familiar to them (e.g. a birthday month or a month in which they celebrate a special event). If learners are unsure which months certain events occur in, write the event on the board and the appropriate month next to it. Ask the learners to point to each month on the calendar in turn and say the months together as a class.
- **[TWM.05]** Remind learners that there are seven days in a week and tell them that there are four weeks in one month. Ask: **Can you have a week with only four days in it? How about a year with 20 months? Why/why not?** Allow learners time to discuss this, drawing out that a week can only be made up of 7 days and a year can only be made up of 12 months. Now ask: **Can Friday sometimes come before Tuesday in a week? Can the order of the months of the year change?** Elicit that the order of days and months is fixed, much like the numbers that we say when we count.
- Discuss the Guided practice example in the Student's Book.

Practise [WB]

- Workbook

Title: Months of the year

Page: 83

- Refer to Activity 1 (variation) from the Additional practice activities.

Apply 👥 🖥

- Display **Slide 1**.
- Give pairs of learners a shuffled set of months of the year cards.
- Learners work together to order the cards.
- Learners who are unable to order the months without support could use a calendar to help them.

Review

- Stick one set of months of the year cards on the board and add in five months from a second months of the year cards set. Mix the months up so that they are not in order.
- 🗣 Ask: **How many months of the year are there?** Elicit that there are 12 months, so there are too many cards on the board.
- Ask volunteers to remove the duplicate cards.
- Count the remaining cards to check that there are 12 and read the names of the months together before finally placing the cards in order.

Assessment for learning

- How many months are there in a year?
- How many weeks are there in a month?
- Tell me the names of some of the months.
- Does the month name change every week?
- Tell me something you might do in September?

Same day intervention
Support

- Concentrate on the current month and any months particularly significant to the learners.

Geometry and Measure – Time

Unit **20** Time

Lesson 3: **O'clock**

Geometry and Measure – Time

Learning objectives

Code	Learning objective
1Gt.01	Use familiar language to describe units of time.
1Gt.03	Recognise time to the hour [and half hour].

Resources

1–12 number cards from Resource sheet 1: 0–20 number cards (per pair); paper plate or circle of paper (per pair); one long straw and one short straw (per pair)

Revise

Use the activity *Stop the clock* from Unit 20: *Time* in the Revise activities to make an initial assessment of what learners already know about o'clock times.

Teach SB [TWM.02]

- Direct learners to the Let's learn picture in the Student's Book. Ask: **What is in the picture?** (a clock) **What do we use it for?** (to tell the time) **What numbers can you see on the clock?** (1–12) Count the numbers 1–12 with learners, pointing at each number on the clock as you do so.
- Display the **Clock tool**. Point to the clock hands and explain that they are called hands because they point to the numbers to show us the time.
- Explain that the short hand is called the hour hand and the number it points to is the hour. Ask a learner to hold their arm next to yours to demonstrate the meaning of 'short' and 'long' (one long arm and one short arm). Explain that the long hand is called the minute hand. If the minute hand is on 12, it means no minutes past the hour. We call this an 'o'clock' time.
- Use the **Clock tool** to show that the hour hand does not reach the next number until the minute hand has gone all the way around. Demonstrate that the minute hand then continues around the clock until another hour has passed.
- Use the **Clock tool** to make a variety of o'clock times, discussing what the time is and why with learners.
- Next, say an o'clock time and ask for a volunteer to come to move the hands on the **Clock tool** to show the time. Repeat several times.
- Ask: **When it has been 1 o'clock, 2 o'clock, 3 o'clock and all the way around to 12 o'clock, which o'clock time do you think comes next?** Elicit that it would then start at 1 o'clock again and the cycle begins again. Point out that the hour hand goes all the way around the clock twice in one whole day and night and therefore you can have, for example, 10 o'clock in the morning and 10 o'clock at night.
- Discuss the Guided practice example in the Student's Book.

Practise WB

- Workbook

Title: O'clock

Page: 84

- Refer to Activity 2 from the Additional practice activities.

Apply

- Display **Slide 1**.
- Give each pair of learners a set of 1–12 number cards, two straws (one long, one short) and a circle of paper or a paper plate.
- Learners work together to make a clock face by placing their number cards on their paper plate.
- Learners then take turns to say an o'clock time which their partner shows on the clock using the two straws as the hands.

Review

- Display the **Clock tool**.
- Choose a 1–12 number card at random.
- Invite individuals or pairs to come forward and show the time to the hour on the tool.

Assessment for learning

- What is the name of the long/short hand?
- Is the hour hand shorter or longer than the minute hand?
- What does it mean if the minute hand is on 12?
- What does it mean if the hour hand is on [name a number] and the minute hand is on 12?
- Can you show me which way the hands go around the clock face?

Same day intervention

Support

- Focus on familiar times of day (for example, school starts at 9 o'clock/home time is at 3 o'clock).

Enrichment

- Ask learners to create a visual school timetable with the approximate time of all events of the day shown to the nearest o'clock time. Learners should draw the hands on clock faces to represent the times.

Lesson 4: **Half past**

Learning objectives

Code	Learning objective
1Gt.01	Use familiar language to describe units of time.
1Gt.03	Recognise time to the [hour and] half hour.

Resources

two sets of 1–12 number cards from Resource sheet 1: 0–20 number cards (per pair); paper plate or circle of paper (per pair); one long straw and one short straw (per pair)

Revise

Choose from the activities *Stop the clock!* or *Make the set* (variations) from Unit 20: *Time* in the Revise activities.

Teach [SB] [📊] [TWM.06]

- Direct learners to the picture in the Student's Book. Ask: **Do you know what time the clock shows?** Address any misconceptions that learners may have (e.g. that it is 4 o'clock or 5 o clock) and point out that it can't be 4 o'clock or 5 o'clock because the minute hand isn't pointing to the 12 and the hour hand isn't pointing straight at the 4 or the 5.
- Explain that when the hour hand is half way between two numbers on the clock face and the minute hand is pointing to the 6, we call it a 'half past' time.
- Remind learners of the direction that the hands travel around the clock and explain that we say it is 'half past' the number that the minute hand travelled past last.
- **[TWM.06]** Display the **Clock tool** and make half past 2. Ask learners if they can read the time. Now make several 'nearly' half past times (e.g. twenty past 8). 💬 Ask: **Is this a half past time?** Elicit that it can't be because the minute hand isn't on the 6 and the hour hand isn't exactly half way between two numbers. Invite learners to come to the front to show a half past time on the clock and to describe to the class what they are doing to ensure that it definitely shows a half past time.
- Next, say a half past the hour time and ask for a volunteer to come to move the hands on the **Clock tool** to show the time. Repeat several times.
- Discuss the Guided practice example in the Student's Book.

Practise [WB]

- Workbook

Title: Half past

Page: 85

- Refer to Activity 2 from the Additional practice activities.

Apply 👥 🖥️

- Display **Slide 1**.
- Use the same resources as in Lesson 3: a paper plate or circles of paper, one long straw and one short straw and two sets of 1–12 number cards.
- Learners set the number cards out on the paper plate/circle of paper to make a clock.
- Learner 1 chooses a number from the spare set of number cards.
- Learner 2 makes 'half past' that time with the straws on the clock.
- Learners swap roles and repeat.

Review [📊]

- Shuffle a set of 1–12 number cards.
- Display the **Clock tool**.
- Pick a volunteer to come to the front of the class and choose a number card.
- Another learner chooses whether the volunteer should make a half past time or an o'clock time.
- The volunteer uses the **Clock tool** to show the time.
- Repeat with other learners.

Assessment for learning

- What does it mean if the minute hand is on 6?
- What does it mean if the hour hand is half way between two numbers?
- Make half past 3 on the clock tool.
- Is this a half past time? Why/why not?

Same day intervention

Support

- Focus on differentiating between 'half past' and 'o'clock' times rather than reading the half past times accurately.

Geometry and Measure – Time

Additional practice activities

Activity 1

> **Learning objective**
> • Identify routines and activities that happen during different days or months.

Resources

seven large labels: Monday, Tuesday, Wednesday, Thursday, Friday, Saturday, Sunday (per class); 12 large labels showing the months of the year (per class) (variation only); paper, pencils and coloured pencils (per child)

What to do

• **[TWM.05]** Remind learners that each week is split into seven different days.
• Give each learner a piece of paper and allocate them a day of the week.
• They draw on their paper something they do during that day.
• When everyone has finished, add the pictures to the correct area in the display and talk about

the different things that we do on those days (for example, **we go to school on Monday/some people have a music lesson on a Thursday**).

Variations

3 Learners make pictures for all seven days.

 Stick up labels for each month of the year.

Learners draw a picture of what they do in that month to add to the display.

Activity 2

> **Learning objectives**
> • To identify o'clock times on an analogue clock
> • To identify half past times on an analogue clock (variation)

Resources

toy clock (variation only); Resource sheet 24: O'clock times – cut into cards (per pair); Resource sheet 25: Half past times – cut into cards (per pair – variation only); Resource sheet 26: O'clock and half past times – cut into cards (per pair – use o'clock or half past cards from this sheet to match the focus of the lesson)

What to do

• Learners work in pairs. Learner A has a shuffled set of cards from Resource sheet 24. Learner B has a shuffled set of cards from Resource sheet 26 (use only o'clock or half past cards from this resource sheet depending on the focus of the lesson).
• Learner A holds up one of their cards. Learner B looks through their cards to find the matching time. They put these cards next to each other as a pair.

• Learners swap roles and repeat with learner B holding up one of their cards and learner A finding the matching card.
• Continue until all cards are in matching pairs.

Variations

1 Give pairs of learners a set of o'clock (or half past as necessary) cards and a toy analogue clock. Learner A shows learner B a card. Learner B must make the matching time on the toy clock.

 Use cards from Resource sheet 25: Half past times rather than Resource sheet 24: O'clock times.

Geometry and Measure – Time

Unit 21: 2D shapes

Collins International Primary Maths Recommended Teaching and Learning Sequence: Term 1, Week 7

Learning objectives

Code	Learning objective
1Nc.06	Use familiar language to describe sequences of objects.
1Gg.01	Identify, describe and sort 2D shapes by their characteristics or properties, including reference to number of sides and whether the sides are curved or straight.
1Gg.07	Identify when a shape looks identical as it rotates.

Unit overview

In this unit, learners work with the four common 2D shapes: circles, squares, rectangles and triangles. They learn to name shapes and recognise them in different sizes, colours and orientations. They explore shapes in the world around them. They learn how to sort shapes according to rules, make simple models and patterns, and use key words to describe them. They start to describe the properties of shapes, for example whether the sides are curved or straight, and the number of sides.

Note

For those teachers/schools following the CIPM Recommended Teaching and Learning Sequence this unit is taught prior to Unit 22: 3D shapes. However, some schools may prefer to change the order of these two units and teach Unit 22 before teaching this unit.

Prerequisites for learning

Learners need to:
- be able to observe a picture, with a purpose
- be able to describe in everyday language different properties of objects or pictures that they see (for example, the road is bumpy/the line is spiky)
- understand the concept of putting things into groups, such as when tidying up the classroom.

Vocabulary

shape, 2D shape, circle, triangle, rectangle, square, straight, curved, sides sort, group, rotate, turn

Common difficulties and remediation

Learners may become confused counting the sides of shapes and continue to count the same sides again. Until they can visualise shapes, model counting the sides and encourage learners to share effective ways of counting.

It can be difficult to grasp that any shape with three straight sides is a triangle, even though they do not look the same. Reinforce this by explaining that sometimes triangles look a bit different and identifying as many different triangles as you can in the classroom.

A square is a rectangle with equal sides. Every square is a rectangle but not every rectangle is a square – to be a square its sides must be the same length. Learners do not need to know this at Stage 1; however, using the phrase 'a square has four sides of the same length' will help to reinforce this concept.

Always use the phrase '2D shapes' so that learners don't confuse the 2D classroom shapes they can handle with 3D shapes.

Supporting language awareness

Introduce the key words at the start of each lesson and display them; discuss them with learners. Use and refer to these words throughout the lessons and encourage learners to use them too.

Tone and style of voice can help embed learning. For example, when showing a 2D circle, trace around the circumference with a finger, simultaneously saying slowly: **one curved side**.

Promoting Thinking and Working Mathematically

TWM.01 Specialising
Learners rotate shapes to check if they look the same or different when rotated.

TWM.04 Convincing
Learners discover that all shapes with three straight sides are triangles, even though they look different, and they find examples to prove this. They also discover that a given 2D shape (such as a square) is always the same shape even if it is rotated, or if the colour or size changes.

TWM.05 Characterising
Learners learn the characteristics of 2D shapes, including the number of sides and whether those sides are curved or straight.

TWM.06 Classifying
Learners sort 2D shapes by their characteristics (for example, putting all the shapes with four sides together).

Success criteria

Learners can:
- recognise a circle, square, rectangle and triangle
- describe whether the sides of a 2D shape are straight or curved
- identify how many sides a 2D shape has
- sort 2D shapes according to some of their properties
- identify shapes when they are rotated.

Geometry and Measure – Geometrical reasoning, shapes and measurements

Unit 21 2D shapes

Lesson 1: **Recognising 2D shapes**

Learning objectives

Code	Learning objective
1Gg.01	Identify[, describe and sort] 2D shapes [by their characteristics or properties, including reference to number of sides and whether the sides are curved or straight].

Resources

classroom 2D shapes: circle, square, triangle, rectangle (per class); coloured pencils (per learner) (for the Workbook); mini whiteboard and pen (per pair)

Revise

Use the activity *Hunt the shapes* from Unit 21: *2D shapes* in the Revise activities to make an initial assessment of which 2D shapes learners already know.

Teach SB

- Discuss the picture in the Student's Book. Explain that shapes are all around us. Engineers and builders use shapes when they design bridges, cars and houses. Ask: **Can anybody see any shapes that they recognise in the picture?**
- Make a list of shapes that learners find. Every time a learner names a shape in the picture, hold up a classroom 2D shape that matches it and say the name of the shape together.
- Explain that the shapes you are learning about this week are called 2D shapes – 2D is short for two-dimensional, which means 'flat'.
- Show the **Shape set tool** and together with learners name the shapes.
- Now use the **Shape set tool** to make a picture of a robot as a class, naming each shape as you use it.
- Discuss the Guided practice example in the Student's Book.

Practise WB

- Workbook

Title: Recognising 2D shapes

Page: 86

- Refer to Activity 1 from the Additional practice activities.

Apply 👥 🖥

- Display **Slide 1**.
- Send pairs of learners on a classroom '2D shape hunt'. They can note on their whiteboards any 2D shapes that they spot in the environment (for example, a picture frame could be a rectangle).
- Share some examples as a class.

Review [TWM.04]

- Display the **Shape set tool**, showing rectangles in different sizes and colours.
- Ask: **What shapes can you see?**
- [TWM.04] Ask: **Are these 2D shapes the same shape or different?**
- Discuss how the shapes are all different colours and sizes but they are still all rectangles.

Assessment for learning

- What colour is the circle/triangle/rectangle/ square on the Shape set tool?
- Are all circles/triangles/rectangles/squares the same colour?
- How do you know this shape is a circle/triangle/ rectangle/square?
- What shapes can you see in the classroom/in this picture?

Same day intervention

Support

- Focus on identifying and naming just two shapes each day (such as circle and square).

Enrichment

- Add different types of triangle to the classroom 2D shapes.

Lesson 2: Describing 2D shapes

Learning objectives

Code	Learning objective
1Gg.01	Identify, describe [and sort] 2D shapes by their characteristics or properties, including reference to number of sides and whether the sides are curved or straight.

Resources

set of classroom 2D shapes: circle, square, triangle, rectangle (per pair); coloured pencils (per learner) (for the Workbook)

Revise

Use the activity *Shape reveal* from Unit 21: 2D shapes in the Revise activities to consolidate sight recognition of circles, squares, triangles and rectangles.

Teach [SB] [TWM.04/05]

- Discuss the picture in the Student's Book. Explain to learners that today they will be learning about the sides of shapes.
- [TWM.05] Ask learners to point to the side of the circle and the sides of the triangle. [T&T] Ask: **What is different about the sides of these shapes?** Elicit that the circle has a curved side and the triangle has straight sides. Learners may also notice that the triangle has three sides whereas the circle only has one. Tell learners that by describing and counting the sides of the shape, they are *characterising-* describing its properties.
- Give each pair of learners a set of classroom 2D shapes (circle, square, triangle and rectangle).
- [TWM.05] Ask learners to find a square. Ask: **How many sides does the square have?** Model counting the sides with learners.
- Now ask: **Are the square's sides curved or straight?**
- [TWM.04] Ask: **Look at other squares on your table – do all the squares have four straight sides?**
- Explore the other shapes in the classroom shape set in this way.
- Discuss the Guided practice example in the Student's Book.

Practise [WB]

- Workbook

Title: Describing 2D shapes

Page: 87

- Refer to Activity 2 from the Additional practice activities.

Apply 👥 🖥

- Display **Slide 1**.
- Using their set of classroom 2D shapes, learner 1 secretly chooses a 2D shape and hides it under their Student's Book.

- [TWM.05] Learner 2 asks questions to guess the shape, for example: 'How many sides does it have?' 'Does it have any curved sides?'
- Learners swap roles and repeat.

Review [|.|] [TWM.04]

- Use the **Shape set tool** to show a variety of triangles (equilateral, isosceles and right-angled) on the board.
- [TWM.04] Ask: **What are these shapes?** Some learners may be confused as the triangles look so different from each other.
- [TWM.04] Ask: **How can we check if these shapes are triangles?** Elicit that you could check if they all have three straight sides.
- Count the sides of each triangle with learners, reminding them that any shape with three straight sides is a triangle.
- Use the string from Additional practice activity 2 to make a variety of triangles for learners.

Assessment for learning

- Is this side straight or curved?
- How many sides does a circle/square/triangle/ rectangle have?
- What is the same/different about a square and a rectangle?
- What is the same/different about squares, rectangles and triangles?

Same day intervention

Support

- If learners are struggling to count the sides of shapes accurately, show them how to mark each side as they count it (on paper) or to put a small sticker on each side as they count (concrete objects).

Enrichment

- Ask learners to describe the shape of their choice with 'It hasn't...' statements (e.g. 'It hasn't got any curved sides'/'It hasn't got 4 sides')

Geometry and Measure – Geometrical reasoning, shapes and measurements

217

Unit **21** **2D shapes**

Lesson 3: **Sorting 2D shapes**

Learning objectives

Code	Learning objective
1Gg.01	Identify, describe and sort 2D shapes by their characteristics or properties, including reference to number of sides and whether the sides are curved or straight.

Resources

set of classroom 2D shapes: circles, squares, triangles, rectangles (per class); three sorting hoops (per class); four classroom 2D shapes (per learner); three large labels: 1 side, 3 sides and 4 sides (per class); coloured pencils (per learner) (for the Workbook)

Revise

Use the activity *The shape shop* from Unit 21: *2D shapes* in the Revise activities to remind learners about the different properties of common 2D shapes.

Teach [SB] [TWM.06]

- Direct learners to the picture in the Student's Book. Ask: **What is the boy doing with the pieces of paper?** Elicit that he could be sorting them into piles according to which shape or which colour they are.
- [TWM.06] Show learners a set of classroom 2D shapes: circles, squares, triangles, rectangles and Ask: **How could we sort these 2D shapes?** Elicit that they could be sorted by size, colour, shape, number of sides or into shapes with curved sides and shapes with straight sides. Say: **When you sort shapes according to their properties, you are** *classifying*.
- Select 6–10 shapes from the box and ask volunteers to come to the front and sort them by colour, placing them in the hoops as appropriate.
- [TWM.06] Return all the shapes to the box. [T&T] Ask: **Can you think of a different sorting rule for these shapes?** Take suggestions for different sorting rules/criteria, writing each one on the board. Decide on one of the sorting criteria (such as whether the shapes have curved or straight sides) and, using the sorting hoops, ask volunteers to help you to sort the shapes. Repeat for the other suggested sorting criteria.
- Discuss the Guided practice example in the Student's Book.

Practise [WB]

- Workbook

Title: Sorting 2D shapes

Page: 88

- Refer to Activity 1 from the Additional practice activities.

Apply 👤 🖥 [TWM.06]

- Display **Slide 1**.
- Each learner takes four shapes from the classroom 2D shape box at random.
- Put up three large labels around the classroom (for example, on tables): **1 side**, **3 sides** and **4 sides**.
- [TWM.06] Learners count the sides of each of their shapes and put them on the correct table.

Review 📊

- Display the **Shape set tool** and use it to show circles, squares, triangles and rectangles in a variety of sizes and colours.
- Ask volunteers to come and sort the shapes according to their own secret sorting rules.
- The rest of the class compete to work out the sorting rule as quickly as possible.

Assessment for learning

- Why does this 2D shape belong in that group?
- What is the same/different about the shapes in this group?
- Why can't I put this 2D shape in this group?
- What is the sorting rule?
- Can you sort these shapes by colour/size/ number of sides/curved or straight sides?

Same day intervention
Support

- Focus on sorting by shape name, size or colour.

Enrichment

- Can learners sort using two different rules (such as red shapes with four sides, red shapes with three sides)?

Lesson 4: **Rotating shapes**

Learning objectives

Code	Learning objective
1Nc.06	Use familiar language to describe sequences of objects.
1Gg.01	Identify, describe and sort 2D shapes by their characteristics or properties, including reference to number of sides and whether the sides are curved or straight.
1Gg.07	Identify when a shape looks identical as it rotates.

Resources

coloured pencils (per learner) (for the Workbook); one circle, square, triangle and rectangle from the class 2D shapes set (per learner); sheet of A3 paper (per learner); dishes of paint in different colours (per group); four rectangles from the class 2D shapes set (per class)

Revise

Use the activity *Shape reveal* from Unit 21: *2D shapes* in the Revise activities.

Teach SB ▃▃

- Direct learners to the first picture in the Student's Book. Tell them to point to the first triangle and ask: **What is happening in this pattern? What shape is this?** Elicit that it is a triangle. Ask: **What about the next shape?** Some learners will know that it is a triangle but some may be unsure. Hold up an equilateral triangle from the classroom 2D shapes and rotate it until it matches the second triangle in the pattern. Say: **It is still the same triangle, but I have rotated it.** Draw learner's attention to the second Let's learn diagram in the Student's Book.
- Explain that 'to rotate' means to turn and that some shapes look different when they are rotated but some look the same. With the triangle, demonstrate that you can rotate a shape a small amount (rotate it through 180°) or a large amount (rotate it through 360°).
- Show the **Shape set tool** and ask a learner to choose a shape to be rotated. **[T&T]** Ask: **Do you think it will look the same or different when we have rotated it?** Repeat with other shapes.
- Model making a repeating pattern with a rectangle and a circle using the **Shape set tool** (rectangle, circle, rectangle, circle, rectangle, circle). Discuss the pattern with learners and invite volunteers to come to the front to make different repeating patterns. Now alter your original repeating pattern (rectangle, circle, rectangle, circle) by rotating each shape half a turn each time you use it. Show the learners that a different pattern is made when the shapes are rotated and that some shapes look different when rotated (e.g. rectangles) but some look the same (e.g. circles).
- Discuss the Guided practice example in the Student's Book.

Practise WB

- Workbook

Title: Rotating shapes

Page: 89

- Refer to Activity 1 (variation) from the Additional practice activities

Apply 👤🖥 [TWM.01]

- Display **Slide 1**. Give each learner a circle, square, triangle and rectangle from the classroom 2D shape set, and a sheet of A3 paper. Provide dishes of paint in different colours per table of learners.
- **[TWM.01]** Learners dip the shapes in the paint and make a pattern, exploring which shapes look different when rotated by rotating them every time they print a shape. Encourage learners to describe and discuss the patterns they make with a partner.

Review

- Give three volunteers a rectangle each to hide behind their backs.
- Show the rest of the class a rectangle and keep it in the same position throughout this activity.
- Count to 3. On 3, the volunteers show their rectangles to the class. They can choose to keep them in the same position or rotate them before taking them from behind their backs.
- The rest of the class work out which of the rectangles have been rotated and which haven't. Encourage class discussion – how do learners know that a shape has or has not been rotated?

Assessment for learning

- Which 2D shapes look the same/different when they have been rotated? Show me.
- Are these shapes (e.g. two rectangles of same colour and size, one rotated 180 degrees) the same or different?

Same day intervention

Enrichment

- Ask: **Why does the triangle/rectangle looks different when rotated? Why does the square/ circle look the same rotated?**

Geometry and Measure – Geometrical reasoning, shapes and measurements

219

Additional practice activities

Activity 1  or 👥 or 👪 ⚠2

Learning objectives
- Identify and label common 2D shapes
- Use 2D shapes to make a picture
- Describe sequences of objects

Resources
up to 20 paper 2D shapes in different colours and sizes (per learner, pair or group); one sheet of A4 or A3 paper (per learner, pair or group); glue stick (per learner, pair or group)

What to do [TWM.05/06]
- Each learner, pair or group decides on a picture to make from their 2D shapes. Give them some ideas such as a robot, rocket or house.
- Explain to learners that they don't have to use all of the shapes and that they can use each shape more than once.
- Once the picture is finished, learners label each shape that they have used.

- [TWM.05/06] When they are tidying the leftover shapes away, ask learners to sort them into piles by shape, colour, size or whether they have curved or straight sides.

Variations [TWM.01]

⚠2 Ask learners to rotate some of the shapes as they use them to create different effects for their pictures.

⚠2 Ask learners to use their 2D shapes to make a repeating pattern. They choose whether to vary the shape, size or colour throughout the pattern. Learners then swap patterns with a partner and take turns to describe what they see in their partner's pattern.

Activity 2 👥 ⚠2

Learning objective
- Understand straight and curved lines and how these make up the properties of 2D shapes

Resources
3 metres of string or ribbon (per pair); classroom 2D shapes: circle, square, triangle, rectangle (per pair) (variation only)

What to do
- Show learners how to make a curved line and a straight line by using the string on the floor.
- Demonstrate making a 2D shape, pointing out the rules: it must be on the floor and the two ends of the string must be touching when it is finished.

- Learners can explore making irregular 2D shapes and then move on to making a circle, a square, a triangle and a rectangle, counting the sides of each of these shapes to self-check their work.

Variations

1 Halve the length of string and use classroom 2D shapes as templates.

3 Ask Challenge 3 learners to make various different shapes using string. You could ask them to make specific shapes (e.g. a triangle, a square and a circle) or let them experiment with making three different irregular 2D shapes.

Unit 22: 3D shapes

Collins International Primary Maths Recommended Teaching and Learning Sequence: Term 1, Week 8

Learning objectives

Code	Learning objective
1Gg.03	Identify, describe and sort 3D shapes by their properties, including reference to the number of faces, edges and whether faces are flat or curved.
1Gg.06	Differentiate between 2D and 3D shapes.
1Gg.07	Identify when a shape looks identical as it rotates.

Unit overview

In this unit, learners work with the five common 3D shapes: spheres, pyramids, cylinders, cuboids and cubes. They explore 3D shapes in the world around us. They use their sorting skills from Unit 21: 2D shapes, to investigate the properties of 3D shapes and start to describe them in terms of the number of curved and flat faces. They learn to differentiate between 2D and 3D shapes, and they use 3D shapes to make simple models and patterns.

Note

For those teachers/schools following the CIPM Recommended Teaching and Learning Sequence this unit is taught after Unit 21: 2D shapes. However, some schools may prefer to change the order of these two units and teach this unit prior to teaching Unit 21.

Prerequisites for learning

Learners need to:
• recognise circles, squares, triangles and rectangles
• know how to use a sorting rule
• understand how to make a simple model or pattern.

Vocabulary

3D shape, sphere, pyramid, cylinder, cuboid, cube, faces, edges, flat, curved, sort, group, rotate, turn, model

Common difficulties and remediation

Learners may become confused counting the faces of a 3D shape. Put the shape on the floor and count all of the visible faces and then reveal the hidden face to add to the total.

All 3D shapes are represented in the Student's Book, Workbook and **Shape set tool** as illustrations, photos or images in 2D. Always combine teaching with the classroom geometric 3D shapes. If possible, pairs should have one of each 3D shape, giving them time to clarify the learning by handling the shapes.

Some learners may not understand the difference between 2D and 3D shapes (for example, thinking that a circle and a sphere are the same). Remind learners often that 2D shapes are flat and 3D shapes are solid, using classroom 2D and 3D shapes to reinforce this.

Supporting language awareness

Introduce the key words at the start of each lesson and display these and discuss them with learners. Use and refer to these words throughout the lessons and encourage learners to use them too. Tone and style of voice can help embed learning.

Promoting Thinking and Working Mathematically

TWM.01 Specialising
Learners rotate shapes to check if they look the same or different when rotated.

TWM.05 Characterising
Learners learn the characteristics of 3D shapes, including the numbers of edges and faces, and whether those faces are curved or flat. They also differentiate between 2D and 3D shapes.

TWM.06 Classifying
Learners sort 3D shapes by their characteristics (for example, all the shapes with curved faces together).

Success criteria

Learners can:
• recognise a sphere, cylinder, pyramid, cuboid and cube
• describe whether the faces of a 3D shape are straight or curved
• identify how many faces and edges a 3D shape has
• sort 3D shapes according to some of their properties
• identify 3D shapes when they are rotated.

Unit **22** **3D shapes**

Lesson 1: **What is a 3D shape?**

Learning objectives

Code	Learning objective
1Gg.03	Identify[, describe and sort] 3D shapes [by their properties, including reference to the number of faces, edges and whether faces are flat or curved].
1Gg.06	Differentiate between 2D and 3D shapes.

Resources

set of classroom 2D shapes: circle, triangle, square, rectangle (per pair); set of classroom 3D shapes: cuboid, cube, sphere, pyramid, cylinder (per pair); nine sticky labels (optional) (per pair); one 2D or 3D shape (per learner); coloured pencils (per learner) (for the Workbook)

Revise [TWM.05]

[TWM.05] Use the activity *Shape feely bags* (variation) from Unit 22: *3D shapes* in the Revise activities to introduce learners to both 2D and 3D shapes and to prompt discussion between learners about the differences between them.

Teach [SB] [📊] [TWM.05]

• Discuss the picture in the Student's Book. Explain that it shows a shape sorter but that the shapes are different from the ones that learners looked at in Unit 21. **[TWM.05]** Ask: **How are the shapes in the picture different from 2D shapes?** Show the learners a few plastic 2D and 3D shapes to highlight the difference.

• Explain that the shapes in Unit 21 are 2D shapes, which means that they are flat and that this week learners will be learning about 3D shapes, which are solid.

• Show learners the five common 3D shapes and the 2D shapes that are similar to them (for example, a cuboid and a rectangle) to demonstrate the difference. Tell learners that 3D shapes have faces and that some of the faces are made up of 2D shapes (for example, a cylinder has two circular faces, a cube has six square faces).

• Show the **3D shape tool**, first displaying a cube. Tell learners: **This shape is called a cube.** Pass around a selection of cubes from the class 3D shapes set so that learners can manipulate the shape as well as see a visual representation of it.

• Repeat this with the other four 3D shapes.

• Discuss the Guided practice example in the Student's Book.

Practise [WB]

• Workbook

Title: What is a 3D shape?

Page: 90

• Refer to Activity 1 (variation) from the Additional practice activities.

Apply [👥] [🖥] [TWM.05]

• Display **Slide 1**.

• Give each pair of learners a tray containing the five 3D shapes and the four 2D shapes from Unit 21.

• Learners sort the shapes into 3D shapes and 2D shapes, explaining why they have put each shape in each set.

• You could extend the activity by giving them sticky labels so that they can label each shape with its name.

Review

• Give each learner a 2D or 3D shape.

• Ask: **Who has a sphere?** Learners who are holding spheres stand up.

• Address any misconceptions (for example, if a learner with a circle stands up).

• Repeat, calling out different shapes.

• Learners can swap shapes with one another and repeat the activity.

Assessment for learning

• Is this a 2D or 3D shape? How do you know?

• What is the difference between a cube and a square?

• Can you sort these shapes into 2D and 3D?

• What is this 3D shape called?

Same day intervention

Support

• Focus on differentiating between 2D and 3D shapes rather than naming the 3D shapes.

Enrichment

• Ask learners to name and label any 3D shapes they see around the classroom.

Lesson 2: **Making models with 3D shapes**

Learning objectives

Code	Learning objective
1Gg.03	Identify[, describe and sort] 3D shapes [by their properties, including reference to the number of faces, edges and whether faces are flat or curved].
1Gg.07	Identify when a shape looks identical as it rotates.

Resources

set of class 3D shapes: cuboids, cubes, spheres, pyramids, cylinders (per pair); set of large 3D shape building blocks or a selection of cardboard boxes, tubes and packaging in the five 3D shapes (per class); coloured pencils (per learner) (for the Workbook)

Revise

Use the activity *The same or different?* from Unit 22: *3D shapes* in the Revise activities.

Teach [SB]

- Direct learners to the picture in the Student's Book. Explain that today we are going to make models and patterns with 3D shapes.
- Point to the house model in the Student's Book and ask: **What shapes can you see in this model of a house? What other models do you think you could make from 3D shapes?** Make a list of suggestions on the board.
- Choose one of the suggestions and ask volunteers to help you to use the class 3D shape set to make it. Demonstrate that some 3D shapes look different when they are rotated so therefore could be used for different purposes in the model. Point out the shapes that look the same when they are rotated and explain: **These shapes look the same when rotated, as all their edges and faces are the same size/shape/length.**
- Focus on using the names of the shapes while you are building the model and encourage learners to do so too.
- Discuss the Guided practice example in the Student's Book.

Practise [WB]

- Workbook

Title: Making models with 3D shapes

Page: 91

- Refer to Activity 2 from the Additional practice activities.

Apply 👥 🖥 [TWM.01]

- Display **Slide 1**.
- Give each pair of learners a selection of 3D shapes.
- They use the shapes to make a model from the suggestions given in **Teach**.
- [TWM.01] They tell another pair which 3D shapes they used in their model and whether they rotated any of them, demonstrating this rotation with the classroom shapes.

Review [TWM.01]

- Use a set of large building 3D shapes or a selection of old cardboard boxes, tubes and packaging in the five 3D shapes to make a class model (such as a rocket).
- Take suggestions of which shapes to use from the class and ask for volunteers to help you to put them together.
- [TWM.01] Write a list on the board of the shapes used and whether they were rotated. Tell learners that by identifying which shapes have been rotated, they are *specialising*.

Assessment for learning

- Which 3D shapes look the same/different when they have been rotated?
- Can you rotate this 3D shape?
- Are these shapes (such as two cuboids of equal colour and size, one rotated 180°) the same or different?
- How would you use these 3D shapes to make a model of a castle?

Same day intervention

Support

- Ask learners to make a model in a small group, as an adult-led activity, so that the adult can prompt discussion about the shapes used and whether they have been rotated.

Geometry and Measure – Geometrical reasoning, shapes and measurements

223

Unit 22 3D shapes

Lesson 3: **Describing 3D shapes**

Learning objectives

Code	Learning objective
1Gg.03	Identify, describe [and sort] 3D shapes by their properties, including reference to the number of faces, edges and whether faces are flat or curved.

Resources

cylinder, cube, cuboid, pyramid and sphere (per pair); coloured pencils (per learner) (for the Workbook)

Revise

Choose an activity from Unit 22: *3D shapes* in the Revise activities.

Teach [SB]

- Look at the picture in the Student's Book. Point to each of the shapes, saying the names of the shapes with learners.
- Tell learners that you can describe 3D shapes by counting how many faces and edges they have and by describing what the faces look like.
- Show learners a cylinder and explain that it has three faces. Show them the two flat faces and then clarify that the curved part of the cylinder is its third face (surface). Discuss what the faces look like (two flat faces and one large curved face). Count the faces together. Repeat this with the other 3D shapes, identifying that a cube and cuboid each have 6 faces, a square-based pyramid has 5 faces (a triangular-based pyramid has 4 faces) and a sphere has one face (surface).
- Pay particular attention to whether the faces are curved or flat and encourage learners to manipulate the class 3D shapes to reinforce this.
- Now show learners a cube and ask: **Where do you think the edges are?** Take suggestions and then show learners the edges by running your finger over each edge, where two faces meet. Slowly and carefully count the 12 edges of the cube with learners. Repeat this with the other 3D shapes, identifying that a cuboid has 12 edges, a square-based pyramid has 8 edges (a triangular-based pyramid has 6 edges), a sphere has no edges and a cylinder has 2 edges.
- ⏱ Show the learners a cube and a cuboid and ask: **What is different about the cube and the cuboid?** Discuss the similarities (12 edges, 6 faces, straight edges) and elicit that the cube's edges are all the same length and its faces are all the same shape, but the cuboid has some long edges and some short edges and some of its faces are square and some are rectangular.
- Discuss the Guided practice example in the Student's Book.

Practise [WB]

- Workbook

Title: Describing 3D shapes

Page: 92

- Refer to Activity 1 from the Additional practice activities.

Apply 👥 🖥 [TWM.05]

- Display **Slide 1**.
- Give each pair a set of five 3D shapes.
- Learner 1 names a 3D shape.
- **[TWM.05]** Learner 2 finds the 3D shape.
- They count the faces and edges together.
- Learners swap roles and repeat.

Review [TWM.05]

- Play the 'What shape am I?' game. Pick a learner to describe a 3D shape, using key vocabulary such as: 'face', 'edge', 'curved' or 'flat'. Remind learners that when they describe a shape's properties, they are *characterising*.
- The other learners guess which shape they are describing.

Assessment for learning

- Does this shape have flat or curved faces or both?
- How many faces does this shape have?
- How many edges does this shape have?
- Describe this shape.

Same day intervention

Support

- Pick one feature for learners to describe shapes by: number of edges, number of faces or whether the faces are curved or flat.

Enrichment

- Encourage learners to describe shapes by mentioning all of the features discussed in the lesson (number of faces and edges and whether the faces are curved or flat or both).

Geometry and Measure – Geometrical reasoning, shapes and measurements

Lesson 4: Sorting shapes

Learning objectives

Code	Learning objective
1Gg.03	Identify, describe and sort 3D shapes by their properties, including reference to the number of faces, edges and whether faces are flat or curved.

Resources

set of classroom 3D shapes: cuboid, cube, sphere, pyramid, cylinder (per pair); two sorting hoops (per pair); 2–4 sheets of A4 paper (per class); coloured pencils (per learner) (for the Workbook); one classroom 3D shape: cuboid, cube, sphere, pyramid or cylinder (per learner)

Revise

Use the activity *3D shape sequences* from Unit 22: *3D shapes* in the Revise activities.

Teach [SB] [image] [TWM.06]

- Direct learners to the picture in the Student's Book. Explain that the shapes have been sorted. **[TWM.06]** Ask: **What do you think the sorting rule was for these shapes?** Take suggestions (for example, shapes with six faces and shapes with a different number of faces/shapes with 12 edges and shapes with fewer edges/shapes with all flat faces and shapes with some curved faces).

- Put a variety of 3D shapes on the **Shape set tool** and say that you are going to sort them by the number of edges. Shapes with eight or more edges will go on one side of the board and shapes with fewer than eight edges on the other side. Ask volunteers to come to the front, find a matching shape in the class 3D shape set and count its edges, before moving the shape on the shape tool to the correct side. Repeat with the other shapes.

- Repeat for other sorting rules, such as shapes with more than five faces and shapes with five or fewer faces / shapes with only flat faces, shapes with only curved faces, faces with flat and curved faces.

- Demonstrate to learners that you can use a sorting rule to make a repeating pattern, for example, shape with a curved face, shape with flat faces, shape with a curved face or red pyramid, blue pyramid, red pyramid, and so on. Invite learners to come to the front to make repeating patterns with the 3D shapes using one sorting rule. Encourage the class to talk about the sequences that they can see.

- Discuss the Guided practice example in the Student's Book.

Practise [WB]

- Workbook

Title: Sorting shapes

Page: 93

- Refer to Activity 1 from the Additional practice activities.

Apply [icons] [TWM.06]

- Display **Slide 1**.

- **[TWM.06]** Pairs of learners sort the five 3D shapes by the sorting rule of their choice. Remind them that when they are sorting the shapes according to their properties, they are *classifying*.

- They then swap places with another pair, who have to work out what their sorting rule is.

Review

- Give each learner one 3D shape.

- Ask a volunteer to give you a sorting rule for the shapes (such as curved and flat faces). Use A4 paper to label areas of the classroom according to this rule (for example, the left side could be for flat faces and the right side for curved faces).

- Learners check their shape's properties and stand in the correct area with their shape.

Assessment for learning

- Why does this 3D shape belong in that group?
- What is the same/different about the shapes in this group?
- Why can't I put this 3D shape in this group?
- What is the sorting rule?
- Can you sort these shapes by colour/size/ number of edges/number of faces/curved or flat faces?

Same day intervention

Support

- Ask learners to sort 3D shapes by whether they can or can't roll. Encourage them to try to roll the shapes across the floor before sorting them and reinforce that if a shape can roll, it must have a curved face and that shapes with flat faces cannot roll.

Enrichment

- Can learners sort according to two different rules? For example, large shapes with 6 faces?

Additional practice activities

Activity 1 👥 ⚠2

Learning objectives
- Become aware of the properties of the five common 3D shapes: cube, cuboid, pyramid, cylinder and sphere
- Discover the differences between 2D and 3D shapes (see variation)
- Sort 3D shapes by some of their characteristics (see variation)
- Make and describe repeating patterns using 3D shapes (see variation)

Resources
cube, cuboid, pyramid, cylinder and sphere (per pair); thick piece of modelling clay or wet sand to make imprints and patterns (per pair)

What to do
- Pick up a cylinder and model how to make an imprint in the clay/sand. Show the different imprints from the shape, one flat and one curved. **[T&T]** Ask: **What do you see?**
- Show learners how to feel the indent left by the shape by running their finger around and inside it. Tell them to turn the shape in their hands, looking for faces that might make a different imprint. Learners continue with all six 3D shapes.
- Ask: **What happens if we put the shape back in its imprint? How many different imprints can one shape make?**

Variations

⚠2 Make a repeating pattern with the imprints or the shapes themselves. Encourage learners to describe the pattern that they have made with reference to the shapes used or the size or colour of each shape.

⚠2 Sort the shapes into those with flat faces and those with curved faces and make imprints with the flat-faced shapes only.

1 Provide 2D shapes as well as 3D shapes to demonstrate how you can only make one imprint from a 2D shape as it isn't solid and doesn't have different faces.

Activity 2 👥 ⚠2

Learning objectives
- Make models of cubes and cuboids out of smaller 3D shapes
- Recognise that shapes come in all sizes

Resources
9, 18, 24 or 36 interlocking cubes (per pair)

What to do
- Give each pair nine interlocking cubes. Ask: **How many cuboids can you make? One or more?**
- Give learners time to investigate the different options. Ask: **How do you know it's a cuboid?**

- Give more interlocking cubes as the activity progresses, providing the numbers listed in Resources.
- Ask: **Why can't you make a cylinder/sphere/pyramid out of the cubes?**

Variation

3 What is the smallest cuboid you can make? What is the largest?

Unit 23: Length and mass

Collins International Primary Maths Recommended Teaching and Learning Sequence: Term 3, Week 8

Learning objectives

Code	Learning objective
1Gg.02	Use familiar language to describe length such as long, longer, longest, thin, thinner, thinnest, short, shorter, shortest, tall, taller and tallest.
1Gg.04	Use familiar language to describe mass, including heavy, light, less and more.
1Gg.08	Explore instruments that have numbered scales, and select the most appropriate instrument to measure length, mass, [capacity and temperature].

Unit overview

In this unit, the concepts of length, height and width are introduced, along with the concept of mass. Learners compare two objects and consider which is longer, shorter, higher, taller, wider, heavier or lighter. They learn how to order more than two objects according to length or mass and how to measure the length and mass of objects, using non-standard units. They learn that the length or mass of an object can change when the object changes (for example, a pencil gets shorter when it is sharpened; a baby gets heavier as it grows). Learners do not use standard units of measurement or numbered scales to measure length or mass in Stage 1, but a selection of rulers, metre sticks and scales should be placed in the classroom (perhaps in the role-play area) for learners to explore during play.

Prerequisites for learning

Learners need to:
• know that comparing two objects means finding out what is different/the same
• understand directional language (e.g. up and down).

Vocabulary

length, long/longer/longest, short/shorter/shortest, height, width, tall/taller/tallest, wide/wider/widest, thin/thinner/thinnest, measure, mass, heavy/heavier/heaviest, light/lighter/lightest, scales, balance, estimate, the same

Common difficulties and remediation

When comparing two objects, the language is longer, shorter (when comparing length) and taller, shorter (when comparing height).

Some learners may struggle to differentiate between taller/longer and bigger. Something that appears 'big' may be wider than another object but not taller/longer than it. Try to include short, wide objects when comparing length/height to address this misconception.

Some learners may be confused by the term 'mass' as they may only be familiar with the term 'weight' which is often used at home and in society in general when mass is what is actually being referred to. Explain the difference as follows: Mass is a measure of how much matter an object has. Weight is a measure of

how strongly gravity pulls on that matter. In space, an astronaut would be weightless because there is no gravity. However, if they could weigh themselves on the spaceship they would still get a measure of their mass.

Supporting language awareness

Introduce the key words at the start of each lesson and display these and discuss them with learners. Use and refer to these words throughout the lessons and encourage learners to use them too.

The key words 'width' and 'wider' will be new to learners so model their use, holding your hands apart and describing this as 'wide', then moving them apart for 'wider'. Ask learners to repeat the gesture to boost understanding of these terms.

When displaying the vocabulary for this unit, it would be useful to have visual references on the wall next to the words (for example, a picture of a long pencil next to 'long' and a short pencil next to 'short').

Promoting Thinking and Working Mathematically

TWM.05 Characterising
Learners describe the length and mass of everyday objects.

TWM.06 Classifying
Learners compare length and mass of objects and arrange them in groups according to these properties.

TWM.07 Critiquing
Learners measure objects using non-standard units and evaluate how well these work for the purpose.

Success criteria

Learners can:
• use non-standard units to measure length and mass
• use language associated with length and mass (such as longer, shorter, heavy, light)
• compare the lengths and masses of two objects
• order objects according to length/height/mass
• say which measuring instruments are suitable for measuring length and mass
• use balance scales.

Unit **23** Length and mass

Lesson 1: **Length, height and width**

Learning objectives

Code	Learning objective
1Gg.02	Use familiar language to describe length including long, longer, longest, thin, thinner, thinnest, short, shorter, shortest, tall, taller and tallest.

Resources

one long piece of string and one short piece of string (per pair); other objects to compare by height, length and width (per class); two crayons of different lengths (per learner)

Revise

Choose an activity from Unit 23: *Length and mass*.

Teach [SB] [TWM.05]

- Discuss the picture in the Student's Book. Ask: **What differences do you see between the two pencils?** Discuss learners' suggestions, including the difference in length. Say: **One pencil is longer than the other.** Ask: **Which is the longer pencil?** Now ask: **Can you spot another difference between the pencils?** Elicit that the long pencil is thinner than the short pencil. Pass around two pieces of string of differing lengths to each pair of learners and encourage them to compare, asking: **Which piece of string is longer?**
- Tell learners that they will be comparing the lengths of objects today. Show them one long and one short piece of string and say: **This piece of string is longer than the other piece.** Now refer to the other piece of string and say: **This piece of string is shorter than the other.** Use hand gestures to reinforce the meaning of 'shorter' and 'longer'.
- Tell learners that you can compare the lengths of most things and demonstrate this with other objects such as two pencils of different lengths. Then order the height of three objects using the words: longest and shortest.
- Explain to learners that you can also compare the lengths of people or buildings, but we call this comparing height, and we use the words 'taller' and 'shorter' instead of 'longer' and 'shorter'.
- Explain that we can also compare how wide something is (such as the width of a window or a doorframe) and in this case, we use the words 'wider' and 'thinner' instead of 'longer' and 'shorter'.
- **[TWM.05]** Ask learners to describe the height, length or width of various classroom objects, asking them questions to extend their descriptions such as: **The plant is tall, but how would you describe its height compared to a house?** Remind learners that when they describe the length, width or height of something, they are *characterising*.
- Explain that sometimes the length, height or width of an object can change. For example, a pencil will get

shorter when it is sharpened and a child grows taller as they get older.
- Discuss the Guided practice example in the Student's Book.

Practise [WB]

- Workbook

Title: Length, height and width

Page: 94

- Refer to Activity 1 from the Additional practice activities.

Apply 👤 and 👥 🖥

- Display **Slide 1**.
- Give each learner two crayons.
- **[TWM.05/06]** They each compare the length of their own crayons, finding the shorter and the longer.
- They then put their crayons with a partner's and order their four crayons from shortest to longest before working together to order the crayons by height – shortest to tallest.

Review [TWM.06]

- Ask pairs of learners to compare their heights to find out who is shorter and who is taller.
- Now challenge small groups of 3–6 learners to put themselves into height order. Encourage discussion involving comparison and using the appropriate vocabulary: taller/shorter/tallest/shortest.

Assessment for learning

- Which is longer/shorter?
- Order these objects tallest/longest to shortest.
- Which is the widest object?

Same day intervention
Support

- Ask learners to compare the lengths and heights of only two objects at a time.

Enrichment

- Give learners larger amounts of objects to order by height/length/width.

Geometry and Measure – Geometrical reasoning, shapes and measurements

Lesson 2: **Measuring length**

Learning objectives

Code	Learning objective
1Gg.02	Use familiar language to describe length including long, longer, longest, thin, thinner, thinnest, short, shorter, shortest, tall, taller and tallest.
1Gg.08	Explore instruments that have numbered scales, and select the most appropriate instrument to measure length[, mass, capacity and temperature].

Resources

selection of rulers and metre sticks (per class); tray of interlocking cubes (per group then per pair for Apply); box of paper clips (per group); selection of other items to use as uniform non-standard units for measuring, (per group then per pair for Apply); interlocking cubes, glue stick, crayon, pencil, notebook (per learner) (for the Workbook); mini whiteboard and pen (per pair)

Revise

Use the activity *Comparing length and height* from Unit 23: *Length and mass* in the Revise activities.

Teach SB [TWM.07]

- Show learners a selection of rulers and metre sticks. Ask: **What do we use these for?** Elicit that we use them to measure length. Model measuring a pencil against a ruler, showing how the pencil lines up with zero. Leave rulers and metre sticks in the role-play area for learners to explore throughout the week.

- Direct learners to the Student's Book. Explain that the paper clips are used to measure the feet. The longer foot measures 8 paper clips and the shorter foot 6 paper clips.

- Explain that there are lots of objects that can be used to measure length and we call these 'non-standard units'. Draw learners' attention back to the picture in the Student's Book and remind them that paper clips have been used to measure the length of the feet. [TWM.07] Ask: **What else in the classroom could we use to measure length?** Take suggestions and write a list on the board. Discuss the suitability of the items suggested for measuring a pencil and the classroom. ⤷ Ask: **Would it be easier to use books or interlocking cubes to measure the length of the classroom? Why?** Also discuss how the non-standard units must all be the same (uniform), i.e. all interlocking cubes or all identical paper clips, and not objects of different lengths.

- [TWM.07] Give all but one group a tray of interlocking cubes and ask them to measure the Workbook with them. Model how to do this before they begin. Give the remaining group paper clips for the same purpose. Compare the groups' measurements at the end. Ask: **Why were the measurements the same for everyone who had cubes? Why was the paper clip measurement different?** Explain that the results were different because they were using different units. Also discuss how the answer may not be an exact number (e.g. 'a 'bit more than 3' or 'almost 4').

- Discuss the Guided practice example in the Student's Book.

Practise WB

- Workbook

Title: Measuring length

Page: 95

- Refer to Activity 1 from the Additional practice activities.

Apply 👥 🖥 [TWM.07]

- Display **Slide 1**. Learners draw around each other's hands. Then they draw a line up the middle to help them to measure and use interlocking cubes to measure the length of the hands, recording how many cubes long the hands are.

- [TWM.07] They experiment with using different non-standard units to measure the hands, recording each measurement on their mini whiteboards and discussing which units worked well for this purpose and which did not to help them to choose which units to use next.

Review [TWM.08]

- Ask learners to share their **Apply** measurements.

- ⤷ Ask: **What did you use to measure your hands? Which worked best? Which didn't work? Why? Why couldn't you use books to measure the length of your hand?**

- [TWM.08] Take suggestions for something that you could measure with books instead. Pick a suggestion and use books to measure it, evaluating the suitability of the books as a unit of measure in class discussion.

Assessment for learning

- What could you measure this pencil with?
- Do you think the longer/shorter object will measure more or less interlocking cubes?
- If we used a different unit of measure, would the answer be the same? Why/why not?

Same day intervention

Enrichment

- Encourage learners to compare two different non-standard units when measuring and explain which was the best choice and why.

Lesson 3: **Mass**

Learning objectives

Code	Learning objective
1Gg.04	Use familiar language to describe mass, including heavy, light, less and more.

Resources

large book (per class); feather (per class); balance scales (per group); tray of three pairs of objects to compare by mass, e.g. toys, fruit, shoes, feathers, coins, books (per group); large, empty cardboard box (per class); paperweight or similarly small, heavy object (per class)

Revise

Choose an activity from Unit 23: *Length and mass* in the Revise activities.

Teach [SB] [TWM.05/06]

- Look at the picture in the Student's Book. Tell learners that one of the children is heavier than the other. Ask: **Which child is heavier? How can we tell?**

- Tell learners that they will be comparing the mass of objects and discovering whether they are heavy or light. Pick up a heavy object (such as a large book), tell learners that it feels heavy. Pass the book around the class for learners to feel.

- Now pick up a light object (such as a feather). Tell learners that it feels light. Pass the feather around so that learners can compare its mass to the book.

- Hold the book in one hand and the feather in the other hand and hold your arms out like balance scales, with the book lower than the feather and say: **The book is heavier than the feather. The feather is lighter than the book.**

- Show learners a set of balance scales and explain that we can use them to compare the mass of objects. Put the book in one side and the feather in the other. Ask: **What happened to the scales?**

- **[TWM.05/06]** Ask volunteers to put different objects in each side of the balance scales to compare their mass. Learners can estimate which item will be the heavier/lighter by holding them first. Discuss what happens each time. **[T&T]** Ask: **What do you think will happen if we put two objects that are the same onto the scales?** Try this and point out how the scales balance when the objects have the same mass.

- Finally, move on to showing the class three (or four) objects that differ greatly in mass and order the objects introducing the language: heavy/heavier/heaviest, light/lighter/lightest.

- Discuss the Guided practice example in the Student's Book.

Practise [WB]

- Workbook

Title: Mass

Page: 96

- Refer to Activity 2 from the Additional practice activities.

Apply [TWM.06]

- Display **Slide 1**.

- Learners work in pairs to compare the mass of three different pairs of objects.

- They compare the objects first by handling them and then by using balance scales (one set per group).

- **[TWM.06]** Encourage learners to extend this activity by discussing the mass of the objects using the appropriate vocabulary and arranging them into groups according to this – for example, heavy objects and light objects or objects that are heavier than a book and objects that are lighter than a book.

Review [TWM.05]

- Show learners a large, empty cardboard box and a paperweight (or similar). Ask: **Which object do you think is heavier?**

- **[TWM.05]** Pass the objects around the class so that learners can compare the mass by holding them. Encourage learners to describe the mass of the objects using the words heavy, light, heavier, lighter, and so on.

- Put the objects on the balance scales and discuss how larger doesn't necessarily mean heavier. The box is lighter than the small paperweight even though it is larger.

Assessment for learning

- Which of these objects do you think is heavier?
- Show me how you would compare the mass of these objects on a balance scale.
- Which object feels heavier/lighter?

Same day intervention

Support

- Give learners objects that differ greatly in mass so that they can see and feel the difference between their mass clearly.

Enrichment

- Allow learners to compare the mass of four objects by comparing two at a time on the balance scales, then comparing the two heaviest objects to find the heaviest object overall.

Lesson 4: **Measuring mass**

Learning objectives

Code	Learning objective
1Gg.04	Use familiar language to describe mass, including heavy, light, less and more.
1Gg.08	Explore instruments that have numbered scales, and select the most appropriate instrument to measure [length,] mass[, capacity and temperature].

Resources

selection of measuring scales, such as bathroom scales and kitchen scales (per class); balance scales (per group); tray of different objects that could be used as uniform non-standard units, e.g. interlocking cubes, plastic counters, identical toy cars, identical coins (per group)

Revise

Use the activity *Being balance scales* from Unit 23: *Length and mass* in the Revise activities.

Teach [SB] [TWM.07]

- Show the learners a selection of measuring scales (such as bathroom scales and kitchen scales) and explain that we use them to measure the mass of objects. Pass the scales around for learners to look at and place them in the role-play area for learners to explore through play.
- Discuss the picture in the Student's Book. Ask: **Is the boy measuring the mass of the teddies or the toy car?** Explain that the teddies are non-standard units of measurement, like the interlocking cubes and paper clips that were used to measure length in Lesson 2. This time it is mass that is being measured. Ask: **How many teddies is the mass of the toy car?**
- Demonstrate how to use non-standard units to measure mass by placing a soft toy on one side of the balance scales and adding interlocking cubes to the other side of the scales until the scales balance. Now count the cubes to find out the mass of the soft toy. Repeat this with other objects to measure the mass and different non-standard units. If appropriate, also discuss how the answer may not be an exact number (e.g. 4). The answer may be 'bit more than 3' or 'almost 4'.
- [T&T] Discuss the suitability of non-standard units. Ask: **Why would it be difficult to measure the mass of a person with interlocking cubes?** Elicit that a lot of cubes would be needed and it would take a long time to count them. Also discuss with the class how the non-standard units must all be the same, i.e. all interlocking cubes or the same type of toy car, and not objects of different mass.
- Discuss the Guided practice example in the Student's Book.

Practise [WB]

- Workbook

Title: Measuring mass

Page: 97

- Refer to Activity 2 from the Additional practice activities.

Apply 👥 and 🖥 [TWM.07]

- Display **Slide 1**.
- Pairs of learners measure the mass of a shoe on the balance scales, using a non-standard unit of their choice.
- [TWM.07] Provide a set of balance scales per group and trays of different objects that could be used as non-standard units, such as interlocking cubes, plastic counters, identical toy cars, identical coins.
- Learners discuss them which units they think would be most suitable for this task and why, then try using the suggested units. They should discuss the outcomes.

Review

- As a class, measure the mass of a pencil case and a book with interlocking cubes to discover which is heavier.
- Take suggestions from learners of something in the classroom that they think will be heavier.
- Measure the mass of the suggested objects with cubes to discover if they were correct.

Assessment for learning

- If it is heavy/light will I need a lot/a few cubes?
- How many cubes does it take to measure the mass of the...?
- What does balanced look like?
- What objects would you choose to measure the mass of the...?

Same day intervention
Support

- Give learners lighter objects to measure so that they have fewer units to count.

Geometry and Measure – Geometrical reasoning, shapes and measurements

Additional practice activities

Activity 1

Learning objective
* Order and compare lengths

Resources
modelling clay (per learner); base for rolling modelling clay (per learner); interlocking cubes (per learner) (variation)

What to do
* Model rolling a snake shape with modelling clay. Ask a volunteer learner to also roll a snake.
* Put both snakes side by side to compare their lengths. Ask: **Which is longer/shorter?**
* [TWM.05] Pairs of learners roll their own snakes. They describe the length of their snakes, then compare their lengths and describe them again in comparison to their partner's snake.
* Ask: **Whose snake is longer/shorter? Do any pairs have snakes of the same length?**

Variations
3 Learners roll out four snakes and order them by length.

3 Learners measure their snakes with interlocking cubes.

Activity 2 and

Learning objective
* Understand the meaning of heavier and lighter in a practical context

Resources
2 kg bag of rice or fine gravel (per class); plastic jug, tray, two yoghurt pots (per pair); balance scales (per group); tray of objects to compare by mass, e.g. toys, fruit, shoes, feathers, coins, books (per group)

What to do
* Model pouring rice from the plastic jug into the yoghurt pots, making one heavier and one lighter and explain what you are doing.
* Invite some learners to compare the mass of the pots, first by feeling and estimating, and then by placing them on the balance scales.
* Learners work in pairs and take turns to pour the rice into the yoghurt pots, with both learners estimating which will be heavier/lighter. Learners then swap roles.
* In groups, learners measure the mass of and compare the pairs of pots on the scales. Ask: **Was your estimate correct? Whose pot was the heaviest/lightest overall?**

Variations
 Give Challenge 1 learners two yoghurt pots containing rice or sand. Ensure that on is a lot heavier than the other. ask the learners to tell you which is heaviest and which is lightest.

 Estimate heavier/lighter by holding two objects (such as toys, fruit, shoes, feathers, coins, books) then place them on the balance scales to see if the estimate was correct.

Unit 24: Capacity and temperature

Collins International Primary Maths Recommended Teaching and Learning Sequence: Term 3, Week 9

Learning objectives

Code	Learning objective
1Gg.05	Use familiar language to describe capacity, including full, empty, less and more.
1Gg.08	Explore instruments that have numbered scales, and select the most appropriate instrument to measure [length, mass,] capacity and temperature.

Unit overview

This unit is an introduction to the topic of capacity. It starts with the basic description of 'full', 'half full' and 'empty' so that learners can recognise and name them. They learn how to order capacities by finding the container that holds the most and the least, then comparing additional containers as 'more' or 'less'. They develop motor skills by actively measuring capacity, using uniform non-standard units such as empty yoghurt pots and plastic cups.

Learners are also introduced to temperature in this unit, with lots of discussion about what feels hot and cold and the opportunity to touch warm and cold objects. There are cross-curricular links with Science as learners think about hot and cold weather and how to keep warm or cool. They continue by differentiating between different numbered measurement scales, although they still use non-standard units in practical work.

Prerequisites for learning

Learners need to:
• understand the concept of measuring
• be able to use comparative language
• recognise what is the same/different.

Vocabulary

capacity, full, empty, half full, most, least, more, less, temperature, warm, hot, hotter, hottest, cool, cold, colder, coldest, measure

Common difficulties and remediation

Capacity is how much a container holds, as opposed to how much the container is holding, or the amount of space taken up by an object, which is volume. That said, the unit involves learning the basic descriptions full, half full and empty. Learners need to recognise and name these measures and understand what each description means and looks like, in order to progress to estimating and measuring capacity. It also contributes to the initial understanding of more or less, most and least.

There will be differences in learners' experiences of hot and cold weather depending on geographical location. You may need to adapt some of the material in Lesson 4 to take account of this.

Supporting language awareness

Introduce the key words at the start of each lesson and display these and discuss them with learners. Use and refer to these words throughout the lessons and encourage learners to use them too.

When displaying the vocabulary for this unit, it would be useful to have visual references on the wall next to the words (e.g. a picture of a full glass, a half empty glass and an empty glass).

Promoting Thinking and Working Mathematically

TWM.05 Characterising
Learners describe the capacity of containers and how much a container is holding.

TWM.06 Classifying
Learners compare objects and weather by temperature and group them as such.

TWM.07 Critiquing
Learners learn to compare capacities and describe them in different ways.

Success criteria

Learners can:
• measure capacity in terms of non-standard units
• use language associated with capacity (such as 'full', 'empty', 'half full', 'more' or 'less')
• compare the capacity of two containers
• compare the amount that is inside two containers
• classify objects and weather as hot, cold or warm
• say which measuring instrument is appropriate for measuring capacity and temperature.

Geometry and Measure – Geometrical reasoning, shapes and measurements

Lesson 1: **Full, half full or empty?**

Learning objectives

Code	Learning objective
1Gg.05	Use familiar language to describe capacity, including full, empty, less and more.

Resources

large bottle, cup, jug of water and large bowl (per class); teaspoon, container of rice or lentils, egg cup, teacup and plastic beaker (per group); blue coloured pencil (per learner) (for the Workbook)

Revise

Choose an activity from Unit 24: *Capacity and temperature* in the Revise activities.

Teach [SB] [.il]

- Discuss the picture in the Student's Book. Ask: **Can you tell which glass is full, which is half full and which is empty?** Explain to learners that they will be learning about capacity, which is a measure of how much something contains. **[T&T]** Ask: **There is water in the glasses. Can you think of anything else that we could use the glasses to measure?**
- Display the **Capacity tool**, showing 'full' for different containers, starting with the bucket, then other containers chosen by learners. Say: **'Full' means to the top but not overflowing.** Give a practical demonstration of this, using various different containers and water.
- Return to the **Capacity tool** and show 'half full' for the containers that you have previously looked at. Ask volunteer learners to point to where they think 'half full' will be on each container. Ensure that they understand that 'half full' means that the liquid comes half way up the container and does not refer to any amount of fluid in the container. Give a practical demonstration of 'half full' as you did for 'full'.
- Ensure that learners understand that when a container is empty it means that there is nothing inside it at all.
- Discuss the Guided practice example in the Student's Book.

Practise [WB]

- Workbook

Title: Full, half full or empty?

Page: 98

- Refer to Activity 1 (variation) from the Additional practice activities.

Apply 👥 🖥 [TWM.05]

- Display **Slide 1**.
- Give each group of learners a spoon, a teacup, an egg cup and a plastic beaker, and some lentils or rice.
- They use the spoon to fill the containers to half full, then full. Learners should describe and discuss how much each container is holding as they fill them, for example, 'This container is nearly half full – I need to add a bit more'/'This container is full now'.

Review

- Discuss the findings of the **Apply** activity.
- Ask: **Which container took the longest to fill/half fill?**
- Check this by using the spoon to half fill each container with lentils.
- 📖 Ask: **Why did it take longer to fill [name of container] than the egg cup?** Discuss the size of the containers in reference to this to prepare for Lesson 2.

Assessment for learning

- Which container is full/half full/empty?
- Is this container full?
- What does empty mean?
- If a cup contains this much (show cup approximately one-fifth full), is it half full? Why/why not?

Same day intervention

Enrichment

- Ask learners to order containers containing a variety of amounts of liquid/rice/lentils from empty to full.

Lesson 2: **Estimating and comparing capacity**

Learning objectives

Code	Learning objective
1Gg.05	Use familiar language to describe capacity, including full, empty, less and more.

Resources

washing - up liquid bottle, bucket, plastic cup, large jug, marker pen (per class); jug of water, yoghurt pot, plastic beaker, marker pen (per group); egg cup (per class)

Revise

Choose an activity from Unit 24: *Capacity and temperature* in the Revise activities.

Teach [SB]

- Direct learners to the picture in the Student's Book. Ask: **What is the same/different about the containers. Which do you think holds the most? Which do you think holds the least?**
- Explain to learners that even when containers look different, it is still possible to estimate which holds the most and which holds the least. Remind learners that estimating means making your best guess.
- Show learners a variety of containers (such as a washing - up liquid bottle, a bucket and a plastic cup). Ask: **Which of these containers do you think will hold the most liquid?** Discuss the variables such as the size and width of the containers and take estimates from learners.
- Use a large jug to fill each container, returning the water to the jug after each time and remembering to mark the level of the water left in the jug after you fill each container.
- With reference to the marks on the jug, ask: **Were our estimates correct?**
- Discuss the Guided practice example in the Student's Book.

Practise [WB]

- Workbook

Title: Estimating and comparing capacity

Page: 99

- Refer to Activity 1 from the Additional practice activities.

Apply 👥 🖥 [TWM.05]

- Display **Slide 1**.
- Learners work in groups. Give each group a jug of water, a yoghurt pot and a plastic beaker.
- As a group, they estimate whether the beaker or the yoghurt pot will hold more water.
- They fill the containers with water as you did in **Teach**, marking the level of water left in the jug both times.

- Encourage learners to discuss which container has the greatest capacity and which has the least. Do any of the containers have a similar capacity? Remind learners that by describing the capacity of the containers, they are *characterising*.

Review [TWM.07]

- Discuss the results of the **Apply** activity.
- Add another container (such as an egg cup). 👥 Ask: **Do you think this will hold more or less water than the beaker/yoghurt pot?**
- Take estimates from learners then check by filling the egg cup.
- Ask the learners to describe the capacity of each container to help you to order them from which holds the most to which holds the least. Then try grouping the containers in different ways according to their suggestions (e.g. containers with a large capacity and containers with a small capacity).

Assessment for learning

- Which container do you think will hold the most/least?
- Put the containers in order of which holds the most to which holds the least.
- Name a container that you think would hold more than this one.
- This container is full and this container is half full – but which one could hold more if they were both full?

Same day intervention

Support

- Spend extra time looking at long, thin containers and short, wide containers with learners to reinforce the concept that a tall container doesn't necessarily hold the most if it is thin.

Enrichment

- Ask learners to look at the selection of containers in the classroom. They select one that they estimate will hold the water from both the beaker and yoghurt pot from **Apply** at once.

Geometry and Measure – Geometrical reasoning, shapes and measurements

Lesson 3: **Measuring capacity**

Learning objectives

Code	Learning objective
1Gg.05	Use familiar language to describe capacity, including full, empty, less and more.
1Gg.08	Explore instruments that have numbered scales, and select the most appropriate instrument to measure [length, mass,] capacity [and temperature].

Resources

variety of measuring jugs and containers with scales up the side (per class); spread tub (per class); small bucket, jug, yoghurt pot, container of water or rice; (per group); mini whiteboard and pen (per group); spoon or egg cup (per class); access to a yoghurt pot, jug, small bucket and water (per learner) (for the Workbook)

Revise

Choose an activity from Unit 24: *Capacity and temperature* in the Revise activities.

Teach [SB]

• Show learners a variety of measuring jugs and containers with scales up the side and explain that they are used to measure capacity. Pass them around the class and then place them in the role-play or water/sand area for learners to explore through play.

• Discuss the picture in the Student's Book. Explain that we can use everyday containers (non-standard units) to measure capacity and in this picture, yoghurt pots are being used. Ask: **How many yoghurt pots equal one jug?** Elicit that it would take four yoghurt pots of water to fill the jug.

• Explain that today when we measure, the container will need to be full to the top and the units we use (such as yoghurt pots) also need to be full or the answer may not be correct. Also discuss with the class how the non-standard units must all be the same (uniform), i.e. all identical yoghurt pots or all identical paper cups, and not different types of containers.

• Model filling a spread tub with water (or rice) from a yoghurt pot. Count how many times you fill the yoghurt pot with water and empty it into the tub. Ask: **How many yoghurt pots did it take to fill the tub?**

• Note that the answer may not be an exact number (e.g. 4). The answer may be 3 and a half or more than 3 but less than 4.

• Discuss the Guided practice example in the Student's Book.

Practise [WB]

• Workbook

Title: Measuring capacity

Page: 100

• Refer to Activity 1 from the Additional practice activities.

Apply 👥🖥

• Display **Slide 1**.

• Learners work in groups to fill a small bucket and a jug with water or rice, using yoghurt pots.

• They write down on their mini whiteboards how many yoghurt pots it took to fill each container.

Review

• [T&T] Ask: **What units could we use to measure capacity instead of yoghurt pots?**

• Write a list of suggestions on the board (such as spoons, egg cups).

• Use one of the suggestions (such as a spoon) to fill a small cup, counting with learners how many spoons it takes.

• Now use a yoghurt pot to do the same and compare the answers. 🗣 Ask: **Why did it take more/fewer yoghurt pots compared to the other unit?**

Assessment for learning

• Where do I need to fill the container to?

• If it takes two yoghurt pots to half fill the container, how many yoghurt pots will it take to fill the whole container?

• Which of these two containers holds more/less?

• How many yoghurt pots of water does the container hold?

Same day intervention

Support

• Give learners who struggle with motor control larger containers to fill and larger or sturdier units of measurement (for example, they could fill a bucket rather than a smaller container and use plastic beakers rather than yoghurt pots to measure).

Lesson 4: Temperature

Learning objectives

Code	Learning objective
1Gg.08	Explore instruments that have numbered scales, and select the most appropriate instrument to measure [length, mass, capacity and] temperature.

Resources

selection of thermometers (per class); hot water bottle (per class); bag of ice cubes (per class); mini whiteboard and pen (per learner); bowl of hot (safe temperature) water, bowl of warm water and a bowl of cold water (per class or group); red and blue coloured pencils (per learner) (for the Workbook)

Revise

Use the activity *Which scale?* from Unit 24: *Capacity and temperature* in the Revise activities.

Teach [SB]

- Discuss the pictures in the Student's Book. Ask: **What is the biggest difference between these two holidays?** Elicit that the weather is different and that on one holiday it would be hot, whereas the other would be cold. [T&T] Ask: **What clothes would you need to pack to keep you feeling comfortable in the weather for the skiing holiday? What about the beach holiday?** Explain that it is important for us to know the temperature so that we can wear the right clothing and do activities to keep us warm or cool.

- Show learners a selection of thermometers and explain that we use them to measure the temperature. Pass the scales around for learners to look at and place them in the role-play area for learners to explore through play.

- Ask: **How else can we tell if something is hot or cold?** Elicit that we can tell by touching it. Pass a hot water bottle and bag of ice cubes around for learners to feel and discuss. Explain that something is 'warm' if it is between hot and cold. Give the example of a bath – it would not be comfortable to have a hot bath or a cold bath but a warm bath is just right.

- Ask: **Can you think of something hot and something cold?** Make a list on the board of hot and cold objects. Then, referring to the list of hot and cold objects, introduce learners to the terms: 'hotter', 'hottest', 'cool', 'colder' and 'coldest'.

- Discuss the Guided practice example in the Student's Book.

Practise [WB]

- Workbook

Title: Temperature

Page: 101

- Refer to Activity 2 from the Additional practice activities.

Apply 👤🖥 [TWM.06]

- Display **Slide 1**. Learners think about things that they eat and drink and whether they are hot or cold. They write (or draw) these items in two lists on their mini whiteboards: 'hot' and 'cold'. Say: **You have classified the food and drink by sorting it into hot and cold items.**

- Share some of the items on the lists, when learners have completed their lists and describe and compare the temperatures of the items on the list. Can learners think of anything to eat or drink that does not belong on the 'hot' list but does not belong on the 'cold' list either? How would they describe its temperature?

Review

- Play the temperature game with the learners:

- Tell a story about a day. Learners must act as though they are hot or cold depending on whether you mention hot or cold objects:
It was a sunny day. [pause] **I had a big glass of iced water before school.** [pause] **In school, the windows were closed so it was getting very warm in the classroom.** [pause] **I had a big bowl of soup for lunch,** [pause] **then I had an ice cream on my way home.** [pause] **I had a bath before I went to bed.** [pause] and so on.

Assessment for learning

- If it is snowing, are you hot or cold?
- Which of these objects is the hottest/coldest?
- Tell me one hot object and one cold object.

Same day intervention
Enrichment

- Ask: **Will a bath always be warmer than a glass of lemonade? Why/why not?** Ask learners to discuss this with reference to the things that may change (the lemonade may not contain ice cubes, it may have been sitting out in the sun getting warm for a while, the bath water may have cooled down).

Geometry and Measure – Geometrical reasoning, shapes and measurements

Additional practice activities

Activity 1

Learning objective
• Estimate and measure capacity in a practical context, using non-standard units of measure

Resources
spread tub, toy bucket, yoghurt pot, beaker, plastic jug, water funnel (per group) or, if outside, paddling pool (per class)

What to do
• Learners work in groups under adult supervision. Class/school rules for water play apply.
• Learners estimate, then measure the capacity of the tub with the pot, then the beaker. Repeat for the toy bucket.
• At each stage, share the results using less than/more than [number of] cups/pots.

Variations
2 Learners use the jug to half fill (or fill) each of the containers. Re-fill the jug before you use it to half fill each container and discuss how much liquid is left in it each time. Ask: **Which container used the most liquid from the jug?**

3 Give Challenge 3 learners extra containers to measure the capacity of. Encourage them to order them from greatest capacity to least afterwards.

Activity 2

Learning objective
• Order objects from coldest to hottest

Resources
bowl of hot water (safe level of heat, not boiling), bowl of warm water, bowl of cold water and bowl of iced water (per group)

What to do
• Give each group a bowl of hot (a safe level of heat) water, a bowl of warm water, a bowl of cold water and a bowl of iced water.
• **[TWM.06]** Invite learners to dip their hands into each bowl and decide whether the water feels hot or cold. Learners should describe the temperatures in comparison to each other and to other hot or cold objects that have been discussed in the lesson.
• Learners order the bowls from coldest to hottest.

Variations
1 Give learners a bowl of warm water and a bowl of cold water and ask them to describe the temperature of each when they dip their hands in, then say which is the coldest or warmest.

2 Write the names of a variety of hot and cold objects on the board (ice, a bath, a boiling kettle, a bowl of fruit, a book) and ask learners to order them from coldest to hottest.

Geometry and Measure – Geometrical reasoning, shapes and measurements

Unit 25: Position and direction

Collins International Primary Maths Recommended Teaching and Learning Sequence: Term 1, Week 9

Learning objectives

Code	Learning objective
1Gp.01	Use familiar language to describe position and direction.

Unit overview

In this unit, learners use everyday language for position and direction to describe how to move objects or how they have been moved. At Stage 1 the focus is on speaking and listening skills, using the key words to describe the movement of any object, with the simultaneous action of moving that object. The unit does not involve counting skills, making maps or quarter, half or three-quarter turns (Stage 2) or co-ordinates (Stage 4), but is the necessary foundation for future learning related to position and transformation.

Prerequisites for learning

Learners need to:
- understand the concept of following and giving simple instructions
- use gesture and spoken word simultaneously
- use some basic positional language in their everyday conversations (such as **over there**, **next to me**).

Vocabulary

up, down, forwards, backwards, around, left, right, direction, turn, inside, outside, under, over, below, on top of, in front of, behind, next to beside, on

Common difficulties and remediation

Learners will be familiar with the instruction **turn around** from playground games and will probably associate it with turning all the way around. In Lesson 1, make it clear that you can turn around on the spot or go around something and that turning all the way around is different from **turn around**, which means to turn to face the other way.

Supporting language awareness

Introduce the key words at the start of each lesson, display them and discuss them with learners. Use and refer to these words throughout the lessons and encourage learners to use them too.

Encourage learners to talk through what they are doing when they are making up or following directions (for example, **I am walking forwards and now I am going behind the door**). Model this to learners by doing it yourself.

If a learner makes the correct gesture for the position of an object (such as gesturing as if to show that the direction is forwards) say: **That's correct – it has gone forwards** to reinforce use of the correct terminology.

Promoting Thinking and Working Mathematically

TWM.08 Improving
Learners add to their vocabulary to describe position and movement and use it to give clearer directions and instructions, which they can then evaluate and improve on.

Success criteria

Learners can:
- follow and give instructions to move an object, using everyday directional language
- describe the position of an object, using everyday positional language
- explain how they are moving an object or where they have moved an object to
- use the words 'left' and 'right' accurately, to describe position and direction.

Lesson 1: **Describing direction**

Learning objectives

Code	Learning objective
1Gp.01	Use familiar language to describe position and direction.

Resources

toy car (per class); teddy or other soft toy (per pair); box/table/chair for teddy to 'climb' (per pair)

Revise

Use the activity *Teacher says...* from Unit 25: *Position and direction* in the Revise activities to make an initial assessment of which directional words learners are already familiar with.

Teach [SB] [TWM.08]

- Discuss the picture in the Student's Book. Ask: **What is the boy doing?** Explain that he is about to come down the slide. He will have climbed up to the top of the slide.
- Explain that there are special words for directions. Use a toy car to demonstrate the words 'up' (move it up the wall), 'down' (move it down the wall), 'around' (move it around in a circle and around an obstacle), 'forwards' (move it forwards) and 'backwards' (move it backwards).
- [TWM.08] Referring to the Key words in the Student's Book, ask for volunteers to come to the front of the class and 'drive' the car. Other learners take turns to give the volunteer learner directions to follow. Intervene to address any misconceptions or misunderstood words when necessary. Encourage learners to use the vocabulary that they have learned today to describe the directions. If a direction is not clear, ask: **How else can you describe that direction? Can you** *improve* **it so it is easier to understand?**
- Repeat the *Teacher says...* Revise activity that you used at the start of the lesson.
- Discuss the Guided practice example in the Student's Book.

Practise [WB]

- Workbook

Title: Describing direction

Page: 102

- Refer to Activity 1 from the Additional practice activities.

Apply 👥 💻

- Display **Slide 1**.
- Provide each pair with a teddy bear or other soft toy and a box/table/chair for him to 'climb' up and down or similar.
- Learner 1 tells learner 2 a story about Ted while moving the teddy bear up, down, around, forwards and backwards, using the key words for this lesson.
- Learners swap roles and repeat.

Review

- Stand in front of the class and move your arms to show up and down. Walk around in a circle and then forwards and backwards.
- With each movement, ask learners to call out the correct directional word.

Assessment for learning

- How would you get the car to the other side?
- Can the car go backwards around the box?
- Is the car moving forwards or backwards?
- Is my hand going up or down?

Same day intervention

Support

- Ask learners to move the car, while you give a commentary about the direction that the car is going in, until they have a firmer grasp of the correct vocabulary.

Enrichment

- Ask learners to 'perform' their story about Ted to the whole class or small groups.

Lesson 2: **Left and right**

Learning objectives

Code	Learning objective
1Gp.01	Use familiar language to describe position and direction.

Resources

soft toy or toy car (per pair); set of direction cards from Resource sheet 17: Direction cards – left, right, forwards, backwards, around (per pair)

Revise

Choose an activity from Unit 25: *Position and direction* in the Revise activities.

Teach [SB]

- Direct learners to the picture in the Student's Book. Explain that we have both a left hand and a right hand and that knowing which way is left and which way is right can help us to describe and follow directions.

- Ask learners to hold their hands in the same position as the picture of the hands in the Student's Book. Show learners that when in that position, one of the hands looks like it is making a capital L. Explain that this is your left hand and you can remember that by checking your hands and finding which one makes the L for left. Explain that the other hand is the right hand.

- Ask learners to wave at you with their left hand and then their right hand. Use this opportunity to check if they have identified the correct hands and correct any learners who have mixed up their hands.

- Show learners a teddy and make him 'walk' to the right (the learners' right, not your right). Ask: **Which direction is Ted walking in?** Ask for a volunteer to make him walk to the left.

- Now practise walking Ted forwards then saying **turn right** or **turn left** and turning him in the correct direction. Ask: **When have you heard somebody say turn right/left before?** (in a car when someone is reading a map/when somebody is giving directions for how to get somewhere).

- Discuss the Guided practice example in the Student's Book.

Practise [WB]

- Workbook

Title: Left and right

Page: 103

- Refer to Activity 2 (variation) from the Additional practice activities

Apply 👥 🖥

- Display **Slide 1**.
- Give pairs of learners a shuffled set of direction cards (left, right, forwards, backwards, around) and a soft toy or toy car.
- Learner A reads the instructions on the cards one by one.
- Learner B follows the directions with the soft toy/ toy car.
- Learners swap roles and repeat.

Review [TWM.08]

- Ask learners to stand up and march on the spot.
- Give learners directions including left and right (such as left turn/march forwards, right turn/march around the table).
- Remind learners to check their hands if they forget which way is left and which way is right.
- [TWM.08] Invite learners to take it in turns to give orders while the class marches. They should use the vocabulary from this unit including 'left' and 'right', making improvements to their directions if they are unclear.

Assessment for learning

- Show me your left hand.
- What can you see to your right?
- Can you make the teddy turn to the right?

Same day intervention

Support

- Give learners L and R stickers to put on the appropriate hands to help them to see which way is left and which way is right.

Geometry and Measure – Position and transformation

241

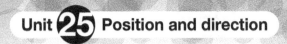

Lesson 3: **Describing position**

Learning objectives

Code	Learning objective
1Gp.01	Use familiar language to describe position and direction.

Resources

teddy or other soft toy (per class); a box that the teddy will fit inside (per class); sheet of paper (per learner); coloured pencils (per learner)

Revise

Use the activity *Treasure hunt* from Unit 25: *Position and direction* in the Revise activities.

Teach [SB]

- Look at the picture in the Student's Book. Discuss where each child is in the picture in relation to the other children and the objects in the picture.
- Explain that there are special words to describe the position that somebody or something is in. Refer learners' attention to the Key words in the Student's Book.
- Put the teddy from Lesson 2 **on** the box, **inside** the box, **behind** the box, **under** the box and hold him **over** the box, using each word to describe where he is to learners.
- Model the teddy's position again, asking learners to call out which position he is in each time.
- Now call volunteers to the front of the class to move the teddy into a given position (such as **under the box**). Repeat for each positional word.
- Discuss the Guided practice example in the Student's Book.

Practise [WB]

- Workbook

Title: Describing position

Page: 104

- Refer to Activity 2 from the Additional practice activities.

Apply 👤🖥

- Display **Slide 1**.
- Give each learner a sheet of paper and coloured pencils.
- Learners follow the instructions on the slide to draw items on, under, next to, inside and outside a box.

Review [TWM.08]

- Ask learners to work in pairs.
- They instruct each other to move an object, using the positional language from this lesson.
- Conclude by asking pairs of learners to share with the rest of the class their set of instructions for moving the object. Encourage them to emphasise the positional language.
- [TWM.08] Ask the rest of the class: **Were their instructions clear? How could they improve them? Which words would you use instead?**

Assessment for learning

- If the teddy is behind the box, is he outside the box too?
- Can we see the teddy when he is inside the box?
- What things do we like to sit under? (such as tree/umbrella/sun)
- Can you put the teddy under the box?

Same day intervention

Support

- Choose pairs of opposites (inside and outside or under and over) to demonstrate repeatedly to learners with the teddy and the box so that they can recall the two positions easily.

Enrichment

- Describe more complex positions to learners when asking them to draw or place something somewhere (e.g. put the cup on top of the book but under the table).

Lesson 4: **Following directions**

Learning objectives

Code	Learning objective
1Gp.01	Use familiar language to describe position and direction.

Resources

set of direction cards from Resource sheet 17: Direction cards – forwards, backwards, left, right (two of each card per pair); mini whiteboard and pen (per learner); blindfold (per class)

Revise

Choose an activity from Unit 25: *Position and direction* in the Revise activities.

Teach SB

- Direct learners to the picture in the Student's Book. Ask: **Why does the blindfolded child need to be given directions?** (because they can't see where they're going) Explain that it is important that their friend gives the correct directions because they could bump into something and hurt themselves if the directions are incorrect.
- Discuss other occasions when it's important to give the correct directions (such as when somebody needs to find a place to meet a friend or when you are asking somebody to find something in your house that you need).
- Explain that it is also important to listen carefully to any directions given to us so that we can find what we need or get to the right place.
- Practise this skill by asking the class to stand up and giving them a series of directions to follow: **walk forwards, STOP, turn right, walk backwards, STOP, turn left, walk around the room, STOP, turn right again** and so on.
- Address any issues that come up, such as the difference between walking or moving an object forwards and moving your pen forwards on a piece of paper (the pen can only move left or right, so forwards tends to mean right in this case and backwards tends to mean left).
- Discuss the Guided practice example in the Student's Book.

Practise WB

- Workbook

Title: Following directions

Page: 105

- Refer to Activity 1 from the Additional practice's activities.

Apply 👥 🖥

- Display **Slide 1**.
- Give pairs of learners two piles of direction cards (one pile consisting of two 'forwards' and two 'backwards' cards, the other pile consisting of two 'left' and two 'right' cards). Give each learner a mini whiteboard and pen.
- Learners place the two piles of cards face down on the table.
- Each learner starts with their pen in the middle of their whiteboard.
- One learner takes the top card from one pile of cards and both learners move their pen in the direction shown. (Show learners a rough amount of distance to move their pen by indicating distance with your finger and thumb.)
- The other learner then takes a card from the other pile and both learners move their pen in that direction.
- Continue alternating between the two piles of cards until all eight cards have been used.

Review

- Ask for a volunteer to be blindfolded.
- Ask other learners to take turns to give the volunteer directions to get from one side of the classroom to the other.

Assessment for learning

- Show me how you would turn left and then walk forwards.
- Which direction is right in?
- Walk up the steps and then down the steps.
- Why is it important to listen carefully to directions?

Same day intervention

Support

- Give learners one direction at a time and allow them plenty of time to think about the instruction before they attempt to follow it.

Geometry and Measure – Position and transformation

243

Additional practice activities

Activity 1 👥 ▲2

Learning objective
• Follow instructions in positional and directional language

Resources

variety of obstacles that can be found in the classroom, such as tables and chairs (per class); mini whiteboard and pen (per learner – see variation)

What to do

• Set up a simple obstacle course in the classroom.
• Divide learners into groups; one learner volunteers and the rest give directions to guide them around the course.
• **[TWM.08]** If the volunteer learner completes the course easily, the directions given were obviously clear. If they struggle to complete the course,

encourage the group to share ideas for how to make their directions clearer and easier to follow. What words should they use? What could they change?
• Learners swap roles and repeat.

Variations

1 Challenge 1 learners should do this activity in an adult-led group if possible so that the adult can encourage and extend their use of directional language.

3 Learners draw an obstacle course on a mini whiteboard and direct a partner through the course using directional language.

Activity 2 👥 ▲2

Learning objective
• Follow instructions in positional and directional language

Resources

Resource sheet 18: Direction and position (per pair); mini whiteboard and pen (per pair)

What to do

• Give the following instructions to learners to follow either by saying them to their partner or writing them on the whiteboard (adapt the instructions as you see fit):
  – **Draw a tree near the house.**
  – **Draw a ball near the tree.**
  – **Draw a kite above the house.**
  – **Draw a person inside the circle.**
  – **Draw a tick under the square.**

Variations

1 and **3** Adapt the instructions for Challenge 1 and 3 learners. Give Challenge 1 learners three instructions only. Give Challenge 3 learners more complex instructions, for example, **Draw a ball near the house but nearer to the tree.**

▲2 Add left and right to the instructions, for example: **Draw a star to the left of the moon.**

Unit 26: Statistics (A)

Collins International Primary Maths Recommended Teaching and Learning Sequence: Term 2, Week 9

Learning objectives

Code	Learning objective
1Ss.01	Answer non-statistical questions (categorical data).
1Ss.02	Record, organise and represent categorical data using: - practical resources and drawings - lists and tables [- Venn and Carroll diagrams - block graphs and pictograms].
1Ss.03	Describe data, using familiar language including reference to more, less, most or least to answer non-statistical questions and discuss conclusions.

Unit overview

This unit introduces the concepts of collecting, sorting and drawing conclusions from data. The unit begins by refreshing learners' knowledge of sorting, allowing them to work on this skill, as it is required later in the unit when organising data that they have collected. It progresses to show learners how to collect data in a variety of simple ways and how to present it in lists and tables. The question 'What do we want to find out?' is asked throughout (e.g. We want to find out which fruit the Hungry Caterpillar ate most of/We want to find out which sport is most popular with learners in this class) to give learners purpose when collecting, sorting and recording their data.

Prerequisites for learning

Learners need to:
- have prior experience of sorting
- recognise what is the same/different by comparison
- be able to formulate a question with two or more potential answers.

Vocabulary

statistics, data, collect, sort, most, least, the same, different, list, table, row, column

Common difficulties and remediation

This unit requires a lot of speaking and listening, peer collaboration and a hands-on approach, as is the nature of data handling. Some learners who may otherwise be identified as being more able in other strands in mathematics may struggle in this area and find that it does not come as naturally to them as maths concepts, so be aware that they may need more support than usual in these lessons. On the other hand, the approach can build confidence in learners who find other aspects of maths, particularly mental maths a challenge. To challenge learners, try asking them to sort objects in a different way or look more deeply for the facts or what the data tells us, rather than giving them more statistics to process.

Help learners to make the connection that each lesson builds on a series of skills that are required together. These skills continue to be developed in Unit 27: Statistics (B).

Supporting language awareness

Introduce the key words at the start of each lesson.

Learners need to know what the key words mean and when to use them in a maths statement, but they do not need to know how to spell them in English.

Promoting Thinking and Working Mathematically

TWM.03 Conjecturing
Learners ask questions and collect data to find out the answers.

TWM.04 Characterising
Learners present data and statistics to provide evidence for an answer.

TWM.06 Classifying
Learners sort data according to different variables.

TWM.07 Critiquing
Learners learn a variety of ways to present data and begin to evaluate these regarding how easy they are to use and interpret.

Success criteria

Learners can:
- sort information to find an answer
- explain the information they have sorted
- collect simple data
- present data in lists and tables
- organise information to answer a question.

Unit 26 Statistics (A)

Statistics and Probability – Statistics

Lesson 1: **Sorting data**

Learning objectives

Code	Learning objective
1Ss.01	Answer non-statistical questions (categorical data).
1Ss.02	Record, organise and represent categorical data using: - practical resources and drawings [- lists and tables - Venn and Carroll diagrams - block graphs and pictograms].
1Ss.03	Describe data, using familiar language including reference to more, less, most or least to answer non-statistical questions and discuss conclusions.

Resources

crayons of mixed sizes and colours (per pair)

Revise

Choose an activity from Unit 26: *Statistics (A)* in the Revise activities.

Teach [SB] [TWM.06]

- Discuss the picture in the Student's Book. Ask: **How have the buttons been sorted?** and **[T&T] Can you think of any other ways that we could sort the buttons?** Take suggestions.
- Explain to learners that how we decide to sort the buttons depends on what we want to find out about them. If we want to find out how many buttons have two holes and how many buttons have four holes, we can sort them according to how many holes they have. If we want to find out how many of each colour button we have, we could sort them according to colour, and so on.
- Tell learners that when the buttons have been sorted, we can count the buttons in each group to give us the answers that we were looking for. Count the buttons in each group in the Student's Book and say: **There are six buttons with two holes and five buttons with four holes.** Explain that this tells us that we have more buttons with two holes. It answers the question, 'Do most of our buttons have two or four holes?'
- **[TWM.06]** Repeat this process by sorting shapes on the **Shape tool**. Ask learners to formulate a question (for example, 'What is the most common shape in this group?') to dictate how the shapes will be sorted. Invite the learners to sort the shapes, and discuss the outcome and whether it answers the question.
- Discuss the Guided practice example in the Student's Book.

Practise [WB]

- Workbook

Title: Sorting data

Page: 106

- Refer to Activity 1 from the Additional practice activities.

Apply 👥 🖥️

- Display **Slide 1**.
- Give pairs of learners some crayons of mixed colours and sizes. Learners think of a question about the crayons, to which they want to find the answer, and devise their own sorting rules based on this.

Review [TWM.06]

- **[TWM.06]** Share and compare results of the **Apply** activity as a whole class. Ask learners to discuss how they sorted their crayons, whether there were any crayons that did not fit into either group clearly and what the outcomes were. Consider the results of pairs with the same sorting rule but different data.
- Now sort one set of crayons with learners according to different questions and sorting rules.
- **[T&T]** Ask: **What do we know about the crayons now?** Make a list of all the facts that you have learned from sorting them in different ways.

Assessment for learning

- What question would you like to find the answer to about these objects?
- What sorting rule will you need to use to find the answer?
- Does this object belong in this group? Why/why not?
- What has sorting these objects told us?

Same day intervention

Enrichment

- Ask learners to consider whether the data has told us 'why' as well as 'how many', for example: **Do we know why there are more red crayons than any other colour?**

Lesson 2: Collecting data

Learning objectives

Code	Learning objective
1Ss.01	Answer non-statistical questions (categorical data).
1Ss.02	Record, organise and represent categorical data using: - practical resources and drawings [- lists and tables - Venn and Carroll diagrams - block graphs and pictograms].
1Ss.03	Describe data, using familiar language including reference to more, less, most or least to answer non-statistical questions and discuss conclusions.

Resources

trays of red, blue, green and yellow paint (per class); large roll of paper or two very large sheets of paper for a display (per class); mini whiteboard and pen (per learner)

Revise

Use the activity *Would you rather…* from Unit 26: *Statistics (A)* in the Revise activities.

Teach SB

- Direct learners to the picture in the Student's Book. Tell learners that the children in the picture have been asked what their favourite fruit is Ask: **What can this data tell us?** Elicit that from the data shown we can work out the most popular fruit for that group of children. Ask: **What is the most popular fruit? What is the least popular fruit? How can you tell?**

- Tell learners that they will be collecting data to answer questions today. Tell them that you would like to find out which is the most popular colour with the class. **[T&T]** Ask: **How could we collect the data?** Take suggestions and add some of your own (e.g. 'Learners could put their hand up when you say their favourite colour and we could count the hands'/'Learners could colour a piece of paper with their favourite colour and hold it up'/'Learners who like blue could stand in one corner of the room, red in a different corner' and so on).

- Provide trays of red, blue, green and yellow paint and ask learners to choose which of those colours is their favourite. Attach a large sheet of paper to the wall (a roll of wallpaper is ideal) and ask each learner to dip one hand into their favourite colour and make one handprint on the paper. 🖐 Ask: **Why must you only make one handprint?** Elicit that the data would be wrong if some people made more than one handprint because it would look as though more people had chosen that colour.

- Note: You will return to this in **Apply** and **Review** and in Lesson 4.

- Discuss the Guided practice example in the Student's Book.

Practise WB 🖵

- Workbook

Title: Collecting data

Page: 107

- Refer to Activity 2 from the Additional practice activities

Apply 🧑 🖵

- Display **Slide 1.** Learners count up the handprints for each colour on the favourite colours display made in **Teach**. They record on their mini whiteboards how many of each colour handprint there are.

- They tick the most popular colour and circle the least popular colour.

Review

- Discuss the results of the **Teach/Apply** activity.

- Did all learners get the same answers? Did anybody count inaccurately?

- Discuss the problems with the way the data is presented (e.g. it isn't organised clearly and all the different coloured handprints are mixed up).

- Put a second large piece of paper on the wall and repeat the handprint activity, but this time divide the paper into clear areas and label them **red**, **blue**, **green** and **yellow**. Learners must put their handprint in the designated area.

- Count up the handprints for each colour with learners to show how much easier it is to do so when the data is clearly organised.

Assessment for learning

- What is the most/least popular colour in the class?
- What does this data tell us?
- How could you organise it to make it clearer?
- How would you collect data to answer the question?

Same day intervention

Support

- Ask learners to count the different colours as a group, to ensure one-to-one correspondence.

Statistics and Probability – Statistics

Unit 26 Statistics (A)

Statistics and Probability – Statistics

Lesson 3: **Using lists**

Learning objectives

Code	Learning objective
1Ss.01	Answer non-statistical questions (categorical data).
1Ss.02	Record, organise and represent categorical data using: - practical resources and drawings - lists [and tables - Venn and Carroll diagrams - block graphs and pictograms].
1Ss.03	Describe data, using familiar language including reference to more, less, most or least to answer non-statistical questions and discuss conclusions.

Resources

mini whiteboard and pen (per group)

Revise

Choose an activity from Unit 26: *Statistics (A)* in the Revise activities.

Teach [SB]

- Discuss the picture in the Student's Book. Explain that it is a shopping list. A list is a way to show information. This list tells us how many of each fruit and vegetable to buy.
- Point out the way that the list is organised. Each fruit and vegetable is on a new row so that we don't get confused when we look at it. The number of each fruit or vegetable that we need to buy is written in numbers rather than words to make it quicker and easier to check.
- Tell learners that lists can be used to organise information or data. Ask: **Can you see any lists in the classroom?** Discuss the lists that learners find.
- Say: **I want to make a salad.** With learners, write a shopping list for ingredients. Give the list a heading ('Shopping list' or 'Salad ingredients') and explain that the heading tells us what the list is about.
- Discuss the Guided practice example in the Student's Book.

Practise [WB] 🖵

- Workbook

Title: Using lists

Page: 108

- Refer to Activity 1 or 2 from the Additional practice activities.

Apply 👥🖵 [TWM.03]

- Display **Slide 1**.
- Ask learners to find out the activities that everybody in their group does after school (such as dancing, football, swimming, singing).
- Learners write this information as a list on a mini whiteboard, showing each activity and the number of people who do each of the activities.

- [TWM.03] Ask learners to discuss the data outcomes as a group so that they can feed back to the class in **Review**. What is the most and least popular activity? Is there any other information that the data tells them?

Review [TWM.04]

- Ask each group to share the lists that they made in the **Apply** activity. Discuss the most and least popular activities for each group, pointing out how easy it is to work this out because the lists are clearly set out and easy to read.
- Use the group lists to make a big class list of after-school activities on the board, with the help of volunteer learners.
- [TWM.04] Discuss the data from the list and whether it has answered the question 'What is the most popular after-school activity for this class?'

Assessment for learning

- Why does a list need a title?
- What is the information that you are organising?
- What is the question?
- Show me how you would set out a list of favourite colours.

Same day intervention

Support

- Give learners scaffolded lists to fill in, indicating where the heading should go and with bullet points for each item.

Enrichment

- Encourage learners to make their own lists on blank paper so that they practise their organisational skills when setting these out. These learners may be able to make longer lists with more categories.

Lesson 4: **Using tables**

Learning objectives

Code	Learning objective
1Ss.01	Answer non-statistical questions (categorical data).
1Ss.02	Record, organise and represent categorical data using: - practical resources and drawings - [lists and] tables - Venn and Carroll diagrams - block graphs and pictograms].
1Ss.03	Describe data, using familiar language including reference to more, less, most or least to answer non-statistical questions and discuss conclusions.

Resources

Resource sheet 19: Table (per group and then per pair); the handprint data collection display from Lesson 2 (per class)

Revise

Choose an activity from Unit 26: *Statistics (A)* in the Revise activities.

Teach [SB] [TWM.03]

- Look at the picture in the Student's book. Ask: **What data do you think is shown here?** Take suggestions and elicit that it shows how many of each type of bike have been sold. Ask: **Is it a list?** Discuss and explain that it is similar to a list but is called a table and is another way of organising data.
- Show learners the rows and columns and their headings. Explain that every new piece of data is presented on a new row.
- Ask: **Which type of bike has sold the most?**
- [TWM.03] Ask learners to suggest some data that you could record in a table. Choose one of the suggestions (e.g. favourite foods or mode of transport to school), collect the data using one of the methods from Lessons 1 and 2 and present it in a table with learners' help.
- Discuss the Guided practice example in the Student's Book.

Practise [WB] 🖥️

- Workbook

Title: Using tables

Page: 109

- Refer to Activity 1 or 2 from the Additional practice activities.

Apply 👥 🖥️ [TWM.03/04/07]

- Display **Slide 1**.
- Give each group a copy of Resource sheet 19: Table.
- Learners find out how many siblings everybody in their group has and record the data in a table. Ask the learners to copy the table header from the slide 'Number of brothers and sisters'.

Review [TWM.07]

- Refer back to the handprint data class display that you made in Lesson 2.
- Working in pairs, ask learners to put the data from this into a table (use photocopies of Resource sheet 19: Table), using the title 'Paint colour' for the left - hand column and 'Favourite paint colour' for the title of the table.
- Make your own version of the table on the board and talk through each feature of the table to ensure that learners are confident with how to input data into a table.
- [TWM.07] Ask: **How has presenting the data in a table made it easier to read?** Discuss this as a class, comparing this to the methods used to collect and present data in previous lessons. Encourage learners to talk about what the data tells them.

Assessment for learning

- Where are the rows in this table?
- Point to the column heading. What does it tell us?
- Look at the table. Can you tell me how many people have two brothers and sisters?
- Where do you write the numbers on the table?

Same day intervention
Support

- Give adult support to learners when they first attempt to put data into a table, to remind them of where each piece of information goes.

Statistics and Probability – Statistics

Additional practice activities

Activity 1

Learning objectives
• Learn the reasoning behind investigating and presenting data without having to collect it
• Develop the skills to work on an answer by a process of steps, using peer support and assessment

Resources
copies of Resource sheet 20: Bugs!, enough for four cards per learner (per learner)

What to do
• Working in pairs, each pair has eight cards, shared equally and turned face down. Learners then turn over their cards and study the patterns on each bug.
• Lesson 1: Pairs sort their cards into groups to find out how many of each different type of bug there are, first sorting their own four, then sorting all eight.

Variations
1 Give Challenge 1 learners 4 bugs only or limit the amount of patterns in their 8 bugs.

2 Lessons 3 or 4: Pairs work together to make a list or table to show how many of each type of bug they have.

Activity 2 and

Learning objectives
• Collect data from a book
• Present data in a list or table

Resources 🖵
copy of *The Very Hungry Caterpillar* by Eric Carle (per class); plastic counters in a variety of colours (per pair); Additional practice activity 2 slide

What to do
• Read the story of *The Very Hungry Caterpillar* to the class and display Additional practice activity 2 slide showing images of all the fruit he eats on the board (one apple, two pears, and so on).
• Each pair counts the amount of each fruit that the caterpillar eats (one apple, two pears, three plums, four strawberries and five oranges). To help them keep track of these, they could use counters in different colours to represent each fruit (for example, four red counters = four strawberries).
• For Lesson 2 the activity can stop here, with learners counting up their counters and sharing with the class which fruit the caterpillar ate the most and least of. For lessons 3 and 4, learners can present the information in list or table format.

• Ask: **Why do you think he ate more oranges than apples?** Take suggestions for conclusions that learners draw from the data, but ensure that you remind them that we only know for sure that he ate more oranges than apples and, although we can try to work out why (for example, he likes oranges better than apples/he was more hungry that day), the data can't tell us why for certain.

Variation
2 Use any fiction book that presents data that can be analysed or, if you do not have one available, tell the class a story about an animal that ate lots of different foods. Use props (such as fake fruit) to give learners something concrete to count rather than just an abstract concept of how many of each food type there were.

Unit 27: Statistics (B)

Collins International Primary Maths Recommended Teaching and Learning Sequence: Term 3, Week 7

Learning objectives

Code	Learning objective
1Ss.01	Answer non-statistical questions (categorical data).
1Ss.02	Record, organise and represent categorical data using: [- practical resources and drawings] - lists and tables - Venn and Carroll diagrams - block graphs and pictograms.
1Ss.03	Describe data, using familiar language including reference to more, less, most or least to answer non-statistical questions and discuss conclusions.

Unit overview

This unit consolidates learners' understanding of the concepts of collecting, sorting and drawing conclusions from data. In this unit the process of handling data moves on to organising the collected data by presenting it in a simple graph or diagram. The unit begins with learners using their sorting skills to make simple Venn and Carroll diagrams, requiring them to sort by one criterion. Learners build on their knowledge of using lists and tables to make a pictogram. Finally, they learn how to present data in a block graph. By the end of the unit learners are able to collect, organise and present their data in a variety of ways.

Prerequisites for learning

Learners need to:
• be able to sort by one criterion
• have experience of collecting data
• have experience of organising data in a list or table
• recognise curved and straight lines.

Vocabulary

statistics, data, sorting, sorting rule, the same, different, Venn diagram, Carroll diagram, pictogram, block graph, title

Common difficulties and remediation

When working with Venn diagrams, some learners may not understand that objects or data that are not members of the set need to be placed outside of the diagram. Learners who experience these difficulties require further practice with sorting objects and discussing rules for placing objects in physical sets.

Supporting language awareness

Display the key words for each lesson and discuss them with learners. Use and refer to these words throughout the lesson and encourage learners to use the words too, prompting them when necessary.

Promoting Thinking and Working Mathematically

TWM.03 Conjecturing
Learners ask questions and collect data to find out the answers.

TWM.04 Characterising
Learners present data and statistics to provide evidence for an answer.

TWM.06 Classifying
Learners sort data according to different variables.

TWM.07 Critiquing
Learners learn a variety of ways to present data and begin to evaluate these regarding how easy they are to use and interpret.

Success criteria

Learners can:
• sort and organise information to find an answer
• explain the information they have sorted
• collect simple data
• present data in Venn diagrams, Carroll diagrams, pictograms and block graphs.

Unit 27 Statistics (B)

Statistics and Probability – Statistics

Lesson 1: **Venn diagrams**

Learning objectives

Code	Learning objective
1Ss.01	Answer non-statistical questions (categorical data).
1Ss.02	Record, organise and represent categorical data using: [- practical resources and drawings] [- lists and tables] - Venn [and Carroll] diagrams [- block graphs and pictograms].
1Ss.03	Describe data, using familiar language including reference to more, less, most or least to answer non-statistical questions and discuss conclusions.

Resources

ten small objects of different types and colours, e.g. counters, pencils, small toys, rubbers, cubes (per pair/group); sorting hoop (per pair/group); approximately ten 2D shapes of different sizes and colours (per class)

Revise

Use the activity *Sorting with a Venn diagram* from Unit 27: *Statistics (B)* in the Revise activities.

Teach 〔SB〕 〔📊〕

- Remind learners that a set is a collection of things in no particular order. Ask learners to form a set according to different criteria, such as boys/girls/hair colour/shoe colour.
- Direct learners to the picture in the Student's Book. Explain that it is a sorting diagram and that we call it a Venn diagram. Explain that we can use Venn diagrams to show objects in a set. Say: **This Venn diagram shows yellow and blue bugs. The blue bugs do not belong in the yellow set, so they cannot go in the ring.**
- Show learners the **Venn diagram tool**. Put a selection of 2D and 3D shapes on the screen and label the circle '2D'. Ask learners to help you to sort the 2D shapes into the circle. 〔🔊〕 Ask: **Can the cube go in the circle? Why/why not?**
- Discuss the Guided practice example in the Student's Book.

Practise 〔WB〕

- Workbook

Title: Venn diagrams

Page: 110

- Refer to Activity 1 from the Additional practice activities.

Apply 〔👥〕 or 〔👨‍👩‍👧〕 〔💻〕 [TWM.06]

- Display **Slide 1**.
- Ask groups or pairs of learners to collect ten different small objects (such as cubes, counters, small toys, pencils, rubbers) from around the classroom. Give each group one sorting hoop.

- [TWM.06] The group chooses one colour (such as red) and puts all of the red objects into the circle.
- Groups feed back to the class about how they sorted their objects, other ways they could have potentially sorted the objects and what question sorting the objects has answered (e.g. How many red objects are there/are there more red objects than any other colour). Remind learners that they have been *classifying* the objects.

Review

- Make a pile of 2D shapes in all colours and different sizes.
- Make a Venn diagram with a sorting hoop.
- Take instructions from the learners to place the shapes correctly, using the criteria of their choice.

Assessment for learning

- What are we sorting?
- What is different/the same?
- Where does this object/information go?
- Does this object belong in the circle?

Same day intervention
Support

- Give learners objects that are easy to sort (for example, red and blue cubes when red cubes are to be sorted into the circle) until they are confident about sorting. Then add in extra items (for example, green and yellow cubes) to reinforce that even when other colours are added, only the red cubes can go into the red cube set.

Lesson 2: Carroll diagrams

Learning objectives

Code	Learning objective
1Ss.01	Answer non-statistical questions (categorical data).
1Ss.02	Record, organise and represent categorical data using: [- practical resources and drawings] [- lists and tables] - [Venn and] Carroll diagrams [- block graphs and pictograms].
1Ss.03	Describe data, using familiar language including reference to more, less, most or least to answer non-statistical questions and discuss conclusions.

Resources

Resource sheet 21: Carroll diagram (per learner); six paper cut-out 2D shapes of different shapes, sizes and colours (per learner); skipping ropes (per class)

Revise

Use the activity *Sorting with a Venn diagram* (variation) from Unit 27: *Statistics (B)* in the Revise activities.

Teach [SB] [◫]

- Direct learners to the picture in the Student's Book. Tell them that the chart is called a Carroll diagram and can be used to sort objects that do or do not belong to different groups, as in a Venn diagram. Explain that its sorting rules are always yes or no questions. Referring to the two groups on the Carroll diagram, ask: **How have the buttons been sorted?** Make it clear that the sorting rule asks 'Are the buttons blue?' If the answer is yes, they go in the blue buttons group. If the answer is no, they go in the not blue buttons group.
- [T&T] Ask: **What other things could we sort using a Carroll diagram?** Take suggestions and use the **Carroll diagram tool** to make a Carroll diagram for some of the suggestions, directed by learners.
- Discuss the Guided practice example in the Student's Book.

Practise [WB]

- Workbook

Title: Carroll diagrams

Page: 111

- Refer to Activity 1 (variation) from the Additional practice activities.

Apply ▪ ▯ [TWM.06]

- Display **Slide 1**.
- Give each learner a set of six paper cut-out 2D shapes in different shapes, sizes and colours and a copy of Resource sheet 21: Carroll diagram.
- **[TWM.06]** Learners use the sorting rule of their choice to make their own Carroll diagrams.
- Each learner shares their Carroll diagram with another learner and they discuss their sorting rules.

Review

- Make a human Carroll diagram.
- Using ten class members and, led by the rest of the class, use skipping ropes to mark two sections and sort the learners by, for example, blonde hair/not blonde hair or boys/not boys.
- 🗩 Ask: **Can anyone be left out of the diagram? Why?**

Assessment for learning

- What is the information?
- What is the sorting rule?
- Does this object belong in this group?

Same day intervention

Support

- Learners work in pairs or small groups for the **Apply** activity.

Statistics and Probability – Statistics

Lesson 3: **Pictograms**

Statistics and Probability – Statistics

Learning objectives

Code	Learning objective
1Ss.01	Answer non-statistical questions (categorical data).
1Ss.02	Record, organise and represent categorical data using: [- practical resources and drawings] - lists and tables - [Venn and Carroll diagrams] - [block graphs and] pictograms.
1Ss.03	Describe data, using familiar language including reference to more, less, most or least to answer non-statistical questions and discuss conclusions.

Resources

coloured pencils (per learner) (for the Workbook); stones, twigs and leaves (ten or fewer of each item) (per group); large piece of paper (per group); marker pen (per group); paper circle (per learner)

Revise

Use the activity *Human pictogram* from Unit 27: *Statistics (B)* in the Revise activities.

Teach [SB] [📊] [TWM.04]

- Direct learners to the picture in the Student's Book. **[T&T]** Ask: **What do you think this chart is showing?** Elicit that it shows the different types of shoes worn in Class 1. Point out that the title of the chart tells us this information. Ask: **Which type of shoe did most children wear? How do you know?**

- Introduce the word 'pictogram' to learners. Explain that a pictogram is a way of showing information. Each pair of shoes shown in the pictogram in the Student's Book represents one child. Demonstrate that by counting all of the pairs of shoes you can find out how many children there were in the class that day in total, and by counting the types of shoe separately, you can find out how many children were wearing each type of shoe. Ask questions that require the learners to read and interpret the data in the pictogram in the Student's Book.

- **[TWM.04]** Now show the **Pictogram tool** and create a pictogram that shows how many days in each of four weeks had sunshine. Use the title 'Sunny days' and the headings 'Week 1', 'Week 2', 'Week 3' and 'Week 4' for the rows. Use a picture of the sun to represent one sunny day, filling in a different number of sunny days in each week. 🗣 Ask: **Which week was the sunniest?** Model how to extract the information, then write which week was the sunniest underneath the chart.

- Explain that representing data allows us to share lots of information and read and understand it quickly.

- Discuss the Guided practice example in the Student's Book.

Practise [WB] [TWM.03/06]

- Workbook

Title: Pictograms

Page: 112
- Refer to Activity 2 (variation) from the Additional practice activities.

Apply [👥] [💻] [TWM.04/06/07]

- Display **Slide 1**.

- Provide groups of learners with stones, twigs and leaves, a large sheet of paper and a marker pen.

- Ask learners to arrange the objects into the rows and columns of a pictogram to show how many of each type of object they have.

- Groups then compare their work.

Review [TWM.04]

- Give each learner a paper circle and ask them to draw their face and write their name on their circle.

- Work together to make a pictogram to show the favourite fruit of learners in the class (or any other variable that you feel is appropriate). Use each learner's face as one vote for that fruit.

- 🗣 Ask: **Which is the least popular fruit? How do you know?**

Assessment for learning

- What does this pictogram show? How can you tell?
- What does each row stand for (represent)?
- Does each pair of shoes mean one person? Or something else?
- What does each paper face equal?

Same day intervention
Enrichment

- Ask learners to collect the data to create a pictogram rather than giving them random data or data that you have already collected.

Lesson 4: **Block graphs**

Learning objectives

Code	Learning objective
1Ss.01	Answer non-statistical questions (categorical data).
1Ss.02	Record, organise and represent categorical data using: [- practical resources and drawings] - lists and tables [- Venn and Carroll diagrams] - block graphs [and pictograms].
1Ss.03	Describe data, using familiar language including reference to more, less, most or least to answer non-statistical questions and discuss conclusions.

Resources

up to ten building blocks or interlocking cubes in green, red, yellow and blue (per class); coloured pencils (per learner) (for the Workbook); Resource sheet 22: Block graph (per group); mini whiteboard and pen (per group); coloured pencils (per group)

Revise

Use the activity *Block graph* from Unit 27: *Statistics (B)* in the Revise activities.

Teach [SB] 🖥

- Direct learners to the picture in the Student's Book. Explain that it shows a block graph. Tell learners that a block graph is another way of showing information. Explain that each block represents one vote. In the case of the block graph in the Student's Book, each block represents one person who likes that flavour of yoghurt. **[T&T]** Ask: **Which are the three most popular yoghurt flavours? How can you tell?** Ask other questions that require the learners to read and interpret the data in the block graph.

- Display **Slide 1**. Explain that the table on the left shows a list of building blocks sorted by colour.

- Provide pairs or groups with building blocks or interlocking cubes in green, yellow, red and blue and ask them to work together to make a block graph from the blocks/cubes to represent the data.

- When pairs/groups have made their block graph, work with the class, to complete the block graph on the slide.

- Model comparison statements such as: **There are more blue blocks than any other colour** and **There are three more green bricks than there are yellow bricks.** Learners make comparison statements to each other about their block graphs.

- Discuss the Guided practice example in the Student's Book.

Practise [WB]

- Workbook

Title: Block graphs

Page: 113

- Refer to Activity 2 from the Additional practice activities.

Apply 👥🖥 [TWM.04]

- Display **Slide 2**.

- Give groups of learners a copy of Resource sheet 22: Block graph.

- Learners make a list of the pets that everyone in their group has on their mini whiteboard. They then use this data to make a block graph with on the resource sheet to show this information.

- Discuss and compare completed graphs as a class. **Talk about the questions that the graphs answer.** Say: **A graph shows evidence to answer a question. That is *convincing*.**

Review [TWM.04]

- Write on the board: Zoo animals: 10 monkeys, 5 parrots, 8 lions, 2 tigers.

- Help learners to build a 3D block graph using classroom building blocks to represent this data.

- Discuss the graph as a class: **Is the data clear? Which type of animal does the zoo have most/least of? What question has the data answered? How else could you have presented the data?**

Assessment for learning

- What information are we organising?
- How does the graph show how many?
- What does each square/block equal?
- How many… are there? How do you know?

Same day intervention

Enrichment

- Give learners jobs such as making labels for the graphs and coming up with appropriate itles.

Statistics and Probability – Statistics

Additional practice activities

Activity 1

Learning objective
• Make Venn or Carroll diagrams and solve a given problem by sorting and organising

Resources
hoop (per pair); six white pebbles and four black pebbles (or any similar resources with one variable) (per pair); black felt-tipped pen (per pair); 20 pebbles in both colours and six skipping ropes (per group) (for variation)

What to do
• Take the learners outside.
• [TWM.06] Pairs of learners make a Venn diagram using one hoop. Give pairs ten pebbles – six white and four black. Learners sort them into two groups according to colour.

• Answer any question with another, for example, for the question 'Do I put that pebble there?' answer with: **Is that pebble different from the ones in the circle?** Check results: black pebbles (4), white pebbles (6).
• As the unit progresses, vary the quantities of pebble type and spend more time peer-reviewing the diagrams.

Variation
Use skipping ropes to make the shape of a Carroll diagram. Give it the headings 'black'/'not black'. Learners sort the pebbles according to the headings.

Activity 2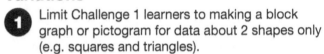

Learning objectives
• Represent data in a block graph
• Represent data in a pictogram (variation)

Resources
interlocking cubes in red, yellow, blue and green – approximately seven of each colour (per pair); Resource sheet 22: Block graph (per pair); paper and scissors (per pair) (for variation); coloured pencils (per pair) (for variation)

What to do
• On the board, write a list showing a number of squares, rectangles, circles and triangles (no number greater than 7). For example, 2 squares, 5 rectangles, 6 circles, 4 triangles.
• Learners make a block graph to represent this data by placing the interlocking cubes on the graph grid resource sheet, for example: 2 red interlocking cubes to represent the 2 squares, 5 yellow interlocking cubes to represent the 5 rectangles.

• Remind learners that one interlocking cube equals one number, so they must read the data carefully and check their numbers of cubes for each column.
• Ensure that learners give their graph a title and label each column with what it represents.

Variations
1 Limit Challenge 1 learners to making a block graph or pictogram for data about 2 shapes only (e.g. squares and triangles).

2 Use the block graph resource sheet and cut out paper shapes to create a pictogram to represent this data.

2 Colour the blocks on the block graph resource sheet rather than using interlocking cubes.

3 Ask Challenge 3 learners to make their own data about shapes by taking handfuls of shapes from a tray rather than using the data presented on the board.

Statistics and Probability – Statistics

0–20 number cards

0	**1**	**2**	**3**
4	**5**	**6**	**7**
8	**9**	**10**	**11**
12	**13**	**14**	**15**
16	**17**	**18**	**19**
20			

Ten frames

Number patterns

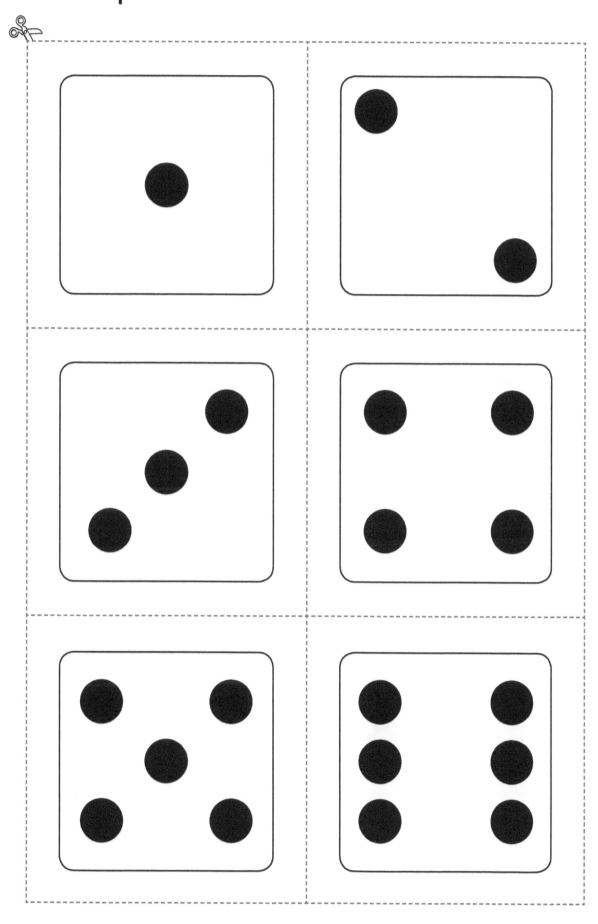

Number patterns

0–10 number track

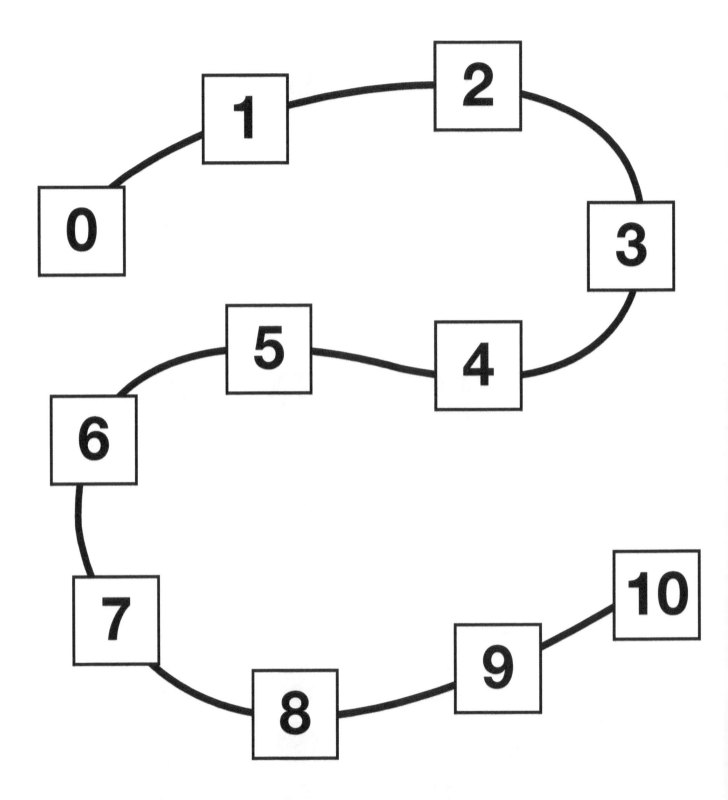

0–20 number track

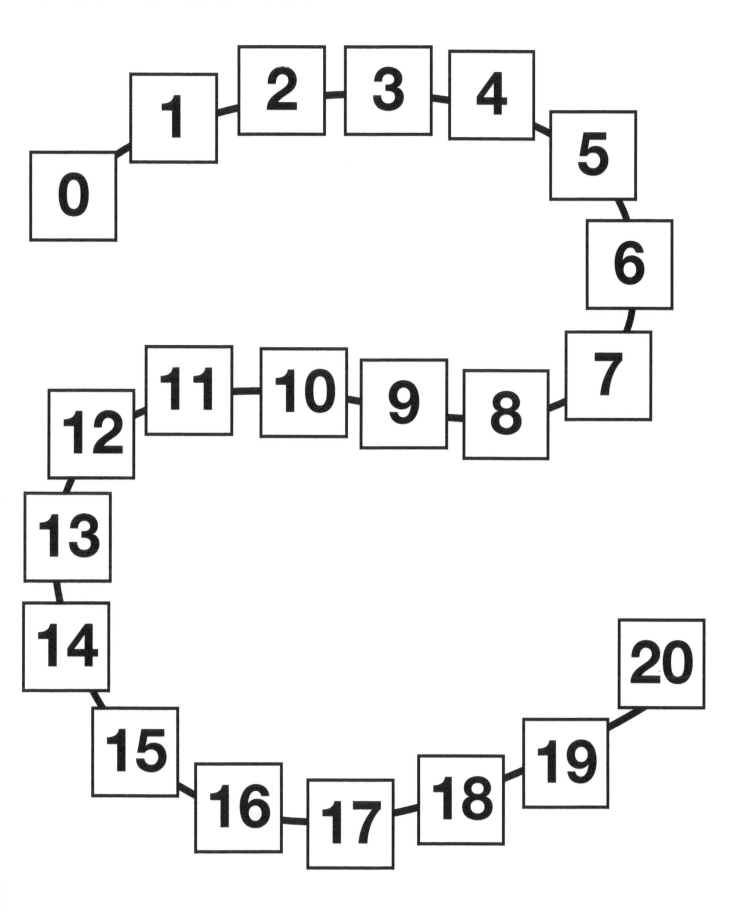

0–20 number word cards

zero	one	two
three	four	five
six	seven	eight
nine	ten	eleven
twelve	thirteen	fourteen
fifteen	sixteen	seventeen
eighteen	nineteen	twenty

Snakes and ladders game

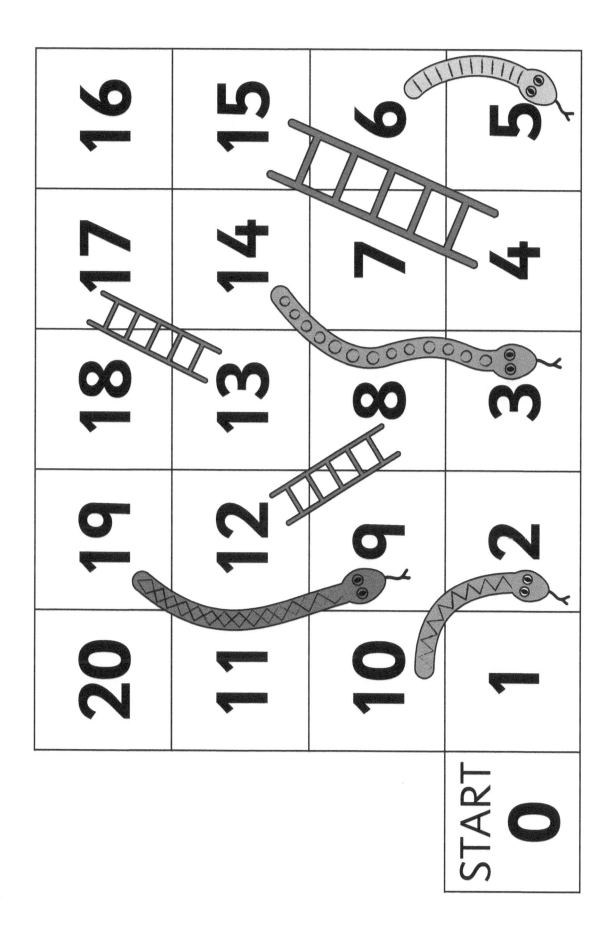

1–5 spinner

How to use the spinner

Hold the paper clip in the centre of the spinner using the pencil and gently flick the paper clip with your finger to make it spin.

You will need:

- pencil
- paper clip

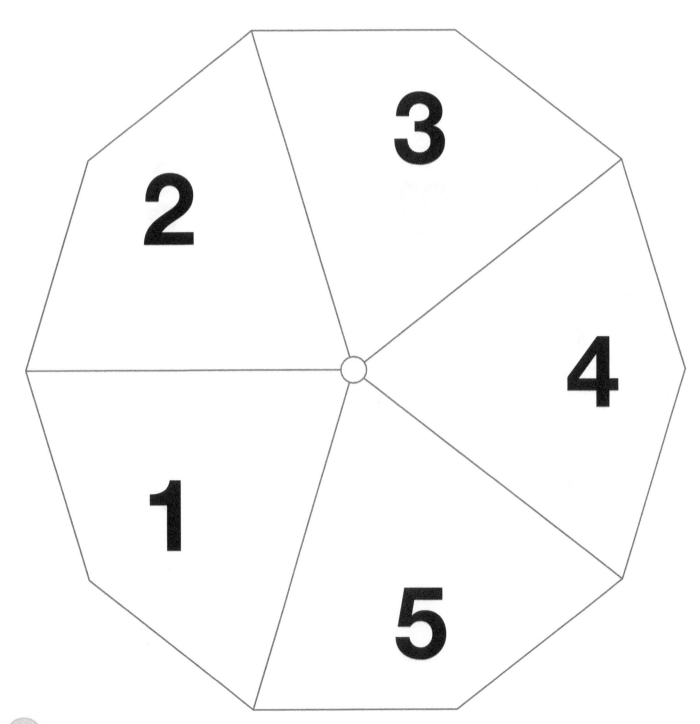

1–6 spinner

How to use the spinner

Hold the paper clip in the centre of the spinner using the pencil and gently flick the paper clip with your finger to make it spin.

You will need:

- pencil
- paper clip

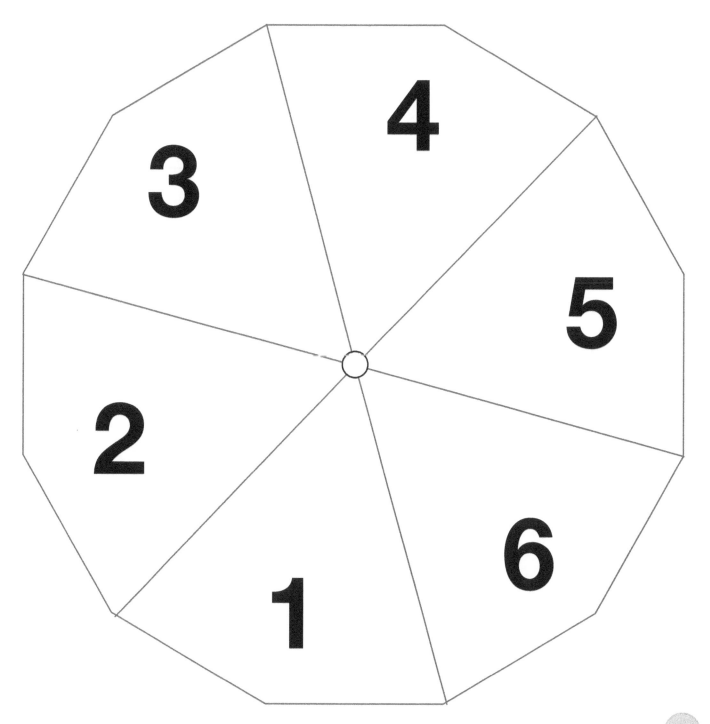

Part–whole diagram

0–10 number line

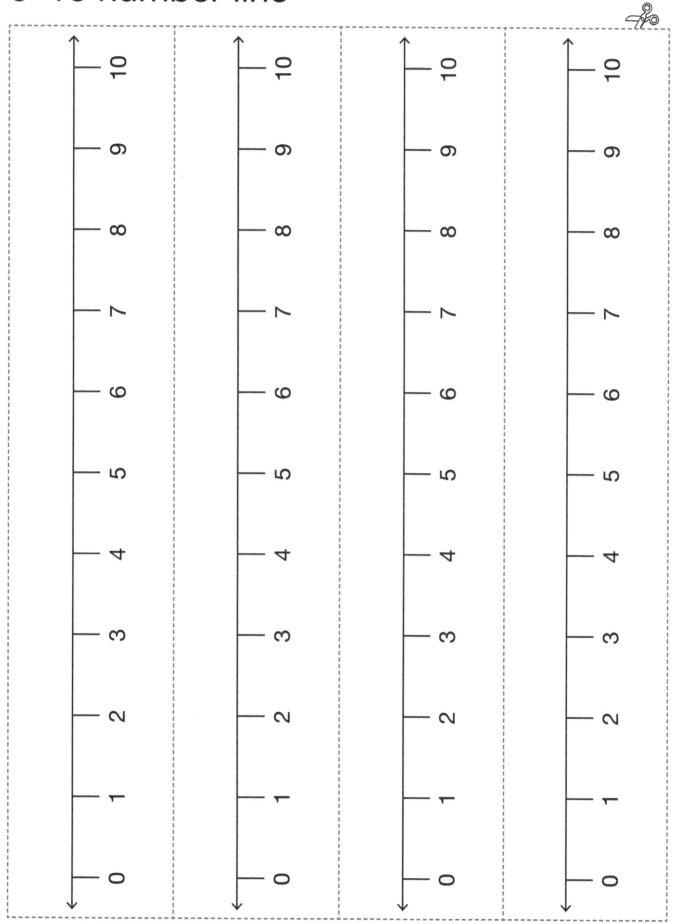

0–20 number line

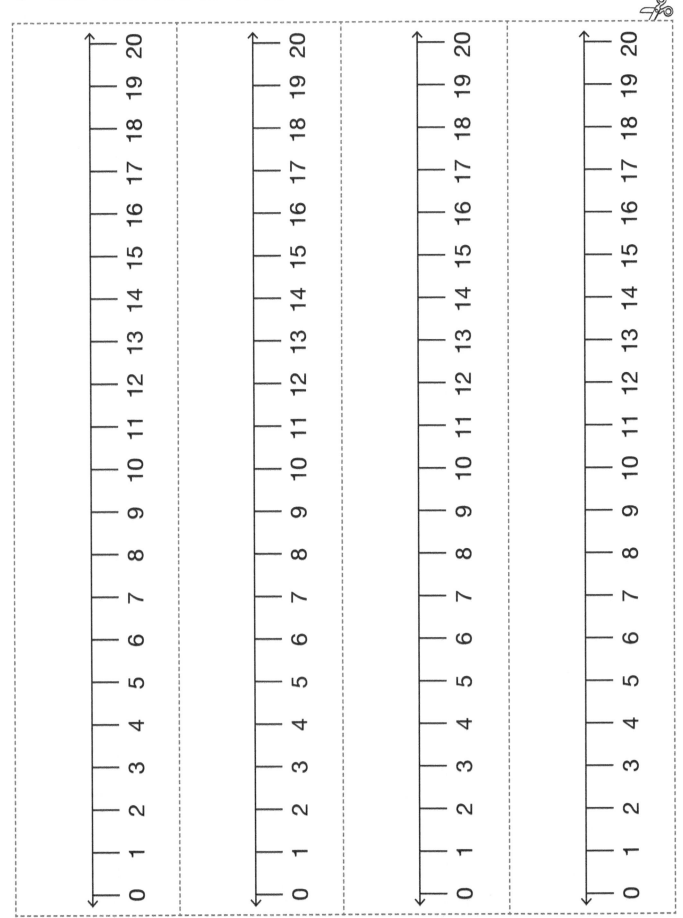

Operations cards

+	+	+	+
−	−	−	−
=	=	=	=

+	−	=

Ordinal number cards

5th

10th

4th

9th

3rd

8th

2nd

7th

1st

6th

Tree

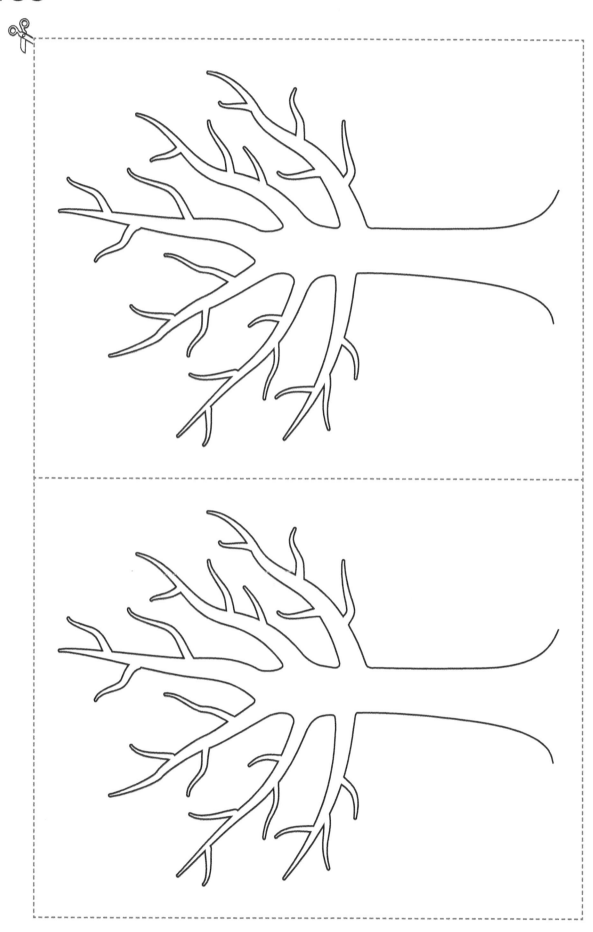

Months of the year cards

✂

January	February	March
April	May	June
July	August	September
October	November	December

Direction cards

up	down	left	right
forwards	backwards	around	under
below	on top of	in front of	behind
next to	over	inside	outside

Direction and position

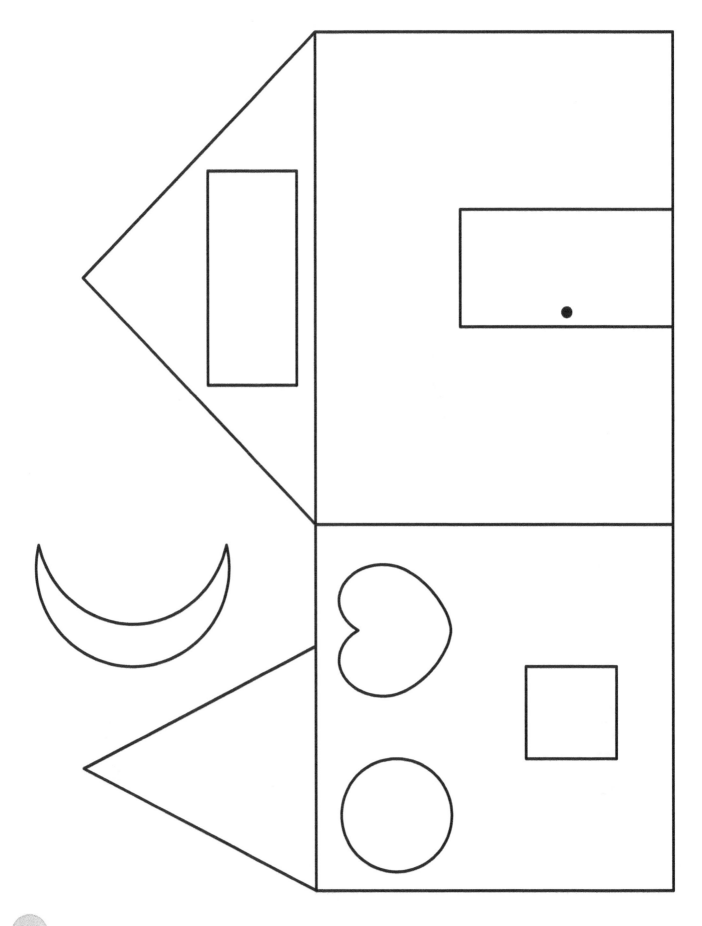

Table

	Total

Bugs!

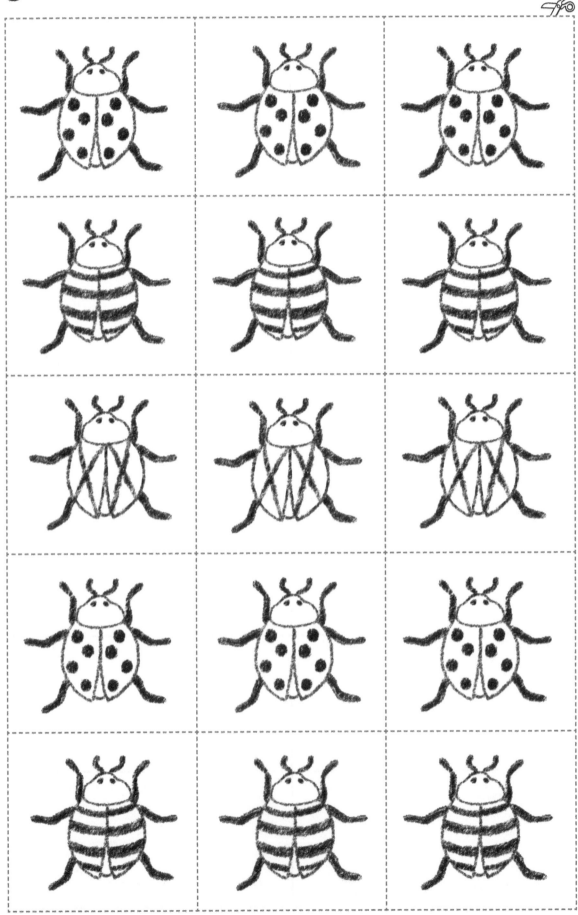

Carroll diagram

Block graph

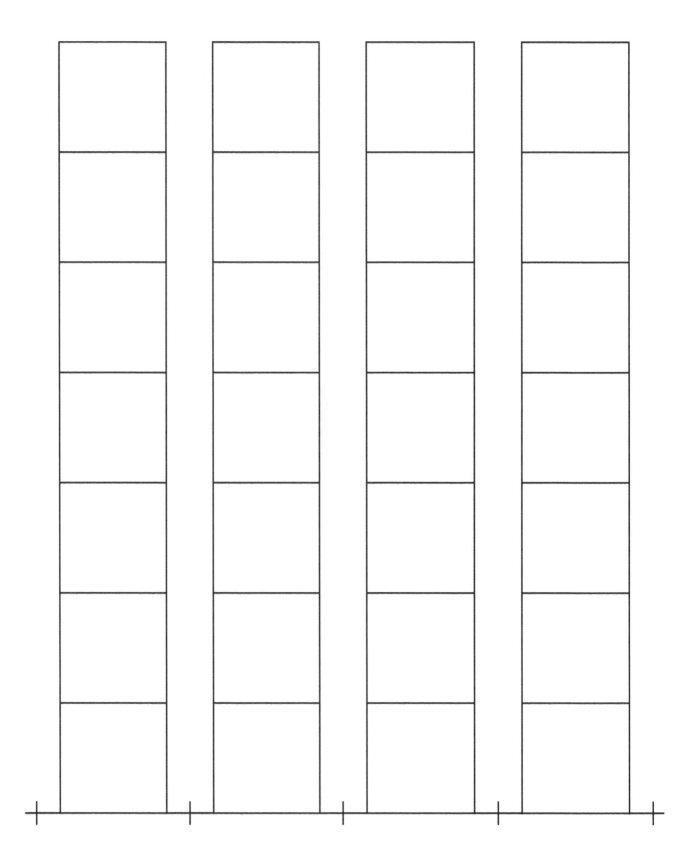

Days of the week cards

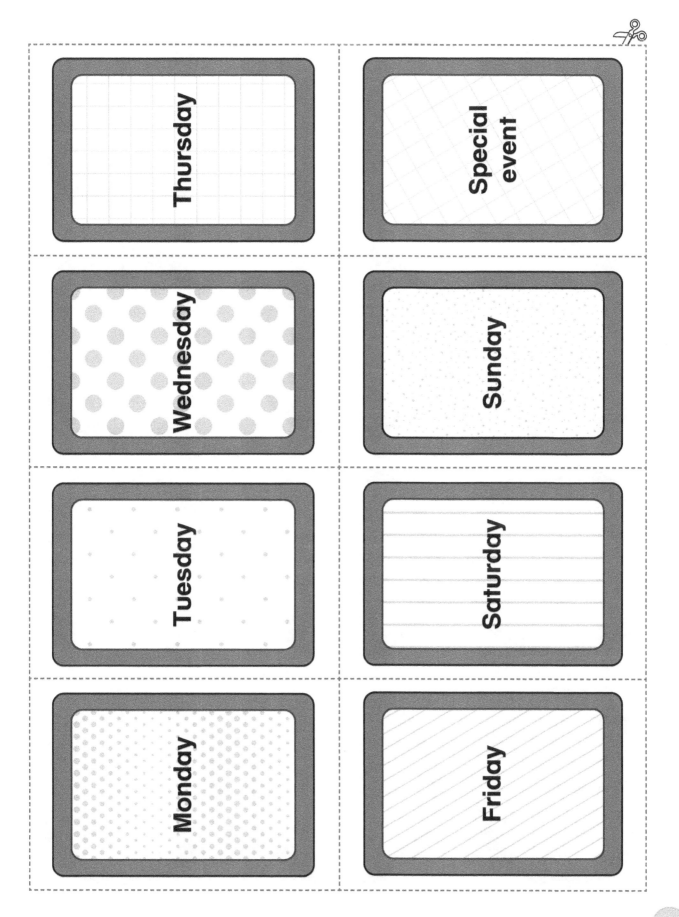

O'clock times

Half past times

O'clock and half past times

12 o'clock	1 o'clock	2 o'clock
3 o'clock	4 o'clock	5 o'clock
6 o'clock	7 o'clock	8 o'clock
9 o'clock	10 o'clock	11 o'clock

Half past 12	Half past 1	Half past 2
Half past 3	Half past 4	Half past 5
Half past 6	Half past 7	Half past 8
Half past 9	Half past 10	Half past 11

Unit 1

Lesson 1: Counting objects
Challenge ❶
1 4   **2** 6

Challenge ⚠
3 8   **4** 8
5 8   **6** 8

Challenge ❸
7 Last group ticked.

Lesson 2: Counting on and back in ones
Challenge ❶
1 4 is coloured in   **2** 5 is coloured in.

Challenge ⚠
3 6   **4** 9
5 7   **6** 4

Challenge ❸
7 Answers will vary.
8 Answers will vary.

Lesson 3: Sequences of objects
Challenge ❶
1

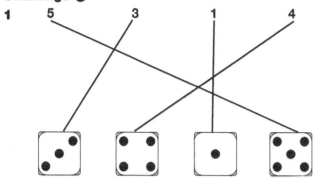

Challenge ⚠
2 8   **3** 6
4 5   **5** 3

Challenge ❸
6 Answers may vary, for example:

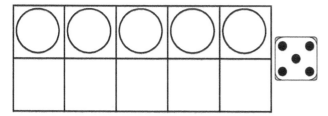

Lesson 4: Estimating to 10
Challenge ❶
1 Estimate: Estimates will vary.
    Count: 4

Challenge ⚠
2 Estimate: Estimates will vary.
    Count: 6
3 Estimate: Estimates will vary.
    Count: 9
4 Estimate: Estimates will vary.
    Count: 3

Challenge ❸
5 Estimate: Estimates will vary.
    Count: 7

Unit 2

Lesson 1: Estimating to 20
Challenge ❶
1 more   **2** less

Challenge ⚠
3a Accept any estimate between 9 and 14, 12
3b Accept any estimate between 17 and 20, 19
3c Accept any estimate between 14 and 18, 16
3d Accept any estimate between 6 and 10, 8

Challenge ❸
4 c

Lesson 2: Counting in twos
Challenge ❶
1a 4, 6, 8, 10   **1b** 5, 7, 9

Challenge ⚠
2a 8, 10, 12, 14   **2b** 3, 5, 7, 9, 11
2c 14, 16, 18   **2d** 11, 13, 15, 17

Challenge ❸
3 18

Lesson 3: Odd and even numbers
Challenge ❶
1 11, odd

Challenge ⚠
2 16, even
3 Any numbers ending in 1, 3, 5, 7 or 9
4 Any numbers ending in 0, 2, 4, 6 or 8

Challenge ❸
5 13, 7, 17, 3, 15, 19, 1, 5, 11

Lesson 4: Counting in tens
Challenge ❶
1 14   **2** 2

Challenge ⚠
3a 17   **3b** 19   **3c** 10
3d 5   **3e** 10   **3f** 1

Challenge ❸
4a 16   **4b** 9

Unit 3

Lesson 1: Counting to 10
Challenge ❶
1 Dotted line correctly traced

Challenge ⚠
2 Dotted line correctly traced

Challenge ❸
3 Dotted line correctly traced

Lesson 2: Reading numbers to 10
Challenge ❶
1 a 2   b 1   c 5

Challenge ⚠
2 Lines drawn to match: 9 ladybirds to 9 on the number track; 4 spiders to 4 on the number track; 8 caterpillars to 8 on the number track

Challenge ❸
3

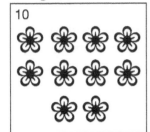 

Lesson 3: Writing numbers to 10
Challenge ❶
1 Numbers correctly traced

Challenge ⚠
2 Numbers copied correctly
3 Answers will vary, ensure that ages are correct

Challenge ❸
4 2, 4, 5, 7, 8, 10

Lesson 4: How many?
Challenge ❶
1 10

Challenge ⚠
2 4   3 8   4 7

Challenge ❸
5 9

Unit 4

Lesson 1: Counting to 20
Challenge ❶
1 Line traced over

Challenge ⚠
2 Line traced over
3 Line traced over

Challenge ❸
4 8, line traced over

Lesson 2: Reading numbers to 20
Challenge ❶
1

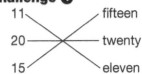

Challenge ⚠
2

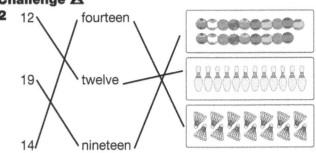

Challenge ❸
3a twelve circled   3b 13 circled

Lesson 3: Writing numbers to 20
Challenge ❶
1a 20   1b sixteen   1c 11   1d nineteen

Challenge ⚠
2a 12   2b nine   2c 14
2d fifteen   2e 10   2f thirteen
2g 8   2h seven

Challenge ❸
3 11, eleven, twelve, 13, thirteen, fourteen, fifteen, sixteen, 17, seventeen, 18, eighteen, nineteen, twenty

Lesson 4: Counting and labelling objects to 20
Challenge ❶
1 13   2 12

Challenge ⚠
3 17, seventeen   4 16, sixteen
5 11, eleven   6 15, fifteen

Challenge ❸
7 nineteen (19 flowers drawn)
8 fourteen (14 leaves drawn)

Unit 5

Lesson 1: Combining sets

Challenge ❶
1 4   **2** 6

Challenge ⚠
3 6   **4** 8   **5** 8   **6** 10

Challenge ❸
7 9

Lesson 2: Part–whole diagrams

Challenge ❶
1 4   **2** 4

Challenge ⚠
3 7   **4** 8
5 8

Challenge ❸
6 9

Lesson 3: Writing addition number sentences

Challenge ❶
1a 6
1b 8

Challenge ⚠
2a 2, 7   **2b** 7, 10
2c 1, 9   **2d** 4, 4

Challenge ❸
3a 6 + 3 = 9   **3b** 4 + 3 = 7

Lesson 4: Using sets of objects to solve additions

Challenge ❶
1 3   **2** 5

Challenge ⚠
3 7   **4** 6
5 7   **6** 8

Challenge ❸
7 9   **8** 9

Unit 6

Lesson 1: Adding more

Challenge ❶
1 3   **2** 5

Challenge ⚠
3 8   **4** 6
5 9   **6** 9

Challenge ❸
7 7   **8** 8

Lesson 2: Adding more to solve additions

Challenge ❶
1a 6   **1b** 6   **1c** 8   **1d** 5

Challenge ⚠
2a 8   **2b** 7
2c 8   **2d** 10

Challenge ❸
3a 7   **3b** 9

Lesson 3: Adding more with a number track

Challenge ❶
1a 9   **1b** 8

Challenge ⚠
2a 10   **2b** 7
2c 4   **2d** 9

Challenge ❸
3a 5 + 3 = 8   **3b** 2 + 4 = 6

Lesson 4: Adding more with a number line

Challenge ❶
1a 5   **1b** 4

Challenge ⚠
2a 8   **2b** 9
2c 6   **2d** 10

Challenge ❸
3a 4 + 3 = 7   **3b** 6 + 2 = 8

Unit 7

Lesson 1: Taking away objects

Challenge ❶
1 3   **2** 3

Challenge ⚠
3 3   **4** 4
5 1   **6** 4

Challenge ❸
7 5   **8** 3

Lesson 2: Taking away to solve subtractions

Challenge ❶
1a 3   **1b** 3

Challenge ⚠
2a 3   **2b** 2
3a 4   **3b** 6

Challenge ❸
4a 10 − 3 = 7   **4b** 9 − 6 = 3

Lesson 3: Taking away with part–whole diagrams

Challenge ❶

1a 2 counters drawn

1b 3 counters drawn

Challenge ⚠

2a 3 counters in each box

2b 2 counters in left hand box; 6 counters in right hand box

2c 1 counter in left hand box; 6 counters in right hand box

2d 4 counters in left hand box
3 counters in right hand box

Challenge ❸

3a 9 – 3 = 6   **3b** 10 – 5 = 5

Lesson 4: Solving subtractions with part–whole diagrams

Challenge ❶

1 1   **2** 3

Challenge ⚠

3 6   **4** 4

5 5   **6** 5

Challenge ❸

7 1   **8** 3

Unit 8

Lesson 1: Counting back in ones to subtract

Challenge ❶

1a 4   **1b** 3

Challenge ⚠

2a 8   **2b** 4   **2c** 3   **2d** 6

Challenge ❸

3a 4   **3b** 3

Lesson 2: Subtracting on a number track

Challenge ❶

1a 2   **1b** 1

Challenge ⚠

2a 6   **2b** 3   **2c** 3   **2d** 0

2e 2   **2f** 4

Challenge ❸

3 2

Lesson 3: Subtracting on a number line

Challenge ❶

1a 5   **1b** 6

Challenge ⚠

2a 5   **2b** 4   **2c** 0   **2d** 1

Challenge ❸

3a 2   **3b** 1

Lesson 4: Counting back to solve subtractions

Challenge ❶

1 8   **2** 7   **3** 5   **4** 9

Challenge ⚠

5 3   **6** 2   **7** 2

8 0   **9** 3   **10** 1

Challenge ❸

11 Answers will vary – accept any two subtractions with 3 as the answer.

12 Answers will vary – accept any two subtractions with 5 as the answer.

Unit 9

Lesson 1: Finding the difference

Challenge ❶

1a 2   **1b** 3

Challenge ⚠

2a 3   **2b** 1   **2c** 2   **2d** 4

Challenge ❸

3 5

Lesson 2: Subtracting by finding the difference

Challenge ❶

1a 2   **1b** 1

Challenge ⚠

2a 3   **2b** 2   **2c** 3   **2d** 8

Challenge ❸

3a 7 – 4 = 3   **3b** 6 – 5 = 1

Lesson 3: Finding the difference on a number line

Challenge ❶

1a 6   **1b** 2

Challenge ⚠

2a 2   **2b** 5   **2c** 2   **2d** 6

Challenge ❸

3 Answers will vary, accept any two numbers with a difference of 3

4 Answers will vary, accept any two numbers with a difference of 5

Lesson 4: Subtraction as difference on a number line

Challenge ❶
1a 3   **1b** 2

Challenge ⚠
2a 1   **2b** 7
2c 3   **2d** 9

Challenge ❸
3 8 − 4 = 4

Unit 10

Lesson 1: Making 10

Challenge ❶
1a 1   **1b** 3

Challenge ⚠
2a 2 sweets drawn, 8 and 2
2b 4 sweets drawn, 6 and 4
2c 7 sweets drawn, 3 and 7
2d 5 sweets drawn, 5 and 5

Challenge ❸
3 Dots shown to indicate any three of the following:
0 and 10, 1 and 9, 2 and 8, 3 and 7, 4 and 6, 5 and 5, 6 and 4, 7 and 3, 8 and 2, 9 and 1, 10 and 0.

Lesson 2: Addition and subtraction facts for 10

Challenge ❶
1 9 + 1 or 1 + 9   **2** 8 = 2 or 2 = 8

Challenge ⚠
3a Accept either: 6 + 4 = 10 or 4 + 6 = 10
Accept either: 10 − 6 = 4 or 10 − 4 = 6
3b 5 + 5 = 10
10 − 5 = 5
3c Accept either: 7 + 3 = 10 or 3 + 7 = 10
Accept either: 10 − 7 = 3 or 10 − 3 = 7
3d Accept either: 10 + 0 = 10 or 0 + 10 = 10
Accept either: 10 − 0 = 10 or 10 − 10 = 0

Challenge ❸
4 Answers will vary.

Lesson 3: Making numbers to 10

Challenge ❶
1 5, 1, 4, 3, 2

Challenge ⚠
2 0, 7, 1, 6, 5, 2, 5, 3, 3, 4

Challenge ❸
3 9, 0, 1, 8, 7, 2, 3, 5

Lesson 4: Addition and subtraction facts to 10

Challenge ❶
1 3, 1 or 1, 3 (for both number sentences)
2 2, 3 or 3, 2 (for both number sentences)

Challenge ⚠
3 (in any order): 5 + 2 = 7, 2 + 5 = 7, 7 − 5 = 2, 7 − 2 = 5
4 (in any order): 6 + 3 = 9, 3 + 6 = 9, 9 − 3 = 6, 9 − 6 = 3

Challenge ❸
5 3 + 5 = 8
8 − 5 = 3
(3 − 8 = 5)
5 + 3 = 8
(8 + 3 = 5)
8 − 3 = 5

Unit 11

Lesson 1: Estimating an answer

Challenge ❶
1 Accept 9 or 10. Answer 10
2 Accept 3–5. Answer 4

Challenge ⚠
3 Accept 4–6. Answer 5
4 Accept 8–10. Answer 10
5 Accept 1–3. Answer 2
6 Accept 1 or 2. Answer 1
7 Accept 8 or 9. Answer 9

Challenge ❸
8 8, nearly 10   **9** Less than 5, 2

Lesson 2: Choosing how to solve an addition

Challenge ❶
1 7   **2** 6

Challenge ⚠
3 10   **4** 9   **5** 6   **6** 9

Challenge ❸
7 8   **8** 9

Lesson 3: Choosing how to solve a subtraction

Challenge ❶
1 5   **2** 8

Challenge ⚠
3 5   **4** 1
5 2   **6** 6

Challenge ❸
7 Answers will vary

Lesson 4: Addition and subtraction in real life

Challenge ❶

1a +   **1b** −

Challenge ⚠

2a 6 + 2 = 8   **2b** 7 + 3 = 10
2c 8 − 4 = 4   **2d** 5 − 2 = 3

Challenge ❸

3 Answers will vary, accept correctly solved additions

Unit 12

Lesson 1: Addition facts to 20

Challenge ❶

1a 15   **1b** 14   **1c** 17
1d 19   **1e** 16   **1f** 17

Challenge ⚠

2a 20   **2b** 18   **2c** 18
2d 17   **2e** 20   **2f** 17
2g 19   **2h** 20

Challenge ❸

3a 18   **3b** 19   **3c** 19   **3d** 20

Lesson 2: Making 10 and adding more

Challenge ❶

1a 7 + 3 + 1 = 11   **1b** 8 + 2 + 3 = 13

Challenge ⚠

2a 8 + 2 + 4 = 14   **2b** 6 + 4 + 2 = 12
2c 7 + 3 + 2 = 12   **2d** 9 + 1 + 2 = 12
2e 8 + 2 + 2 = 12   **2f** 8 + 2 + 6 = 16

Challenge ❸

3a 5 + 5 + 2 = 12   **3b** 7 + 3 + 4 = 14
3c 6 + 4 + 1 = 11   **3d** 8 + 2 + 5 = 15

Lesson 3: Near doubles

Challenge ❶

1a 4, 5   **1b** 8, 9

Challenge ⚠

2b Double 6 = 12, 12 + 1 = 13
2c Double 8 = 16, 16 + 1 = 17
2d Double 9 = 18, 18 + 1 = 19

Challenge ❸

3a Double 7 = 14, 14 + 2 = 16
3b Double 6 = 12, 12 + 2 = 14

Lesson 4: Addition and estimation to 20

Challenge ❶

1 Estimates will vary   12 + 4 = 16
2 Estimates will vary   8 + 3 = 11

Challenge ⚠

3 Estimates will vary   9 + 4 = 13
4 Estimates will vary   14 + 5 = 19
5 Estimates will vary   9 + 10 = 19

Challenge ❸

6 **a** Estimates will vary 9 + 5 = 14
    b Answers will vary

Unit 13

Lesson 1: Subtraction facts to 20

Challenge ❶

1a 15   **1b** 11

Challenge ⚠

2a 3, 13   **2b** 2, 12
2c 1, 11   **2d** 4, 14

Challenge ❸

3a 13   **3b** 10

Lesson 2: Number facts to 20 – part–whole diagrams

Challenge ❶

1 4 dots   **2** 2 dots

Challenge ⚠

3 Answers will vary, accept dots that make 16
4 Answers will vary, accept dots that make 20
5 Answers will vary, accept dots that make 13

Challenge ❸

6 Answers will vary, accept dots that make 18

Lesson 3: Number families

Challenge ❶

1 5 + 4 = 9, 4 + 5 = 9, 9 − 5 = 4, 9 − 4 = 5

Challenge ⚠

2a 6 + 2 = 8, 2 + 6 = 8, 8 − 6 = 2, 8 − 2 = 6
2b 4 + 3 = 7, 3 + 4 = 7, 7 − 4 = 3, 7 − 3 = 4
2c 7 + 5 = 12, 5 + 7 =12, 12 − 7 = 5, 12 − 5 = 7

Challenge ❸

3 Answers will vary.

Lesson 4: Equal statements

Challenge ❶

1a Answers will vary, throughout the lesson, accepts answers that create correct equal statements
1b Answers will vary

Challenge ⚠

2a Answers will vary
2b Answers will vary
2c Answers will vary
2d Answers will vary

Challenge 3

3a Answers will vary

3b Answers will vary

Unit 14

Lesson 1: Doubling amounts to 5

Challenge ❶

1a 3 dots drawn

1b 1 dot drawn

1c 2 dots drawn

1d 5 dots drawn

Challenge ⚠

2a Double 4 = 8

2b Double 5 = 10

2c Double 1 = 2

2d Double 2 = 4

2e Double 3 = 6

2f Double 0 = 0

Challenge 3

3 2 – 4; 3 – 6; 5 – 10

Lesson 2: Doubling amounts to 10

Challenge ❶

1a 12   **1b** 18

Challenge ⚠

2a 14   **2b** 12

2c 20   **2d** 16

Challenge 3

3a 14

3b 18

3c 20

Lesson 3: Doubling on a number line

Challenge ❶

1a 10   **1b** 2

1c 6   **1d** 8

Challenge ⚠

2a 12   **2b** 16

2c 18   **2d** 0

2e 14   **2f** 20

Challenge 3

3 Double 2 = 4, double 7 = 14, double 8 = 16

Lesson 4: Doubling facts to 10

Challenge ❶

1a 2   **1b** 10

Challenge ⚠

2a 12   **2b** 18

2c 6   **2d** 14

2e 8   **2f** 4

2g 16   **2h** 0

Challenge 3

3a 5   **3b** 4   **3c** 8   **3d** 10

3e 7   **3f** 3   **3g** 9

Unit 15

Lesson 1: What is money?

Challenge ❶

1 All coins and notes circled

Challenge ⚠

2 Answers will vary

3 Answers will vary

Challenge 3

4 Answer to state that we use money to pay for things.

Lesson 2: Recognising coins

Challenge ❶

1 Answers will vary, throughout this lesson accept answers that correspond to your local currency

2 Answers will vary

Challenge ⚠

3 Answers will vary

Challenge 3

4 Answers will vary

Lesson 3: Recognising notes

Challenge ❶

1 Paper or plastic

2 Answers will vary, throughout this lesson accept answers that correspond to your local currency

Challenge ⚠

3 Answers will vary

Challenge 3

4 Answers will vary

Lesson 4: Sorting coins and notes

Challenge ❶

1 Answers will vary, throughout this lesson accept answers that correctly use the coins and notes provided

Challenge ⚠

2 Answers will vary

Challenge 3

3 Answers will vary

Unit 16

Lesson 1: Zero

Challenge ①

1

2

Challenge ②

3 three flowers drawn, 6 flowers drawn, no flowers drawn, 8 flowers drawn

4 5 bananas and 2 apples

Challenge ③

5 5   6 9

Lesson 2: Comparing numbers to 10

Challenge ①

1

2

Challenge ②

3 Any more than 4 dots drawn.

4 Any fewer than 6 dots drawn.

5 5

6 7

Challenge ③

7 True   8 False

Lesson 3: Ordering numbers to 10

Challenge ①

1 2, 4, 7, 10

Challenge ②

2a 9   2b 4

2c 2   2d 6

2e 7   2f 8

Challenge ③

3a 3, 4, 5, 6, 7

3b 8, 9

4a 2, 4, 7   4b 3, 5, 9   4c 6, 8, 10

Lesson 4: Ordinal numbers

Challenge ①

1 3rd, 5th, 6th, 9th

Challenge ②

2 2nd, 4th, 7th, 8th, 10th

Challenge ③

3 cat 4th, butterfly 3rd, snake 1st, snail 7th, parrot 2nd, ladybird 5th, rabbit 6th

Unit 17

Lesson 1: Partitioning numbers into tens and ones

Challenge ①

1 Block of ten coloured blue, four ones coloured red.

Challenge ②

2a One ten and five ones drawn.

2b One ten and eight ones drawn.

2c One ten and two ones drawn.

2d One ten and three ones drawn.

Challenge ③

3a 10, 6

3b 10, 1

3c 20, 0

3d 10, 9

Lesson 2: Combining tens and ones

Challenge ①

1a 13   1b 18

Challenge ②

2a 11   2b 17

2c 20   2d 19

Challenge ③

3a 15   3b 10

Lesson 3: Representing numbers in different ways

Challenge ①

1   a 10 + 7

    b 10 + 4 + 3

    c 10 + 5 + 2

Challenge ⚠
2 Accept any of the following:
  10 + 8
  10 + 1 + 7
  10 + 2 + 6
  10 + 3 + 5
  10 + 4 + 4
  10 + 5 + 3
  10 + 6 + 2
  10 + 7 + 1
3 Accept any of the following:
  10 + 9
  10 + 1 + 8
  10 + 2 + 7
  10 + 3 + 6
  10 + 4 + 5
  10 + 5 + 4
  10 + 6 + 3
  10 + 7 + 2
  10 + 8 + 1

Challenge 🔳
4 Accept any answer that successfully regroups 12

Lesson 4: Comparing and ordering numbers to 20

Challenge ❶
1a 3, 10, 15
1b 8, 11, 16

Challenge ⚠
2a 7, 12, 14   **2b** 3, 15, 20
2c 14, 17, 18   **2d** 5, 10, 11

Challenge 🔳
3a 4, 6, 12, 19, 20
3b 3, 8, 9, 10, 16

Unit 18

Lesson 1: Halving objects

Challenge ❶
1 Insect halves correctly matched.

Challenge ⚠
2 Apple, strawberry, sandwich, orange

Challenge 🔳
3 Correct lines that halve the chocolate bars

Lesson 2: Halves of shapes

Challenge ❶
1 Triangle, circle, rectangle

Challenge ⚠
2 Shapes correctly matched
3 Correct lines that halve the shapes (a line of symmetry)

Challenge 🔳
4 A different line of symmetry through each square (two diagonals, one vertical, one horizontal)

Lesson 3: Halves of amounts (1)

Challenge ❶
1 Balls, cars

Challenge ⚠
2a 4 cars each   **2b** 5 soft toys each
2c 6 bowling balls each   **2d** 3 cubes each

Challenge 🔳
3a 6 bananas
3b 8 pineapples
3c 7 peaches

Lesson 4: Halves of amounts (2)

Challenge ❶
1a 4 dots – 4 dots   **1b** 5 dots – 5 dots
1c 3 dots – 3 dots   **1d** 2 dots – 2 dots

Challenge ⚠
2a 4-4   **2b** 6-6
2c 2-2   **2d** 5-5

Challenge 🔳
3 7 cakes each

Unit 19

Lesson 1: Halving numbers to 10

Challenge ❶
1a 4 dots drawn in each half
1b 3 dots drawn in each half
1c 2 dots drawn in each half

Challenge ⚠
2a 5   **2b** 1   **2c** 3
2d 2   **2e** 4

Challenge 🔳
3a 6   **3b** 10

Lesson 2: Halving numbers to 20

Challenge ❶
1a 2
1b 1
1c 3

Challenge ⚠
2a 6   **2b** 10
2c 8   **2d** 9

Challenge ❸
3 7   **4** 9

Lesson 3: Combining halves (1)

Challenge ❶
1 3

Challenge ⚠
2 4   **3** 5

Challenge ❸
4 6

Lesson 4: Combining halves (2)

Challenge ❶
1a 8   **1b** 6

Challenge ⚠
2a 10   **2b** 14
2c 12   **2d** 18

Challenge ❸
3 16   **4** 20

Unit 20

Lesson 1: Days of the week

Challenge ❶
1 Monday, Wednesday, Thursday, Friday, Sunday

Challenge ⚠
2a Thursday   **2b** Monday   **2c** Friday

Challenge ❸
3 Sunday, Tuesday
   Thursday, Saturday
   Tuesday, Thursday

Lesson 2: Months of the year

Challenge ❶
1 Any appropriate drawing of an event labelled with the correct month.

Challenge ⚠
2 Any appropriate events filled in for the correct months.

Challenge ❸
3 January, March, April, June, July, August, September, November

Lesson 3: O'clock

Challenge ❶
1 9 o'clock, 1 o'clock and 6 o'clock ticked

Challenge ⚠
2 4 o'clock, 9 o'clock, 1 o'clock, 10 o'clock

Challenge ❸
3   a   b

Lesson 4: Half past

Challenge ❶
1 Half past 9 ticked, half past 8 ticked, half past 7 ticked

Challenge ⚠
2 a Half past 10   **b** half past 8
    c half past 3   **d** half past 6

Challenge ❸
3   a   b

Unit 21

Lesson 1: Recognising 2D shapes

Challenge ❶
1

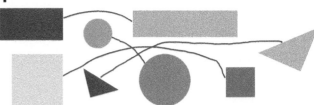

Challenge ⚠
2 The shapes in the house correctly coloured

Challenge ❸
3 6 triangles, 4 squares, 3 circles, 5 rectangles

Lesson 2: Describing 2D shapes

Challenge ❶
1

Challenge 🔺
2

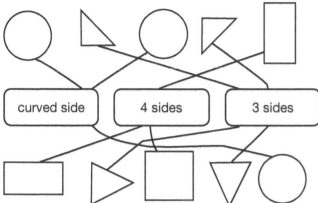

Challenge 🔳
3

Lesson 3: Sorting 2D shapes
Challenge ❶
1a

1b

1c

Challenge 🔺
2a All circles coloured
2b All triangles ticked
2c All quadrilaterals circled

Challenge 🔳
3 Any appropriate sorting suggestions, e.g.: by colour, by size, by number of sides, by straight and curved sides.

Lesson 4: Rotating shapes
Challenge ❶
1 Square: yes
   Triangle: no

Challenge 🔺
2 The circle and the square coloured.

Challenge 🔳
3

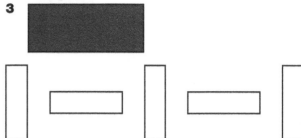

Unit 22

Lesson 1: What is a 3D shape?
Challenge ❶
1 Cube, pyramid and cylinder coloured.

Challenge 🔺
2 Cube joined to image
   Sphere joined to image
   Cylinder joined to image
   Pyramid joined to image
   Cuboid joined to image

Challenge 🔳
3 9 3D shapes
   8 2D shapes

Lesson 2: Making models with 3D shapes
Challenge ❶
1 Cuboid, cylinder and pyramid ticked.

Challenge 🔺
2a cylinders and top cuboid coloured
2b cubes, pyramids, cylinders and cuboids

Challenge 🔳
3 Any appropriate answer explaining that you can't tell if a cube has been rotated because the sides are the same length and the faces are all the same shape.

Lesson 3: Describing 3D shapes
Challenge ❶
1 cube and cuboid coloured

Challenge 🔺
2

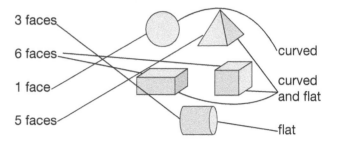

Challenge 3
3 cube or cuboid
4 cylinder

Lesson 4: Sorting shapes
Challenge 1
1 Cylinder and sphere ticked, crosses for cube and cuboid

Challenge 2
2

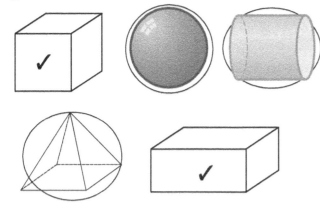

Challenge 3
3 Any 2 suitable answers, e.g. all flat faces and some curved faces/shapes with 5 faces or more and shapes with fewer than 5 faces/shapes with 8 edges or more and shapes with fewer than 8 edges

Unit 23

Lesson 1: Length, height and width
Challenge 1
1

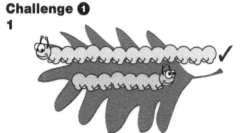

2

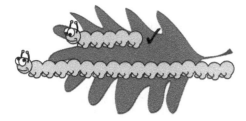

Challenge 2
3 Answers will vary. Check that objects drawn fulfil the criteria.
4 Answers will vary. Check that objects drawn fulfil the criteria.

Challenge 3
5 2, 5, 1, 3, 4

Lesson 2: Measuring length
Challenge 1
1a 6
1b 4
1c 3

Challenge 2
2a Answers will vary depending on the length of classroom objects used.
2b Answers will vary depending on the length of classroom objects used.
2c Answers will vary depending on the length of classroom objects used.
2d Answers will vary depending on the length of classroom objects used.

Challenge 3
3 Answers will vary depending on the length of classroom objects used and non-standard unit chosen.

Lesson 3: Mass
Challenge 1
1

Challenge 2
2a

2b

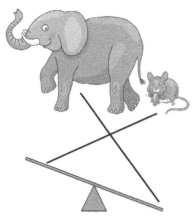

3 Answers will vary. Any appropriately lighter object drawn on the scales.

4 Answers will vary. Any appropriately heavier object drawn on the scales.

Challenge 3

5a

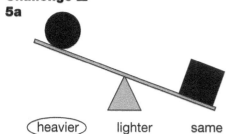

heavier   lighter   same

5b

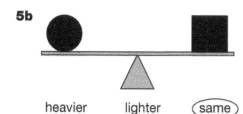

heavier   lighter   same

Lesson 4: Measuring mass

Challenge 1

1a 5   **1b** 3

Challenge 2

2a Take away   **2b** Add

Challenge 3

3a 6

3b 6 cubes drawn on the scale

Unit 24

Lesson 1: Full, half full or empty?

Challenge 1

1

Challenge 2

2

empty   full   half full

Challenge 3

3

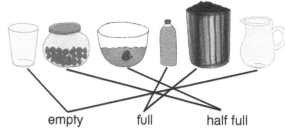

empty   full   half full

Lesson 2: Estimating and comparing capacity

Challenge 1

1 Bucket circled

Challenge 2

2a 4, 1, 2, 3

2b 1, 3, 4, 2

Challenge 3

3 Water bottle, washing up liquid, small vase, glass.

Lesson 3: Measuring capacity

Challenge 1

1 Answers will vary (practical activity)

Challenge 2

2 Answers will vary (practical activity)

Challenge 3

3 Bucket circled

Lesson 4: Temperature

Challenge 1

1 a Cold   **b** Hot   **c** Hot

Challenge 2

2

 COLOURED RED

 COLOURED RED

 COLOURED BLUE

 COLOURED BLUE

 COLOURED RED

COLOURED BLUE

 COLOURED BLUE

COLOURED RED

Challenge 3

3 2, 4, 3, 1

Unit 25

Lesson 1: Describing direction

Challenge 1

1 Line connecting up to the right hand image and down to the left hand image

Challenge 2

2 forwards, around, down

Challenge 3

3

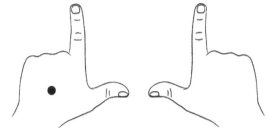

Lesson 2: Left and right

Challenge 1

1

Challenge 2

2 a left

b right

c Cross in the box to the left of the circle.

Challenge 3

3 Answers will vary.

Lesson 3: Describing position

Challenge 1

1 The following drawn on the picture: A fish in the water and a boy on the bridge.

Challenge 2

2

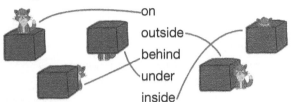

Challenge 3

3 The following added onto the drawing: a boy in the car, a bird over the car and a road under the car.

Lesson 4: Following directions

Challenge 1

1

Challenge 2

2

Challenge 3

3 The last 'Go right' circled.

Unit 26

Lesson 1: Sorting data

Challenge 1

1

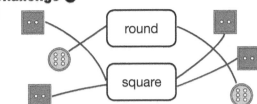

2 Square

Challenge 2

3

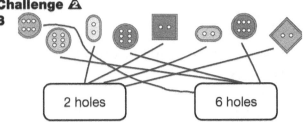

4 the same

Challenge 3

5

Letters				Numbers		
A	B	J		3	4	5
K	L	N		6	7	9
P	T	W		10		

6 Letters

Lesson 2: Collecting data

Challenge 1

1 a 4   **b** 1   **c** 2

Challenge 2

2 a 1

   b 3

   c 5

   d 2

   e ice cream

Challenge 3

3 Any two appropriate suggestions, e.g. people could put their hands up, people could stand in different areas, people could write on a whiteboard and hold it up.

Lesson 3: Using lists

Challenge 1

1 Elephants: 2

Tigers: 4

Zebras: 3

Challenge 2

2a Circles: 2

Rectangles: 2

Triangles: 4

Squares: 5

2b Square

Challenge 3

3a Any appropriate suggestion, e.g. What the weather was like this week?/What weather do people like best?/How many days we had each type of weather

3b Any appropriate suggestion e.g. Type of weather/Weather

Lesson 4: Using tables

Challenge 1

1

Sweet	Total
	3
	5
	4

Challenge 2

2

Vehicle	Total
	4
	1
	3
	2

Challenge 3

3

Button type	Total
	11
	8
	6

Unit 27

Lesson 1: Venn diagrams

Challenge 1

1 numbers

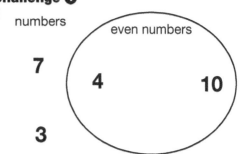

even numbers

7   4   10   1

3

Challenge 2
2 Shapes

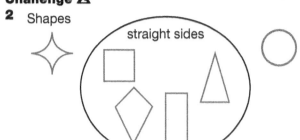

straight sides

Challenge 3
3 Answers may vary, for example: Sea.

Lesson 2: Carroll diagrams
Challenge 1
1

Letters with curves	Letters with no curves
B C D	A E

Challenge 2
2

Letter	Not a letter
H K T C P	8 12 9 4 16 20

5 are letters.
6 are not letters.

Challenge 3
3

Less than 10	More than 10
8 5 9 4 2	11 17 13 16 12 18

Even number	Odd number
8, 4, 2, 18, 16, 12	5, 17, 11, 9, 13

Lesson 3: Pictograms
Challenge 1
1 9, 4, 2

Challenge 2
2 5, 3, 7, 10, 2

Challenge 3
3 Pictures of: two bananas, five apples, three oranges

Lesson 4: Block graphs
Challenge 1
1

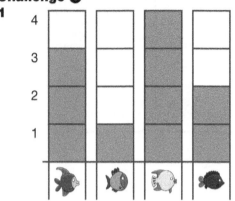

Challenge 2
2

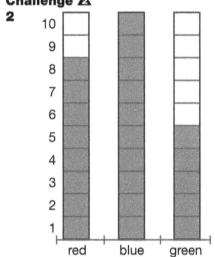

red   blue   green

Statements will vary

Challenge 3
3

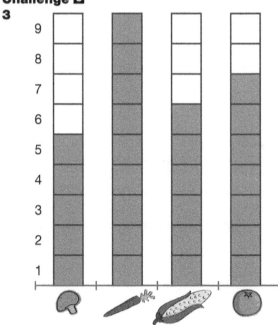

The column for carrots coloured blue.

Stage 1 Record-keeping

Class: _____   Year: _____

KEY

A: Exceeding expectations in this sub-strand	B: Meeting expectations in this sub-strand	C: Below expectations in this sub-strand

Strand: **Number** Sub-strand: **Counting and sequences**		
Code	**Learning objectives**	
1Nc.01	Count objects from 0 to 20, recognising conservation of number and one-to-one correspondence.	
1Nc.02	Recognise the number of objects presented in familiar patterns up to 10, without counting.	
1Nc.03	Estimate the number of objects or people (up to 20), and check by counting.	
1Nc.04	Count on in ones, twos or tens, and count back in ones and tens, starting from any number (from 0 to 20).	
1Nc.05	Understand even and odd numbers as 'every other number' when counting (from 0 to 20).	
1Nc.06	Use familiar language to describe sequences of objects.	

A	B	C

Class: _____   **Year:** _____

Strand: **Number**		
Sub-strand: **Integers and powers**		
Code	**Learning objectives**	
1Ni.01	Recite, read and write number names and whole numbers (from 0 to 20).	
1Ni.02	Understand addition as: – counting on – combining two sets.	
1Ni.03	Understand subtraction as: – counting back – take away – difference.	
1Ni.04	Recognise complements of 10.	
1Ni.05	Estimate, add and subtract whole numbers (where the answer is from 0 to 20).	
1Ni.06	Know doubles up to double 10.	
A	B	C

Strand: **Number**		
Sub-strand: **Money**		
Code	**Learning objective**	
1Nm.01	Recognise money used in local currency.	
A	B	C

Class: _____ **Year:** _____

Strand: **Number**		
Sub-strand: **Place value, ordering and rounding**		

Code	Learning objectives	
1Np.01	Understand that zero represents none of something.	
1Np.02	Compose, decompose and regroup numbers from 10 to 20.	
1Np.03	Understand the relative size of quantities to compare and order numbers from 0 to 20.	
1Np.04	Recognise and use ordinal numbers from 1st to 10th.	

A	B	C

Strand: **Number**		
Sub-strand: **Fractions, decimals, percentages, ratio and proportion**		

Code	Learning objectives	
1Nf.01	Understand that an object or shape can be split into two equal parts or two unequal parts.	
1Nf.02	Understand that a half can describe one of two equal parts of a quantity or set of objects.	
1Nf.03	Understand that a half can act as an operator (whole number answers).	
1Nf.04	Understand and visualise that halves can be combined to make wholes.	

A	B	C

Class: _____ **Year:** _____

Strand: **Geometry and Measure** Sub-strand: **Time**		
Code	**Learning objectives**	
1Gt.01	Use familiar language to describe units of time.	
1Gt.02	Know the days of the week and the months of the year.	
1Gt.03	Recognise time to the hour and half hour.	
A	B	C

Strand: **Geometry and Measure** Sub-strand: **Geometrical reasoning, shapes and measurements**		
Code	**Learning objectives**	
1Gg.01	Identify, describe and sort 2D shapes by their characteristics or properties, including reference to number of sides and whether the sides are curved or straight.	
1Gg.02	Use familiar language to describe length such as long, longer, longest, thin, thinner, thinnest, short, shorter, shortest, tall, taller and tallest.	
1Gg.03	Identify, describe and sort 3D shapes by their properties, including reference to the number of faces, edges and whether faces are flat or curved.	
1Gg.04	Use familiar language to describe mass, including heavy, light, less and more.	
1Gg.05	Use familiar language to describe capacity, including full, empty, less and more.	
1Gg.06	Differentiate between 2D and 3D shapes.	
1Gg.07	Identify when a shape looks identical as it rotates.	
1Gg.08	Explore instruments that have numbered scales, and select the most appropriate instrument to measure length, mass, capacity and temperature.	
A	B	C

Class: _____ **Year:** _____

Strand: **Geometry and Measure** Sub-strand: **Position and transformation**		
Code	**Learning objective**	
1Gp.01	Use familiar language to describe position and direction.	
A	B	C

Strand: **Statistics and Probability** Sub-strand: **Statistics**		
Code	**Learning objectives**	
1Ss.01	Answer non-statistical questions (categorical data).	
1Ss.02	Record, organise and represent categorical data using: – practical resources and drawings – lists and tables – Venn and Carroll diagrams – block graphs and pictograms.	
1Ss.03	Describe data, using familiar language including reference to more, less, most or least to answer non-statistical questions and discuss conclusions.	
A	B	C

Class: _____   Year: _____

Thinking and Working Mathematically	
Code	**Characteristics**
TWM.01	Specialising Choosing *an example* and checking to see if it satisfies or does not satisfy specific mathematical criteria.
TWM.02	Generalising Recognising an underlying pattern by identifying *many* examples that satisfy the same mathematical criteria.
TWM.03	Conjecturing Forming mathematical questions or ideas.
TWM.04	Convincing Presenting evidence to *justify* or challenge a mathematical idea or solution.
TWM.05	Characterising Identifying and describing the mathematical properties of an object.
TWM.06	Classifying Organising objects into groups according to their mathematical properties.
TWM.07	Critiquing Comparing and evaluating mathematical ideas, representations or solutions to identify advantages and disadvantages.
TWM.08	Improving Refining mathematical ideas or representations to develop a more effective approach or solution.

A	B	C

Cambridge Global Perspectives™

Below are some examples of lessons in *Collins International Primary Maths Stage 1* which could be used to develop the Global Perspectives skills. The notes in italics suggest how the maths activity can be made more relevant to Global Perspectives.

Please note that the examples below link specifically to the learning objectives in the Cambridge Global Perspectives curriculum framework for Stage 1. However, skills development in a wider sense is embedded throughout this course and teachers are encouraged to promote research, analysis, evaluation, reflection, collaboration and communication as general best practice. For example, the pair work and group activities suggested throughout this Teacher's Guide offer opportunities to develop skills in communication, collaboration and reflection which build towards the specific Global Perspectives learning objectives.

Cambridge Global Perspectives	Learning Objectives for Stage 1	Collins International Primary Maths Stage 1
RESEARCH	Recording findings • Record information on a given topic in pictograms or simple graphic organisers	• Unit 26, Lessons 3-4 *Focus on recording data collected and talking about information recorded on the various examples provided.* • Unit 27, Lessons 1–4 *Focus on recording data collected and talking about information recorded on the various examples provided.*
ANALYSIS	Interpreting data • Talk about information recorded in pictograms or graphic organisers	
REFLECTION	Personal viewpoints • Talk about what has been learned during an activity with support	• Unit 11, Lesson 2, Review *Focus not only on what has been learned, but also on how it was learned – the strategies used.* • Unit 26, Lesson 3, Review *Discuss how well the lists show the information and develop to reflect on why it is useful to make lists (and other methods of recording information).*
COLLABORATION	Cooperation and interdependence • Share resources with others while working independently or with a partner	*Many of the **Apply** activities in the main TG lesson notes and the **Additional practice activities** following each TG unit require learners to share resources and work collaboratively. From time to time, discuss with them how well they are doing this, and encourage them to think about the benefits of collaboration. Examples of suitable activities are:*
	Engaging in teamwork • Work positively with others	• Unit 2, Lesson 1, Apply • Unit 2, Additional practice activities 1 & 2 • Unit 4, Lesson 3, Apply • Unit 10, Additional practice activities 1 & 2 • Unit 15, Lesson 1, Apply • Unit 20, Lesson 3, Apply • Unit 25, Lesson 1, Apply